Electric Circuit Analysis
Principles and
Applications

ELECTRONIC SYSTEMS ENGINEERING SERIES

Consulting editors   **E L Dagless**
University of Bristol

J O'Reilly
University College of Wales

OTHER TITLES IN THE SERIES

Advanced Microprocessor Architectures   *L Ciminiera and A Valenzano*

Optical Pattern Recognition Using Holographic Techniques   *N Collings*

Modern Logic Design   *D Green*

Data Communications, Computer Networks and OSI (3rd Edn)   *F Halsall*

Multivariable Feedback Design   *J M Maciejowski*

Computer Architecture and Design   *A J van de Goor*

Digital Systems Design with Programmable Logic   *M Bolton*

Introduction to Robotics   *P J McKerrow*

MAP and TOP Communications: Standards and Applications   *A Valenzano,*
     C Demartini and L Ciminiera

Integrated Broadband Networks   *R Händel and M N Huber*

Electric Circuit Analysis

Principles and Applications

K. F. Sander

Emeritus Professor, University of Bristol

Addison-Wesley Publishing Company

Wokingham, England • Reading, Massachusetts • Menlo Park, California • New York
Don Mills, Ontario • Amsterdam • Bonn • Sydney • Singapore
Tokyo • Madrid • San Juan • Milan • Paris • Mexico City • Seoul • Taipei

The programs in this book have been included for their instructional value. They have been tested with care but are not guaranteed for any particular purpose. The publisher does not offer any warranties or representation, nor does it accept any liabilities with respect to the programs.

Many of the designations used by manufacturers and sellers to distinguish their products are claimed as trademarks. Addison-Wesley has made every attempt to supply trademark information about manufacturers and their products mentioned in this book. A list of the trademark designations and their owners appears below.

Cover designed by Hybert Design and Type, Maidenhead and printed by The Riverside Printing Co. (Reading) Ltd.
Printed in Great Britain by Mackays of Chatham, Kent.

First printed 1992.

British Library Cataloguing-in-Publication Data. A catalogue record for this book is available from the British Library.

Library of Congress Cataloguing-in-Publication Data is available from the Library of Congress.

Trademark notice
IBM is a trademark of International Business Machines Corporation
Turbo Pascal is a trademark of Borland International Incorporated
PSpice is a trademark of MicroSim Corporation

◆ *Preface*

Philosophy

This book is about linear circuit analysis and some of its applications. The principles of circuit analysis were laid down in the course of those fundamental experiments with simple wire ciruits which led to the concepts of voltage, current, resistance, capacitance and inductance. Since those days, the many other electrical and electronic devices which have been invented are habitually described in terms of circuits showing equivalent behaviour. This situation means that circuit analysis has become the language of electrical and electronic engineering, and a basic necessity to any practitioner in the very wide range of applications covered by these subjects. Experience shows that students find it difficult to acquire the necessary proficiency. One factor contributing to this situation is perhaps that it seems to be a common philosophy for courses and textbooks to start with a formal theoretical presentation before approaching the simple problems which form the bulk of applications which students are liable to encounter, both in their course and in real life. In this book the alternative approach is adopted: the principles of circuit analysis are presented in the simplest form, followed by application to a wide range of examples of the type that the student will inevitably encounter in his or her course, and indeed in a practising environment. This approach is in no way meant to denigrate the role of formal theory: theory is a vital necessity for the understanding, for example, of filter design and in general for network theory, but not for the analysis of a very wide range of commonly encountered practical situations. The elements of formal theory are presented at a later stage in the book, primarily as a background to computer-aided analysis.

Design

Circuit analysis is the essential first step towards circuit design. Today, it is customary to put great emphasis on computer-aided analysis tools such as SPICE, but it remains true that the great majority of design starts with the algebraic analysis of simple circuits. Refinement follows either by 'breadboard' construction or, if appropriate, with detailed analysis by computer methods. The importance of 'manual' analysis is in no way lessened by the availability of computer methods: the two are complementary. This design aspect has been kept well in mind when choosing appropriate examples.

Mathematical requirements

Some mathematical background is necessary. In its simplest form, circuit analysis uses the mathematical techniques of elementary algebra, requiring manipulation of real and complex quantities. Quantitative formulation of a problem leads to a set of linear algebraic equations, which then have to be solved. Once time variation is considered, it becomes necessary to use the concepts of integration and differentiation. As far as this text is concerned, actual integration extends no further than simple algebraic and trigonometric functions. It is assumed that this background is available.

Layout

The layout of the text, in general terms, is that Chapters 1–5 deal with the basic ideas of circuit analysis and later chapters represent either important applications or topics of importance in extending the range of circuit analysis. The final chapter forms an introduction to the computer analysis program SPICE. In all chapters, numerous worked examples are incorporated in the text. Each chapter ends with a series of examples, numerical answers to which are to be found at the end of the book.

In Chapter 1 the various electrical quantities of interest and the units in which they are measured are defined. Although in the simplest situations voltages and currents in a circuit may not vary with time, in most practical situations some form of time-variation is present. With this fact in mind, in Chapter 2 are considered the various averages which are found to be useful when such variation in time is taking place. Circuit analysis proper is first considered in Chapter 3, in the context of circuits containing only resistances which are driven by constant voltage or current sources. The very important case when sources vary sinusoidally in time is considered in Chapter 4, in which the phasor method of dealing with such variation is described. Chapter 5 also concerns circuits in which sources vary sinusoidally with time, but differs from Chapter 4 in that frequency is regarded as variable rather than fixed. The chapter is built around examples of commonly encountered circuit configurations. In these five chapters analysis has proceeded from first principles; they serve as a self-contained introduction to elementary circuit analysis. Formal methods of analysis using loop and nodal variables are outlined in Chapter 6. Such methods are of interest in the context of this book primarily in relation to computer-aided analysis. The other main use at the elementary level of these formal methods is to provide proofs of network theorems such as those of Thévenin and Norton: these theorems have been amply demonstrated by the worked examples in previous chapters so that formal proofs have been related to Appendix A.

The next two chapters provide extended examples of the application of circuit analysis: Chapter 7 deals with three-phase circuits and Chapter 8 with magnetically coupled circuits and transformers, both important topics in electrical engineering. Chapter 9 introduces the problem of finding the response of a circuit to transient excitation. In it is shown how transient analysis of a circuit leads to a differential equation for the unknown voltage or current, and how in simple cases these equations can be solved. It is also shown how solutions can be obtained using a numerical approach. Chapter 10 continues the topic of transient analysis by describing the Laplace transform method as applied to circuits. The close parallel between the formalism of the Laplace transform and that of the phasor method is emphasized. The concept of signal spectrum has been referred to in earlier chapters: in Chapter 11 this concept is formalized by the development of the Fourier series. The application to circuit problems is illustrated by example. A brief introduction is given to the discrete Fourier transform, the fast Fourier transform and to the Fourier integral transform. Previous to this point in the book, all the circuits considered have been linear. Extension is made in Chapters 12 and 13 to include non-linear devices, specifically semiconductors. Chapter 12 introduces the idea of modelling a diode characteristic by means of linear segments. The use of the technique is illustrated by application to the analysis of various rectifier circuits. Chapter 13 introduces the idea of linearization and small-signal analysis. The use of this technique is illustrated by application to field-effect and bipolar transistors. Results for the small-signal response of simple amplifier circuits are obtained. With this chapter the analytic content of the book is concluded. Chapter 14 is of a different nature, and might be said to form an introduction to real circuits: it considers the question as to how accurately a circuit diagram with its essentially fictional components can predict the behaviour of the real, physical circuit it is supposed to represent. In this chapter the characteristics of real resistors, inductors and capacitors are considered, together with the unexpected, and usually unwanted, effects of wired connections. This is an important and frequently neglected topic, even though the effects are often apparent in the 'simple' laboratory experiments students are required to perform.

Finally, Chapter 15 is devoted to introducing the computer-aided analysis program SPICE. Arguments can be advanced that this topic should be closely integrated with the text, rather than left to a final chapter. However, the professional application of circuit analysis is usually for design purposes, and is in great part concerned with simple circuits of the type which have appeared in this book. For such design purposes results are best formulated algebraically: the essentially numerical analysis by computer methods serves as a verification, not as a substitute for the algebraic analysis. It is the author's belief that a close integration with numerical methods at this elementary stage forms a hindrance to the student trying to acquire the necessary algebraic skills. Nevertheless, the presentation of topics in Chapter 15 follows closely to the order of presentation in earlier chapters, so that the text can easily be used in a more closely integrated mode if desired. Further, the worked examples in the chapter are reworkings of examples from earlier chapters. It should be said that although the presentation is not strongly dependent on which version of SPICE is used, the contents have been entirely worked on the 'class-room' version of the MicroSim PSpice, which is freely available and can be run on an IBM-compatible PC.

The development of the views presented in this book owes much to many discussions with colleagues at Cambridge and Bristol Universities, discussions which have taken place over many years. To name all these colleagues is hardly possible, but they have my thanks. It is also a pleasure to acknowledge gratefully many constructive comments made by referees who read earlier versions of the text. Finally, my thanks are due to Dr Martin Izzard for many useful discussions about Chapter 15.

K. F. Sander

◆ *Contents*

1 ◆ *Components and Sources*

This chapter introduces the various component types associated with linear circuits. Their electrical characteristics are described and characterizing equations given.

1.1 ◆ Introduction

The study of any electric circuit starts with a circuit diagram, which takes the paper form of a number of well defined component types connected together. The ability to translate a particular configuration of hardware into a circuit diagram representing its properties took a long time to develop, and it is of interest to consider briefly the course of this development.

Quantitative understanding first came to electrostatics through the concepts of charge and potential at the hands of Coulomb and Poisson. In these terms, the electrostatic generators used by the early experimenters transferred charge to an isolated electrode which thereby attained a high potential. These experimenters discovered that the 'electric fluid' could be stored in a Leyden jar, consisting of a glass jar coated inside and out with metal foil. In modern terms, the Leyden jar has the property of capacitance.

The discovery of the 'voltaic pile' marked the beginning of experiments with steady electric currents, although realization that these experiments could be analysed in terms of electrostatic charge and potential was not immediate. The ability of the pile to drive current round a circuit can be quantified as an electromotive force equal to the difference in potential between its terminals. The variation of conducting power between substances had been known to the electrostatic experimenters; quantification of this concept followed the work of Ohm, which effectively defined the concept of the resistance of a conductor as the ratio of drop in potential to current. It was observed that a current-carrying resistive wire became hot. Measurements by Joule led to the understanding that electrical energy was being dissipated as heat at a rate equal to the product of potential drop and current and further that electrical energy was equivalent to mechanical energy. This result was clearly in accord with the ideas that potential difference is associated with work done in moving charge and that current is the rate of flow of charge.

The discovery of magnetic induction by Faraday led to the third concept needed to account for circuit behaviour, that of inductance. He investigated the effect that a current-carrying circuit might have on an adjacent circuit and found that interruption of the current produced a transient effect in the adjacent circuit.

This effect Faraday explained in terms of the linkage of the two circuits by magnetic flux. In circuit terms an electromotive force is induced in the adjacent circuit proportional to the rate of change of current in the other. The coefficient of proportionality defines the mutual inductance between the circuits. It was later shown that interruption of the current in a single coil produces a large voltage across that coil, a consequence of the self-inductance of the coil. The three properties of resistance, inductance and capacitance are sufficient to account for all the phenomena found in circuits. It may be noted that each is related to an energy transaction: resistance causes conversion of electrical energy to heat, inductance stores electrical energy in a magnetic field and capacitance stores it in an electrostatic field. The two forms of energy storage can be compared with kinetic and potential energy in mechanics.

As the understanding of circuit behaviour grew, so applications were found, leading to the need to design and engineer circuits for specific purposes. The earliest professional users of circuits were the telegraph engineers engaged in the setting-up and operation of telegraph networks of ever-increasing size and complexity. At the same time the power engineers were developing power sources of continually greater power output. They investigated the problems of distributing electrical power and also developed the use of alternating current. The pioneers of 'wireless' discovered, among many other things, the use of resonant circuits to provide selectivity of reception between stations. Today engineered circuits are found in a vast range of applications, such as power production and distribution, domestic equipment, control systems, TV and radio, terrestrial and satellite communications and computers.

Such circuits all depend on the use of the passive elements of resistance, inductance and capacitance as well as many other devices, of which transistors are just one example. Because of this widespread use of circuits and the power of circuit analysis, the operation of other devices is frequently described by means of equivalent circuits. For the purpose of translating hardware into a circuit diagram, or a designed circuit into hardware, resistance, inductance and capacitance are 'lumped' into individual components: resistors, inductors and capacitors. Ranges of components such as these together with the development of sources and measuring techniques provide the basis for the design and construction of useful circuits.

1.2 ♦ Voltage and current

Voltage is the convenient term in modern usage for the more precise but cumbersome term 'potential difference'. The voltage between two points of a circuit measures the work done in taking unit charge between them. Work has to be done on the charge when it is taken from a low potential to a high potential. The unit of voltage is the volt (V).

In a circuit diagram the current is considered to flow along thin conductors into and out of components. The current in a connecting wire is equal to the rate of passage of charge past any point on the connector. (In a solid conductor it will be distributed in some way throughout the conductor, a fact which has to be taken into account in formulating the properties of devices.) The unit of charge is the coulomb (C) and of current the ampere (A), usually abbreviated to amp. A current of one amp carries a charge of one coulomb per second.

Since voltage represents work per unit charge, the product of voltage and current measures rate of doing work, or power. The unit of work is the joule (J) and of power the watt (W). Both voltage and current may be, and usually are, varying in time.

EXAMPLE 1.1

(i) A torch bulb is rated at 3.5 V, 0.2 A. What is the power rating?

(ii) A car headlight is rated at 12 V, 50 W. Calculate the current. How much charge is passed in 1 h?

(i) Power = $3.5 \times 0.2 = 0.7$ W.

(ii) Current = $50/12 = 4.2$ A. The current is the charge (in coulombs) passed each second. Hence charge in 1 h = $(50/12) \times 3600 = 15\,000$ C

1.3 ♦ Circuit components

A circuit component is considered to have two terminals for connection. The two relevant variables are the voltage, $v(t)$, between these terminals and the current, $i(t)$, flowing through them. The property of the component is defined by the relationship between those two variables. The conventions adopted for direction and sign are illustrated in Figure 1.1. The arrow next to $v(t)$ indicates that the numerical value of $v(t)$ is positive when the terminal at the head of the arrow is more positive than the terminal at the foot. The arrow against $i(t)$ indicates that $i(t)$ will be positive when the net flow of charge is in the direction of the arrow. (In fact, the mobile charge carriers in conductors are electrons, which have a negative charge. The flow of the actual charge carriers is therefore in the opposite direction to the conventional current flow.) It is further conventional that the relative directions of $v(t)$ and $i(t)$ shown in the figure are adopted in the definition of the component. With these conventions positive voltage and positive current will result in power being supplied to the component.

1.3.1 ◆ Resistance

For an ideal resistor of resistance R

$$v(t) = Ri(t) \tag{1.1}$$

This equation is referred to as Ohm's law. With the above conventions an 'ordinary' resistor will have a positive value of resistance. The unit of resistance is the ohm (Ω), corresponding to a voltage drop of one volt when carrying one amp. The circuit symbol for resistance is shown in Figure 1.2(a).

It is useful to have the concept of the conductance of a resistor, defined as $G = 1/R$. This is equivalent to rewriting Equation 1.1 in the form

$$i(t) = Gv(t) \tag{1.2}$$

The unit of conductance is the siemens (S), being the conductance of a one ohm resistance.

It will be noted that, since R is an algebraic constant, in the case of a resistor voltage and current vary in time in the same way: they differ only in magnitude.

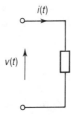

Figure 1.1 Showing the conventional directions for voltage and current with a two-terminal component.

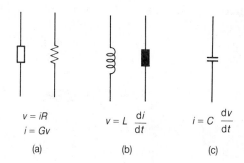

$$v = iR$$
$$i = Gv$$

(a)

$$v = L \frac{di}{dt}$$

(b)

$$i = C \frac{dv}{dt}$$

(c)

Figure 1.2 Circuit symbols for (a) resistance, (b) inductance and (c) capacitance. In (a) and (b), the left-hand symbol is 'preferred', the right-hand symbol 'permissible'.

EXAMPLE 1.2 Calculate (i) the resistance of the light bulbs of Example 1.1 and (ii) the resistance of a 240 V, 1 kW electric heating element.

(i) In Example 1.1 (i), resistance = 3.5/0.2 = 17.5 Ω; in Example 1.1 (ii), resistance = 12/(50/12) = 2.88 Ω.

(ii) The current is 1000/240, so that the resistance is given by 240 × 240/1000 = 57.6 Ω.

1.3.2 ◆ Inductance

For an ideal inductor of inductance L

$$v(t) = L \frac{\mathrm{d}}{\mathrm{d}t} i(t) \tag{1.3}$$

The unit of inductance is the henry (H). A current increasing at the rate of one amp per second flowing through an inductance of one henry will produce a voltage drop of one volt. The symbol for inductance is shown in Figure 1.2(b).

EXAMPLE 1.3 A coil having inductance 0.5 H is connected across a 12 V battery. How long does it take for the current to reach 1 A?

$$0.5 \frac{di}{dt} = 12$$

Integrating, we find

$$i = (12/0.5)t = 24t$$

In this equation i is in amps and t in seconds. Hence 1 A is reached in $1/24 = 0.042$ s.

1.3.3 ◆ Capacitance

For an ideal capacitor of capacitance C

$$i(t) = C\frac{\mathrm{d}}{\mathrm{d}t}v(t) \tag{1.4}$$

In terms of the charge $q(t)$ stored on the capacitor this relation may be written

$$q(t) = Cv(t) \tag{1.5}$$

$$i(t) = C\frac{\mathrm{d}}{\mathrm{d}t}q(t)$$

The unit of capitance is the farad (F) and corresponds to a charge of one coulomb stored at a voltage of one volt. It is a very large unit and capacitors are more likely to be found labelled in microfarads or smaller submultiples. The circuit symbol is shown in Figure 1.2(c).

> **EXAMPLE 1.4** A capacitor of capacitance 0.01 F is charged to 20 V. Calculate the charge stored. If the charge starts to drain away through a circuit taking 0.2 A, calculate the drop in volts over a period of 0.1 s.
>
> Using Equation 1.5
>
> $$\text{charge} = 20 \times 0.01 = 0.2 \text{ C}$$
>
> Using Equation 1.4
>
> $$\frac{\mathrm{d}v}{\mathrm{d}t} = 0.2/0.01 = 20$$
>
> In this equation the right-hand side has units of volts per second. Hence over 0.1 s the drop is
>
> $$20 \times 0.1 = 2 \text{ V}$$

1.3.4 ◆ Mutual inductance

As has been mentioned earlier, the phenomenon of mutual induction was discovered before that of self-induction. The mutual induction between two inductors is characterized by a mutual inductance, M, defined in a manner similar to self-inductance but with the current referred to one inductor and the voltage to the other. This is illustrated in Figure 1.3. As with self-inductance the unit of mutual inductance is the henry. Field theory used to calculate mutual inductance shows that the same value of M applies to both the situations shown. The sign of M depends on the choice of terminals used to define the conventional direction of voltage. It is a common practice to mark terminals with a dot, as in Figure 1.4, to indicate directions with which M will be positive. The equations for this circuit are

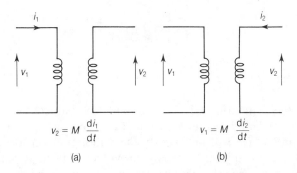

Figure 1.3 Illustrating the significance of the mutual inductance, *M*, between two magnetically coupled inductances.

$$v_1 = L_1 \frac{d}{dt} i_1 + M \frac{d}{dt} i_2$$

$$v_2 = M \frac{d}{dt} i_1 + L_2 \frac{d}{dt} i_2$$

(1.6)

EXAMPLE 1.5 A coil of inductance 0.2 H is coupled to a coil of inductance 5 H to give a mutual inductance of 0.9 H. A current of 1 A flows through the 0.2 H inductor, with the 5 H inductor unconnected. The current is interrupted and falls linearly to zero in 0.01 s. Calculate the voltage appearing across the 5 H inductor.

Taking the moment of interruption as $t = 0$, we have for the interval $0 < t < 0.01$.

$$i_1 = 1.0 - t/0.01, \quad i_2 = 0$$

Hence, referring to Equations 1.6,

$$v_2 = 0.9 \frac{d}{dt} i_1 = 0.9 \times (-100) = -90 \text{ V}$$

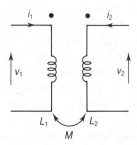

Figure 1.4 The 'dot' notation used to define connections giving a positive value to *M*.

Outside this interval $v_2 = 0$. The sign is only relevant if we have to measure the voltage in the correct sense to make M positive.

1.4 ◆ Linearity

If in any of the Equations 1.1, 1.3, 1.4 and 1.6 characterizing the various components $v(t)$ and $i(t)$ are particular functions satisfying the equation, then $Kv(t)$ and $Ki(t)$ will also satisfy the same equation for any value of the constant K, positive or negative, providing that R, L and C do not depend on current. Such equations are termed linear, and a network for which the components are described by linear equations is termed a linear network. If the network is composed solely of resistance, inductance and capacitance, none of which of themselves produce energy, it is also termed passive.

1.5 ◆ Sources

When a current flows round a circuit, energy is consumed in overcoming resistance to the passage of charge or is transformed into magnetic or electrostatic energy in inductors or capacitors. This energy is provided by the source. Basically, a source is some device which increases the energy of a charge passing through it, measured by the voltage difference between the terminals. This energy can originate in various forms: it might be as a chemical reaction (as in common 'voltaic' cells) or as mechanical work provided by the steam turbine driving a generator. The origin matters little: as far as the circuit is concerned only the increase in energy of a unit charge going from one terminal to the other is relevant. One type of ideal source is easily imagined as one in which a unit charge gains the same amount of energy whatever the rate of passage of charge. Such a source is termed a **constant voltage** source. When such a source is connected in a circuit, the current which flows for a particular source voltage is fixed by the circuit. Another type of ideal source can be imagined which delivers charge at a fixed rate, however much work is required. A practical example illustrating this concept is the electrostatic generator in which charge is 'sprayed' onto a moving belt of insulating material and thus carried to an electrode where the charge is removed from the belt. The source current is literally charge density multiplied by the velocity of the belt. Such a source is termed **constant current**. Neither type is strictly realizable in practice but, just as circuits can be modelled in terms of resistance, inductance and capacitance, so can real cells be modelled by either of these sources in combination with other components. As will be seen, the choice of model depends on convenience. We now consider the formal definitions of these ideal sources.

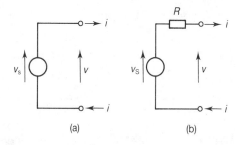

Figure 1.5 The ideal voltage source: (a) symbol and definition, $v = v_s$ independently of the value of i; (b) the addition of a resistance to simulate the behaviour of a real cell.

1.5.1 ♦ Constant voltage source

This is defined by the property that the voltage between the terminals of the source is independent of the current it delivers. In terms of the variables shown in the diagram of Figure 1.5(a), for all values of i

$$v = v_s \tag{1.7}$$

In this definition the word 'constant' merits clarification: the implied constancy refers to variation with respect to the current delivered by the source, not variation in time. The source strength v_s will often be a function of time.

   Such an ideal source is not physically realizable, but in combination with passive components may be used to model the behaviour of a real source. For example, a source such as a dry cell or an accumulator shows a fall of voltage as increasing currents are supplied. Such behaviour can be modelled by a circuit containing a constant voltage source in series with a resistance, as shown in Figure 1.5(b). The resistance is known as the internal resistance of the cell.

1.5.2 ♦ Constant current source

This source is defined by the property that the current delivered is independent of the voltage between the terminals. In terms of the variables in Figure 1.6(a), for all values of v

$$i = i_s \tag{1.8}$$

As with the constant voltage source the 'constant' in the description refers to constancy with variation of voltage, not time: the source strength i_s may well be a function of time.

   It is relevant to note that the cell modelled by a constant voltage source and internal resistance, shown in Figure 1.5(b), can equally well be

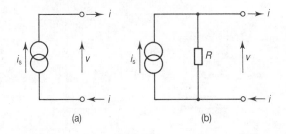

Figure 1.6 The ideal current source: (a) symbol and definition, $i = i_s$ independently of the value of v; (b) the addition of a resistance to simulate the behaviour of a real cell.

modelled using a constant current source as in Figure 1.6(b). When connected to an external circuit the cell is characterized as far as that external circuit is concerned only by the relation between v and i: for the circuit of Figure 1.5(b) this relation is

$$v = v_s - Ri \tag{1.9}$$

For the circuit of Figure 1.6(b) the relation is

$$i = i_s - \frac{v}{R} \tag{1.10}$$

This can equally well be written

$$v = Ri_s - Ri$$

which is identical to Equation 1.9 if the resistance is the same in both cases and

$$v_s = Ri_s$$

If these conditions are satisfied no measurement of behaviour external to the cell could differentiate between the two circuit models: either would describe the behaviour correctly. This freedom to replace one model by the other can often be used to great advantage in the analysis of circuits. It is important to note that what goes on inside the cell is not described by either of the models.

EXAMPLE 1.6 The voltage between the terminals of a power supply is 20 V when no current is drawn, dropping to 16 V for a current of 0.5 A. Determine the internal resistance of the supply. Give models for the supply in terms of (i) a constant voltage source and (ii) a constant current source.

A drop of voltage of 4 V for a current of 0.5 A implies an internal resistance of $4/0.5 = 8\ \Omega$.

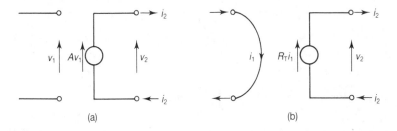

Figure 1.7 Dependent voltage sources: (a) voltage controlled, voltage gain A; (b) current controlled, transfer resistance R_T.

(i) The model of the form of Figure 1.5(b) has $v_s = 20$ V, $R = 8\ \Omega$.

(ii) For the form of Figure 1.6(b), $i_s = 20/8 = 2.5$ A, $R = 8\ \Omega$.

1.5.3 ◆ Dependent sources

The sources described in the previous sections are considered to be of prescribed source strength, independent of any other variable in the circuit of which they are part. A dependent source may be either constant voltage or constant current, but the source strength is controlled by another circuit variable. They therefore have more than two terminals. In Figure 1.7 are shown two forms of constant voltage source, the one controlled by an input voltage v_1 and the other by an input current i_1. The constant A in Figure 1.7(a) is dimensionless, and might be termed **voltage gain**. The constant R_T in Figure 1.7(b) has dimensions of resistance, and is termed a **transfer resistance** since the voltage and current to which it refers are in different parts of the circuit. In Figure 1.8 are shown constant current sources, the one voltage controlled and the other current controlled. The constant G_T in Figure 1.8(a) has the dimensions of conductance and similarly to the transfer resistance is termed **transfer conductance**. The constant B in Figure 1.8(b) is dimensionless and might be termed **current gain**.

As shown, the voltage-controlled source takes zero input current and the current-controlled source causes zero voltage drop in the input circuit to which it is connected. These conditions can be altered by the addition of other components.

In all of the above circuits either the input voltage or the input current is zero so that no power is supplied by the input circuit. The power supplied by the source to the output circuit must therefore have some other origin. Thus although dependent sources are linear they must be classed as active and not passive. When, for example, a dependent source is used to model a transistor in a transistor amplifier, the circuit power would derive ultimately from the battery supplying the amplifier.

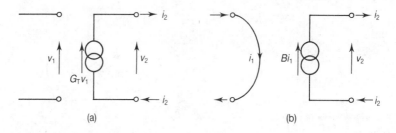

Figure 1.8 Dependent current sources: (a) voltage controlled, transfer conductance G_T; (b) current controlled, current gain B.

1.6 ◆ Energy

As mentioned in Section 1.1 each of the properties of resistance, inductance and capacitance has a connection with energy. For any two-terminal component, with voltage and current directions defined in Figure 1.1, energy is being supplied to the component at a rate

$$p = vi \tag{1.11}$$

If in this expression v and i are varying with time p represents the instantaneous rate of energy input, or the power. The unit of power is the watt (W). The total energy supplied over a given time can be found by integration.

1.6.1 ◆ Resistance

For an ideal resistor of resistance R

$$v = iR$$

so that

$$p = i^2 R \tag{1.12}$$

In terms of conductance this may be written

$$p = v^2 G \tag{1.13}$$

The electrical power dissipated within the resistor appears as heat. Resistance is termed a dissipative element since it causes irrevocable loss of electrical energy.

1.6.2 ◆ Inductance

For an ideal inductor of inductance L

$$v = L\frac{\mathrm{d}i}{\mathrm{d}t}$$

so that

$$p = Li\frac{di}{dt}$$

The total energy supplied over the interval from t_1 to t_2 is given by

$$W = \int_{t_1}^{t_2} p\, dt$$

$$= L \int_{t_1}^{t_2} i\frac{di}{dt} dt$$

Denoting the currents at t_1, t_2 by i_1, i_2 and changing the variable to i, this integral may be written

$$W = L \int_{i_1}^{i_2} i\, di$$

$$= \tfrac{1}{2}L(i_1^2 - i_2^2)$$

Thus the energy which must be provided to increase the current in an inductance from zero to a value i is

$$W = \tfrac{1}{2}Li^2 \tag{1.14}$$

This energy is stored in the magnetic field within the inductor, and is recoverable in electrical form.

EXAMPLE 1.7 For the case described in Example 1.3, calculate the energy stored in the inductor when the current has reached 1 A. Verify that this is the same as the energy drawn from the battery.

For the stored energy we have

$$W = \tfrac{1}{2} \times 0.5 \times 1^2 = 0.25 \text{ J}$$

The battery provides energy at the rate $12 \times iW$, so that the total energy supplied between $t = 0$ and $t = 1/24$ is given by

$$\int_0^{1/24} 12 \times 24t\, dt = 12 \times 24 \times \tfrac{1}{2} \times (\tfrac{1}{24})^2 = 0.25 \text{ J}$$

1.6.3 ◆ Mutual inductance

The magnetic field, and hence the stored energy, in the presence of two mutually coupled inductors depends on the current in each. The value of the stored energy can be found in the same way as for a single inductor, by considering the power provided by the source supplying each inductor. Using Equations 1.6, at a time when the voltages are v_1, v_2 corresponding to currents i_1, i_2, the total power input is given by

$$p = v_1 i_1 + v_2 i_2 = i_1\left(L_1\frac{d}{dt}i_1 + M\frac{d}{dt}i_2\right) + \left(M\frac{d}{dt}i_1 + L_2\frac{d}{dt}i_2\right)$$

$$= \frac{1}{2}\frac{d}{dt}(L_1 i_1^2 + L_2 i_2^2 + 2Mi_1 i_2)$$

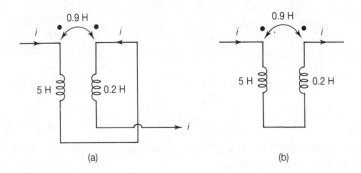

Figure 1.9 Example 1.8, showing two alternative connections with coupled inductances.

Hence

$$\int_{t_1}^{t_2} p\,\mathrm{d}t = [\tfrac{1}{2}(L_1 i_1^2 + L_2 i_2^2 + 2Mi_1 i_2)]_{t_1}^{t_2}$$

If at t_1 both i_1 and i_2 are zero, corresponding to zero stored energy, for other values of the currents

$$W = \tfrac{1}{2}(L_1 i_1^2 + L_2 i_2^2 + 2Mi_1 i_2) \tag{1.15}$$

In this expression M must be given its correct sign in relation to the direction of the currents.

> **EXAMPLE 1.8** The two inductors of Example 1.5 are connected in series and then carry a current of 1 A. Find the two possible values of stored energy.
>
> Using Equation 1.15 we have
>
> $$W = \tfrac{1}{2}(0.2 \times 1^2 + 5 \times 1^2 \pm 2 \times 0.9 \times 1 \times 1) = 2.6 \pm 0.9$$
>
> The possible values are therefore 3.5 J or 1.7 J. Using the dot notation introduced in Section 1.3.4, the value of 3.5 J applies to the connection illustrated in Figure 1.9(a); the lower value applies to the connection in Figure 1.9(b).

1.6.4 ◆ Capacitance

For an ideal capacitor of capacitance C

$$i = C\frac{\mathrm{d}v}{\mathrm{d}t}$$

By a similar process to that used in the previous section, the energy which must be supplied to raise the voltage from v_1 to v_2 is

$$W = C \int_{v_1}^{v_2} v \, dv$$

$$= \tfrac{1}{2}C(v_2^2 - v_1^2)$$

To raise the voltage from zero to a value v requires

$$W = \tfrac{1}{2}Cv^2 \tag{1.16}$$

This energy is stored in the electrostatic field within the capacitor, and is recoverable.

EXAMPLE 1.9 Calculate the energy stored in the capacitor of Example 1.4.

We have

$$W = \tfrac{1}{2} \times 0.01 \times (20)^2 = 2 \, \text{J}$$

1.7 ◆ Summary

Component	Characterizing equation
Resistance	$v = iR$
Conductance	$i = vG$
Inductance	$v = L(di/dt)$
Capacitance	$i = C(dv/dt)$

All units are those of the International System (SI). Specifically electrical units are as follows:

Voltage	volt (V)	
Charge	coulomb (C)	
Current	amp (A)	$C\,s^{-1}$
Work/energy	joule (J)	$C\,V$
Power	watt (W)	$J\,s^{-1}, V\,A$
Resistance	ohm (Ω)	$V\,A^{-1}$
Conductance	siemens (S)	$A\,V^{-1}$
Inductance	henry (H)	$V\,s\,A^{-1}$
Capacitance	farad (F)	$C\,V^{-1}$

The following common multiples are used:

Giga	G	10^9
Mega	M	10^6
Kilo	k	10^3
Milli	m	10^{-3}

Micro	μ	10^{-6}
Nano	n	10^{-9}
Pico	p	10^{-12}

Energy is related to a circuit's properties as follows:

Resistance: energy dissipated at rate $i^2 R = v^2 G$.
Inductance: stored energy is $\frac{1}{2} L i^2$.
Capacitance: stored energy $\frac{1}{2} C v^2$.

PROBLEMS

1.1 Determine the current flowing and the power dissipated when a steady voltage of 12 V is maintained across resistors of the following values: 470 Ω, 2.2 kΩ, 56 kΩ.

1.2 A resistor of 4.7 MΩ forms part of a chain carrying a steady current of 17 μA. Determine the voltage across it and the power dissipated.

1.3 An inductor of 50 mH is connected across a 15 V supply. Calculate the current 1 ms after connection. How long would it take to reach the same current if the inductance were 2 H?

1.4 For both the cases in Problem 1.3 calculate the energy stored after the 1 ms interval.

1.5 An inductor of 100 mH is carrying a steady current of 0.5 A. Interruption of the circuit causes the current to fall according to $i(t) = 0.5 \exp(-1000t)$. Obtain an expression for the voltage appearing across the inductor.

1.6 Calculate the charge and energy stored when a steady voltage of 45 V is maintained across capacitors of the following values: 10 000 μF, 470 μF, 2.2 μF.

1.7 In a sample-and-hold circuit a capacitor of 100 pF is charged over a period of 1.2 μs by a current of 70 μA. Calculate the final voltage appearing across the capacitor.

1.8 The voltage of a dry cell drops from 9 V on open circuit to 8.3 V when a current of 25 mA is drawn. Estimate the internal resistance of the cell.

1.9 A constant current source is simulated in a transistor amplifier by a resistor of 680 kΩ connected to the 9 V power supply line. Express this combination in terms of a constant current source and a resistance.

1.10 One capacitor of capacitance 2 μF is charged to 10 V. A similar but uncharged capacitor is connected in parallel. Assuming that charge is conserved, calculate the final voltage. Calculate also the initial and final values of the total stored energy. By considering the likely consequence of performing the experiment, make suggestions to account for the loss of energy.

2 ◆ Waveforms

The variation in time of a current or a voltage is described by a waveform. In this chapter are considered ways by which waveforms can be expressed in a quantitative fashion, and how the various useful averages are defined and calculated. It concludes by considering the calculation of power.

2.1 ◆ Introduction

The waveforms of voltage and current sources driving circuits vary very widely in characteristics. At one extreme are the highly regular and predictable waveforms associated with test equipment (and with which the majority of calculations are concerned); at the other are the irregular waveforms associated with, for example, speech or video signals. The latter are predictable only in a statistical sense. The way in which the response of circuits to these various waveforms may be calculated will be considered in later chapters; the present chapter looks at some of the waveforms and at various averages which may be usefully employed to describe them. It also looks at the calculation of the power associated with such waveforms. Whereas voltage and current waveforms are the direct outcome of circuit calculations, the calculation of power involves the multiplication of two waveforms, a non-linear operation. Both voltage and current waveforms can be displayed and examined using the cathode ray oscilloscope, but because of the need to multiply waveforms power is more difficult to determine. A related problem arises when the currents from two different sources are combined: whereas the resultant current is a simple sum, the resultant power dissipation is in general not simply related to the powers that each current would produce acting by itself. This again is a consequence of the non-linear operation involved.

2.2 ◆ Some waveforms

In the following sections some features of a number of waveforms are introduced. Basically a waveform can only be described by its shape. For waveforms of any degree of regularity mathematical expressions can usually be found to convey the information, although different expressions may be needed for different time intervals. Firstly are considered cases which are perfectly regular in time, and then some examples of 'one-off', or transient, situations.

2.2.1 ◆ Direct current

When the voltages and currents in a circuit do not vary with time, we are said to be dealing with **direct current**, or DC. Thus an ordinary dry cell or accumulator provides DC. The abbreviation DC is commonly used in an adjectival sense to describe a system operating in this way, or indeed to describe a voltage as in 'DC voltage'. It may be observed that DC is an idealization in that supplies have to be connected at some definite time: the steady state situation takes some time to set up.

2.2.2 ◆ Alternating current

Probably the most common situation exists when the circuit is excited by a waveform which varies sinusoidally in time. As with DC, such a sinusoidal excitation has to be applied at some definite time but, as with DC, a steady state is finally set up. For a linear circuit all voltages and currents in this steady state will vary sinusoidally in time, with waveforms typically of the form

$$v(t) = \hat{V} \cos(\omega t + \phi)$$

In this situation we are said to be dealing with **alternating current**, or AC. This abbreviation too is used as an adjective, as in 'AC voltage'. In this equation the constant ω is the angular frequency, measured in radians per second (rad s^{-1}). It is more usually specified in terms of the frequency, f, given by

$$f = \omega/2\pi$$

Frequency is measured in cycles per second, or hertz (Hz). The reciprocal of frequency is equal to the periodic time, T:

$$T = 1/f$$

The distinction between the various voltages and currents lies in the other two constants, $\hat{V}$ and ϕ. The constant $\hat{V}$ is termed the peak voltage: it is the greatest value, positive and negative, taken by the voltage over the

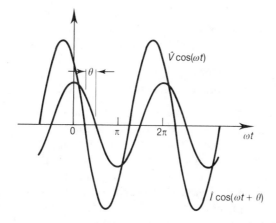

Figure 2.1 Typical steady-state AC voltage and current waveforms: the voltage has peak value $\hat{V}$, the current peak value $\hat{I}$, and the current leads the voltage by the angle θ.

course of one cycle. The constant ϕ cannot be given a value until an origin for time is given. Since no absolute origin for time exists, it is customary to take one voltage, usually the exciting source, as a reference for all other variables with the form

$$v(t) = \hat{V} \cos(\omega t) \tag{2.1}$$

A related current in the circuit would then be described in the form

$$i(t) = \hat{I} \cos(\omega t + \theta) \tag{2.2}$$

These two waveforms are illustrated in Figure 2.1. In Equation 2.2, $\hat{I}$ is the peak current and θ is the phase of $i(t)$ with respect to $v(t)$. On account of the periodicity of the cosine function the value of θ is restricted to the range $\pi > \theta > -\pi$. If $\theta > 0$ the current is said to lead the voltage, if $\theta < 0$ it is said to lag the voltage. Phase is an angle which may be specified in degrees or in radians.

It was said at the beginning of this section that AC was a commonly encountered waveform. This is for two reasons: one the fact that in a linear circuit with AC excitation all voltages and currents have the same basic shape, differing only in peak value and relative phase; the other is that it can be shown that any waveform can be expressed as an appropriate combination of sine waves. This process is known as Fourier analysis, and is studied in Chapter 11. For networks containing non-linear elements sine wave excitation will not normally produce sine wave variation throughout the network.

The range of frequencies encountered is very wide: power generation and distribution takes place at 50 Hz or 60 Hz; radio frequencies extend from 15 kHz to 150 MHz; TV frequencies go up to 1000 MHz (1 GHz); terrestrial and satellite communication operate up to 100 GHz. The techniques of circuit analysis are used over the whole of this range.

2.2.3 ◆ Periodic non-sinusoidal waveforms

This is a very general category whose only common property is periodicity. Some examples are shown in Figure 2.2. Figure 2.2(a) shows a square waveform. Such a waveform has uses for diagnostic purposes. With equal on and off periods of perhaps 0.5 μs it might be found in digital switching circuits. In Figure 2.2(b) is shown a triangular waveform; Figure 2.2(c) shows a recurrent pulse waveform, such as might be found in a radar system. In this application the pulse length might be 0.5 μs with a recurrence period of 1 ms. Two other waveforms likely to be encountered which are derived from sine waves are shown in Figures 2.2(d) and 2.2(e). The first is described as a half-wave rectified sine wave, the second as a full-wave rectified sine wave. These are all fairly common waveforms, but the possibilities are infinite. It is to be noted that, although a linear network may be excited by one of these waveforms, the voltages and currents elsewhere will not in general be of the same shape.

2.2.4 ◆ Transient waveforms

A transient is essentially something of finite, usually short, duration and occurs sufficiently infrequently to be regarded as a single event. A naturally occurring example is a lightning strike in which the peak current flows for only a few microseconds. In radar systems the pulse illustrated in Figure 2.2(c) is so short in comparison with the interval between pulses that it can be considered as a single transient. Another example is provided when an electrical circuit is switched from one steady state to another: the change-over will be accompanied by a transient.

Transients are important in electrical circuits, and certain ideal waveforms are often used in analysis: some examples are shown in Figure 2.3. The waveform shown in Figure 2.3(a) is known as the **unit step** and is typical of a switching situation. In Figure 2.3(b) is shown a **ramp** waveform sometimes used in the analysis of control systems.

2.2.5 ◆ Statistical waveforms

The waveforms associated with signals such as speech have no simple representation by mathematical expressions. Information about them has to be provided by statistical methods and used in conjunction with a knowledge of circuit response to sine wave excitation over a wide frequency range.

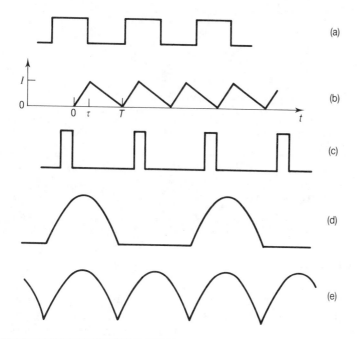

Figure 2.2 Some periodic, non-sinusoidal waveforms: (a) square wave; (b) triangular wave; (c) pulse waveform; (d) half-wave rectified sine wave; (e) full-wave rectified sine wave.

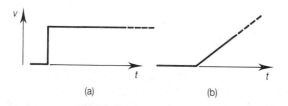

Figure 2.3 Some transient waveforms: (a) step; (b) ramp.

2.3 ♦ Averages related to periodic waveforms

Although complete information about a waveform can only be conveyed by its shape, averages often give useful information. Two averages in particular are relevant, the 'straight' average of the voltage or current and the average of the square of the waveform.

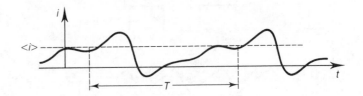

Figure 2.4 Illustrating the significance of the average $\langle i \rangle$ of a periodic current waveform $i(t)$.

2.3.1 ◆ Average value

Consider a current waveform $i(t)$ periodic in time T. The average, denoted by $\langle i \rangle$, is calculated from

$$\langle i \rangle = \frac{1}{T} \int_0^T i(t) \, dt \tag{2.3}$$

In this case a simple physical interpretation of the average is possible: since the integral gives the total charge transferred in the cycle, the average $\langle i \rangle$ is equal to the direct current which would transfer the same charge as $i(t)$ over the period of the waveform. This is illustrated graphically in Figure 2.4: $\langle i \rangle$ is the height of the rectangle whose area is equal to that under the waveform.

2.3.2 ◆ Mean square

The average of the square of the waveform, or the mean square, is denoted by $\langle i^2 \rangle$ and is calculated from

$$\langle i^2 \rangle = \frac{1}{T} \int_0^T i^2(t) \, dt \tag{2.4}$$

One physical example for which the mean square is relevant is the calculation of the average power dissipated in a resistor carrying the current $i(t)$: the instantaneous rate of dissipation of energy in a resistance R is

$$p(t) = Ri^2(t)$$

Thus the average power, which would measure the heating effect, is equal to

$$\langle p(t) \rangle = P = \langle i^2 \rangle R$$

A measure of magnitude for a waveform of which the average $\langle i \rangle$ is zero is provided by the root-mean-square (RMS) value given by

$$i_{\mathrm{RMS}} = (\langle i^2 \rangle)^{1/2} \tag{2.5}$$

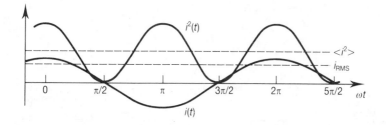

Figure 2.5 Illustrating the significance of the mean-square $\langle i^2 \rangle$ and RMS i_{RMS} for a sinusoidal waveform.

As a measure of magnitude this has the advantages that it is never zero except for a zero waveform, and that it is directly related to average power by

$$P = Ri^2_{RMS}$$

The two averages defined by Equations 2.3 and 2.4 have to be determined by application of those equations to the particular waveform. Some examples will now be given.

2.3.3 ◆ Calculations

Direct current
It requires little calculation to see that for a steady current i_0

$$\langle i \rangle = i_0, \ \langle i^2 \rangle = i_0$$

Hence

$$\langle i \rangle = i_{RMS} = i_0$$

This is the only waveform for which the average and RMS are equal.

Sinusoidal waveform
For this case

$$i(t) = \hat{I} \cos(\omega t), \ \omega = 2\pi/T$$

Consideration of the graph of $\cos(\omega t)$, or evaluation of the integral in Equation 2.3, shows that the average value $\langle i \rangle$ is zero.

The time variation of $i^2(t)$ is shown graphically in Figure 2.5, illustrating the identity

$$\cos(\omega t) = \tfrac{1}{2}[1 + \cos(2\omega t)]$$

Using the fact that the average of $\cos(2\omega t)$ is zero, we find

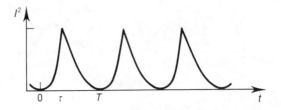

Figure 2.6 Graph of $i^2(t)$ for the triangular waveform of Figure 2.2(b).

$$\langle i^2 \rangle = \frac{1}{T} \int_0^T \hat{I}^2 \cos^2(\omega t)\, dt$$

$$= \tfrac{1}{2} \hat{I}^2 \tag{2.6}$$

From this result we have

$$i_{RMS} = \frac{1}{\sqrt{2}} \hat{I} \tag{2.7}$$

A triangular waveform

Using the notation of Figure 2.2(b) and the formula for the area of a triangle, $(1/2) \times$ base $\times$ height, we have

$$\langle i \rangle = \tfrac{1}{2} \hat{I}$$

The current waveform is described mathematically by the equations

$$i(t) = \hat{I}\frac{t}{\tau} \qquad 0 < t < \tau$$

$$= \hat{I}\left(1 - \frac{t - \tau}{T - \tau}\right) \quad \tau < t < T$$

A graph of $i^2(t)$ is shown in Figure 2.6. The area under the graph between 0 and τ is given by

$$\hat{I}^2 \int_0^\tau \left(\frac{t}{\tau}\right)^2 dt = \hat{I}^2 \frac{\tau}{3}$$

Since the graph between τ and T is the same parabolic shape and has the same peak height, the same result can be used to calculate the area between τ and T, giving

$$\tfrac{1}{3} \hat{I}^2 (T - \tau)$$

Summing, the total area is equal to $\hat{I}^2 T/3$. The mean square is therefore given by

$$\langle i^2 \rangle = \tfrac{1}{3} \hat{I}^2$$

and the RMS value by

$$i_{RMS} = \frac{1}{\sqrt{3}} \hat{I}$$

It will be noticed that this is different from the result obtained for the sinusoidal waveform (Equation 2.7). The RMS has to be evaluated for each waveform.

EXAMPLE 2.1 A current waveform is periodic with a period of 2 s. Its value in amps is given by the following expressions:

$$0 < t < 1 \qquad i = 0.2 + 0.6t$$

$$1 < t < 2 \qquad i = 0$$

Determine the mean and RMS values.

$$\langle i \rangle = \frac{1}{2} \int_0^2 i(t) \, dt$$

$$= \frac{1}{2} \int_0^1 (0.2 + 0.6t) \, dt$$

$$= \tfrac{1}{2} [0.2t + 0.3t^2]_0^1$$

$$= 0.25$$

$$\langle i^2 \rangle = \frac{1}{2} \int_0^1 (0.2 + 0.6t)^2 \, dt$$

$$= \frac{1}{2} \int_0^1 (0.04 + 0.24t + 0.36t^2) \, dt$$

$$= \tfrac{1}{2} (0.04 + 0.12 + 0.12) = 0.14$$

$$i_{RMS} = 0.374$$

2.3.4 ◆ RMS values for a composite waveform

It sometimes happens that a current, through a resistor for example, has two distinct components. In general the mean square for such a composite waveform is not simply related to the mean squares of the separate components, although on occasion this may be the case. Consider the current $i(t)$ expressed in the form

$$i(t) = i_1(t) + i_2(t)$$

We have

$$\langle i^2 \rangle = \frac{1}{T} \int_0^T [i_1(t) + i_2(t)]^2 \, dt$$

$$= \frac{1}{T} \int_0^T [i_1^2(t) + i_2^2(t) + 2i_1(t)i_2(t)] \, dt$$

$$= \langle i_1^2 \rangle + \langle i_2^2 \rangle + 2\langle i_1 i_2 \rangle \qquad\qquad \textbf{(2.8)}$$

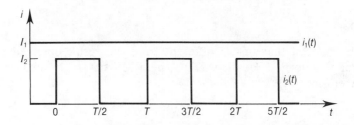

Figure 2.7 Constituents of the composite waveform used in Section 2.3.4.

Only if the average of the product i_1i_2 is zero is the mean square of the composite waveform equal to the sum of the separate mean squares.

As an example consider the waveforms shown in Figure 2.7: $i_1(t)$ is constant at I_1; $i_2(t)$ takes the value I_2 during the interval $0 < t < T/2$ and is zero for $T/2 < t < T$. In this case

$$\langle i_1^2 \rangle = I_1^2$$

$$\langle i_2^2 \rangle = \tfrac{1}{2}I_2^2$$

$$\langle i_1i_2 \rangle = \tfrac{1}{2}I_1I_2$$

Therefore

$$\langle i^2 \rangle = I_1^2 + \tfrac{1}{2}I_2^2 + I_1I_2$$

The waveform $i(t)$ contains an average component of

$$i_2 = \tfrac{1}{2}I_2$$

Consider the following rearrangement: define a current $i_3(t)$ as the constant current $I_1 + I_2/2$ and $i_4(t)$ as a square wave oscillating about zero between $I_2/2$ and $-I_2/2$. The average of $i_4(t)$ is zero. With these definitions

$$i(t) = i_1(t) + i_2(t) = i_3(t) + i_4(t)$$

We have

$$\langle i_3^2 \rangle = (I_1 + \tfrac{1}{2}I_2)^2$$

$$\langle i_4^2 \rangle = \tfrac{1}{4}I_2^2$$

$$\langle i_3i_4 \rangle = \langle (I_1 + \tfrac{1}{2}I_2)i(t) \rangle$$

$$= (I_1 + \tfrac{1}{2}I_2) \langle i_4 \rangle$$

$$= 0$$

Thus in this case i_3i_4 is zero and

$$\langle i^2 \rangle = (I_1 + \tfrac{1}{2}I_2)^2 + \tfrac{1}{4}I_2^2$$

$$= I_1^2 + \tfrac{1}{2}I_2^2 + I_1 I_2$$

in agreement with the previous result.

This result can be generalized: when $i_1(t)$ is a constant (DC) and $\langle i_2 \rangle$ zero the mean square of the composite waveform is the sum of the individual mean squares.

EXAMPLE 2.2   A voltage waveform is periodic with period 0.05 s, and has values given in volts by the following expressions:

$$0 < t < 0.01 \qquad\qquad v = 10 + 500t$$

$$0.01 < t < 0.05 \qquad\qquad v = 16.25 - 125t$$

Determine the RMS value.

We can regard the waveform as composed of a constant voltage of 10 V and a triangular waveform of peak height 5 V, duration 0.05 s. Applying the results of Equation 2.8,

$$\langle v^2 \rangle = 10^2 + \tfrac{1}{3}5^2 + 2 \times 10 \times (\tfrac{1}{2} \times 5) = 158.3$$

Hence

$$v_{\text{RMS}} = 12.6 \text{ V}$$

2.3.5 ♦ RMS values for a sum of sine waves

An important result of general importance concerns waveforms which can be expressed as the sum of sine waves whose frequencies are different integer multiples of a fundamental frequency. Consider a composite waveform

$$i(t) = \hat{I}_1 \cos(\omega t + \theta_1) + \hat{I}_2 \cos(2\omega t + \theta_2) + \ldots$$
$$+ \hat{I}_n \cos(n\omega t + \theta_n) + \ldots$$

For the square of the current waveform we have an expression of the form

$$i^2(t) = \hat{I}_1^2 \cos^2(\omega t + \theta_1) + \hat{I}_2^2 \cos^2(2\omega t + \theta_2) + \ldots$$
$$+ 2\hat{I}_1 \hat{I}_2 \cos(\omega t + \theta_1) \cos(2\omega t + \theta_2) + \ldots$$
$$+ 2\hat{I}_m \hat{I}_n \cos(m\omega t + \theta_m) \cos(n\omega t + \theta_n) + \ldots$$

in which m, n are necessarily different integers. Consider first the average of the typical term in the third row, the average of the expression

$$\cos(m\omega t + \theta_m) \cos(n\omega t + \theta_n)$$

Using the standard trigonometrical formula, this expression can be put into the form

$$\tfrac{1}{2}\{\cos[(m+n)\omega t + \theta_m + \theta_n] + \cos[(m-n)\omega t + \theta_m - \theta_n]\}$$

Since m and n are different the average of each term of this expression is always zero. Hence, using the result derived above that the average of the square of a cosine is $1/2$, we have

$$\langle i^2(t) \rangle = \tfrac{1}{2}(\hat{I}_1^2 + \hat{I}_2^2 + \ldots + \hat{I}_n^2 + \ldots) \qquad (2.9)$$

We can therefore conclude that, for a current composed of a sum of sine waves, the mean square current is the sum of the mean squares of the individual terms.

EXAMPLE 2.3 Find the RMS current for the waveform

$$i(t) = 5\cos(\omega t) + 2\sin(3\omega t)$$

in which the current is given in amps.

Applying the above result, we have

$$\langle i^2 \rangle = \tfrac{1}{2}(25 + 4) = 29/2$$

$$i_{\text{RMS}} = 3.8 \text{ A}$$

2.4 ◆ Decomposition into DC and AC components

The average $\langle i \rangle$ is often referred to as the **DC component** of the current $i(t)$:

$$i_{\text{DC}} = \langle i \rangle$$

Corresponding to this, an **AC component** is then defined by

$$i_{\text{AC}} = i(t) - \langle i \rangle \qquad (2.10)$$

This current is AC only in the sense that $\langle i_{\text{AC}} \rangle = 0$, not in the strict sense of Section 2.2.2. With these definitions the average of the product $i_{\text{DC}} i_{\text{AC}}$ is given by

$$\langle i_{\text{DC}} i_{\text{AC}} \rangle = i_{\text{DC}} \langle i_{\text{AC}} \rangle = 0$$

Hence

$$\langle i^2 \rangle = i_{\text{DC}}^2 + \langle i_{\text{AC}}^2 \rangle \qquad (2.11)$$

2.5 ◆ Power

The calculation of power bears similarities to the calculation of mean squares in that a product of time-varying functions is involved. In the case of power these are the voltage, $v(t)$, across the load and the current, $i(t)$,

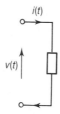

Figure 2.8 Conventional directions of voltage and current associated with a circuit component for which the power vi is dissipated in the component.

through it, as shown in Figure 2.8. At time t energy is being delivered to the load at a rate

$$p(t) = v(t)i(t) \tag{2.12}$$

When the waveforms are periodic then the power is also periodic. In this situation it is frequently the average power that is significant. This average is given by

$$P = \frac{1}{T} \int_0^T v(t)i(t)\, \mathrm{d}t \tag{2.13}$$

The integral in this equation is the total energy supplied to the load during one cycle. For a transient waveform it is likely to be this total energy that matters. In either case the integral has to be determined for the given waveforms.

2.5.1 ♦ Measurement of power

Although this section is concerned with the calculation of the power associated with given waveforms, it is pertinent to comment that power can be found directly from measured waveforms.

To evaluate the integral in Equation 2.13 it is necessary to multiply the two waveforms and then to perform the integration. Given suitable equipment the waveforms can be recorded in digital form and the operation carried out numerically in a computer. For periodic waveforms this can be done in real time. Alternatively, the two waveforms could be fed to an analogue multiplier followed by an analogue integrator: this is the basis of many low-frequency wattmeters.

However, calculation is valuable in many situations as an aid to understanding. It will show, for example, which parts of a waveform in a switching circuit contribute most to power consumption and hence will indicate the steps necessary to minimize that power consumption.

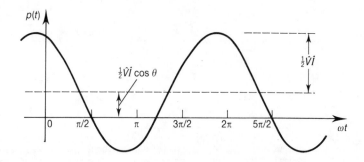

Figure 2.9 Graph showing the instantaneous power associated with voltage $\hat{V}\cos(\omega t)$ and current $\hat{I}\cos(\omega t + \theta)$ (Equation 2.14, Section 2.5.3).

2.5.2 ◆ DC excitation

If both v and i are independent of time then power is also constant, and the average power is given by vi. Thus, if $v(t) = v_0$ and $i(t) = i_0$, then

$$P = v_0 i_0$$

2.5.3 ◆ AC excitation

In this very important case both waveforms are sinusoidal, with representations given by Equations 2.1 and 2.2, repeated below:

$$v(t) = \hat{V}\cos(\omega t)$$
$$i(t) = \hat{I}\cos(\omega t + \theta)$$

The power being delivered to the load at time t is therefore

$$\begin{aligned} p(t) &= \hat{V}\hat{I}\cos(\omega t)\cos(\omega t + \theta) \\ &= \hat{V}\hat{I}\cos(\omega t)[\cos(\omega t)\cos\theta - \sin(\omega t)\sin\theta] \\ &= \tfrac{1}{2}\hat{V}\hat{I}[\cos\theta + \cos(2\omega t)\cos\theta - \sin(2\omega t)\sin\theta] \\ &= \tfrac{1}{2}\hat{V}\hat{I}[\cos\theta + \cos(2\omega t + \theta)] \end{aligned} \tag{2.14}$$

This expression is illustrated in Figure 2.9. The power delivered to the load thus consists of a steady component together with a component varying at twice the frequency of the voltage or current. By use of Equation 2.7 it will be seen that the factor $\hat{V}\hat{I}/2$ is equal to $V_{\mathrm{RMS}}I_{\mathrm{RMS}}$ so that Equation 2.14 can be written

$$p(t) = V_{\mathrm{RMS}}I_{\mathrm{RMS}}[\cos\theta + \cos(2\omega t + \theta)] \tag{2.15}$$

The average rate of delivery of energy is

$$P = \langle p(t) \rangle = V_{\mathrm{RMS}}I_{\mathrm{RMS}}\cos\theta \tag{2.16}$$

This power will be dissipated as either heat or some other form of energy. With a public utility supply it will be what the consumer is charged for. If the current is in phase with the voltage the angle is zero and in this case

$$P = V_{RMS}I_{RMS}$$

This expression is of precisely the same form as for DC, and for this reason it is customary to use the RMS value as the measure for AC voltage and current.

EXAMPLE 2.4 An electric motor is connected to an AC supply of 240 V (RMS). It takes a current of 15 A (RMS) which lags on the voltage by an angle of 15°. Determine the power delivered.

$$P = 240 \times 15 \cos(15°)$$

$$= 3.48 \text{ kW}$$

Although the average power is in many cases the quantity of prime interest the alternating part represented by the second term in Equation 2.15 exists and is sometimes important. If, for example, the light output from an ordinary domestic filament lamp run on AC is observed with a suitable photodetector, it will still be found that a large component fluctuating at a frequency of 100 Hz is present in the output. This is because the filament has low mass and low capacity for retaining heat, so that its temperature varies as the power input varies and so the light output varies.

2.5.4 ◆ Periodic non-sinusoidal variation

The last section considered the very special case when both voltage and current were represented by sine waves. In general the waveforms will not be of the same shape and $v(t)$ and $i(t)$ will exhibit quite different patterns of variation with time. Each case has to be considered on its merits. Consider as an example the waveforms shown in Figure 2.10, which might refer to a switched power supply. The energy delivered in one cycle is calculated from the area under the triangle in the waveform of $p(t)$ and is equal to $EI\tau/2$. Thus the average power is

$$P = \frac{\tau}{2T}EI$$

If this expression is compared with previous results it will be seen that it is quite different from the product of RMS voltage and current: power must be worked out from the product $v(t)i(t)$.

2.5.5 ◆ Transient waveform

In spite of the fact that in the example in Section 2.5.4 some current is flowing all the time, the power is delivered in short bursts. The energy in a single burst (found above to be $EI\tau/2$) is sometimes the important

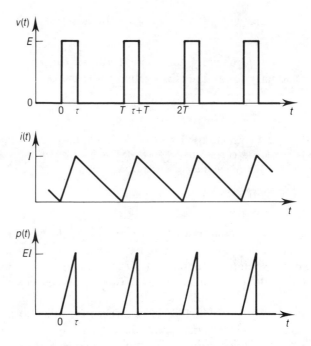

Figure 2.10 Illustrating the calculation of power in the case of periodic, non-sinusoidal waveforms (Section 2.5.4).

quantity. For example, in a small electronic device a sudden influx of energy can cause a transient temperature rise sufficient to destroy the device, whereas the same energy distributed throughout the cycle would have negligible effect.

EXAMPLE 2.5 A semiconductor diode carries a pulse of current of the waveform shown in Figure 2.11. Calculate the energy dissipated in the diode under each of the following assumptions: that the voltage drop occasioned by the current is

  (i) 0.7 V independent of current,
  (ii) $[0.7 + 0.1i(t)]$ V

Since $p = vi$, the energy is

$$w = \int_0^T vi \, dt$$

taken over the length of the pulse, $T = 1$ ms. We have the following.

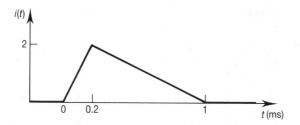

Figure 2.11 Current waveforms pertaining to Example 2.5, Section 2.5.5.

(i) For $v = 0.7$,

$$w = 0.7 \int_0^T i \, \mathrm{d}t$$

$$= 0.7 \times \tfrac{1}{2} \times 2 \times 10^{-3}$$

$$= 7 \times 10^{-4} \text{ J}$$

(ii) For $v = 0.7 + 0.1i(t)$,

$$w = \int_0^T [0.7 + 0.1i(t)]i(t) \, \mathrm{d}t$$

$$= 0.7 \int_0^T i \, \mathrm{d}t + 0.1 \int_0^T i^2 \, \mathrm{d}t$$

The first term has the value obtained in part (i); the second, using the results obtained in Section 2.3.3, is equal to $0.1 \times (1/3) \times 4 \times 10^{-3}$. Hence

$$w = (7 + 1.33) \times 10^{-4} = 8.33 \times 10^{-4} \text{ J}$$

2.6 ♦ Summary

For a periodic waveform $i(t)$ of period T

average value $\qquad \langle i \rangle = \dfrac{1}{T} \displaystyle\int_0^T i(t) \, \mathrm{d}t$

mean square value $\qquad \langle i^2 \rangle = \dfrac{1}{T} \displaystyle\int_0^T i^2(t) \, \mathrm{d}t$

RMS value $\qquad i_{\text{RMS}} = (\langle i^2 \rangle)^{1/2}$

The average power delivered to a load with voltage $v(t)$ and current $i(t)$ both periodic in T is

$$P = \dfrac{1}{T} \int_0^T v(t)i(t) \, \mathrm{d}t$$

If $v(t) = \hat{V}\cos(\omega t)$, $i(t) = \hat{I}\cos(\omega t + \theta)$

$$v_{\text{RMS}} = \frac{1}{\sqrt{2}}\hat{V}$$

$$i_{\text{RMS}} = \frac{1}{\sqrt{2}}\hat{I}$$

$$P = v_{\text{RMS}}\, i_{\text{RMS}}\cos\theta$$

In general, the energy delivered in the interval $t_1 < t < t_2$ is

$$w = \int_{t_1}^{t_2} v(t)i(t)\,dt$$

PROBLEMS

2.1 Determine the average and RMS currents for the following periodic waveforms:

(i) $i(t) = \hat{I}\cos(\omega t)$ $\cos(\omega t) > 0$
      $= 0$ $\cos(\omega t) < 0$

(ii) $i(t) = \hat{I}|\cos(\omega t)|$

(iii) $i(t) = \hat{I}\exp(-t/T)$ $0 < t < T$, period T

2.2 The value in amps of a current waveform of period 20 ms is given as follows:

    $0\text{ ms} < t < 10\text{ ms}$ $i(t) = 1 + 100t$

    $10\text{ ms} < t < 20\text{ ms}$ $i(t) = 0$

Show that the average current is 0.75 A and determine the RMS value. Determine also the RMS value of the current waveform $i(t) - 0.75$.

2.3 The output from a power supply is derived from AC mains of period T. The waveform is periodic with period $T/3$ and is described by the expression

$$v(t) = 240\cos(2\pi t/T) \qquad - T/6 < t < T/6$$

Determine the mean output voltage.

2.4 The DC output from a power supply can be adjusted by control of a 'firing' angle, α. The output is equal to the average of the input waveform, one period of which is described by

$$v(t) = 240\sin(\omega t) \qquad \alpha < \omega t < \pi + \alpha$$

Determine the firing angle to give an output of (i) 100 V and (ii) 30 V.

2.5 In a battery charger the current flowing through the battery is given in amps by

$$i = 2\sin(\omega t) - 1 \qquad \sin(\omega t) > 0.5$$
$$= 0 \qquad\qquad\quad \sin(\omega t) < 0.5$$

in which $\omega = 100\pi$.

The battery can be represented by a constant voltage source of 12 V in series with a resistance of 0.5 Ω. Determine the average power delivered to the battery and that dissipated as heat in the resistance.

2.6 A high-voltage pulse generator supplies a pulse of duration 5 μs once every 2 ms. During the 5 μs interval the pulse has the trapezoidal shape described by the expressions

$$v(t) = \hat{V}(t/\tau) \qquad\qquad 0 < t < \tau$$
$$= \hat{V} \qquad\qquad\qquad \tau < t < 4\tau$$
$$= \hat{V}(5 - t/\tau) \qquad\quad 4\tau < t < 5\tau$$

in which $\tau = 1\,\mu$s, $\hat{V} = 6$ kV. For the remainder of the period the voltage is zero. The generator is connected to a 50 Ω resistive load. Determine (i) the energy delivered to the load in a single pulse, (ii) the average current and (iii) the average power supplied. Determine also the average power during the pulse.

2.7 A radar pulse consists of a burst of sinusoidally varying voltage lasting 4 μs. This burst is repeated with a repetition rate of 1 kHz. During the pulse the voltage is given by the expression

$$v(t) = \hat{V}\sin(2\pi ft) \qquad 0 < t < 4\,\mu\text{s}$$

in which $f = 2.5$ GHz, $\hat{V} = 1$ kV. Determine the RMS voltage of the waveform.

2.8 A transistor in a switching circuit behaves like a load whose characteristics vary with time. In Figure 2.12 are shown observed voltage and current waveforms associated with the turn-on transition and the turn-off transition. Calculate the energy supplied to the transistor during each transition.

A switching cycle consists of a turn-on transition followed by a turn-off. If this repeats at a frequency of 5 kHz calculate the power dissipated in the transistor.

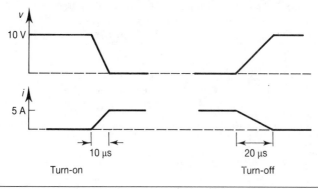

Figure 2.12

2.9 A switching device in a circuit is alternately switched on and off by a 10 kHz timing pulse. At the 'off', the voltage across the device rises from zero according to

$$v(t) = \hat{V}[1 - \exp(-t/2\tau)]$$

while the current falls according to

$$i(t) = \hat{I}\exp(-t/\tau)$$

At the 'on' these expressions become respectively

$$v(t) = \hat{V}\exp(-t/2\tau)$$

$$i(t) = \hat{I}[1 - \exp(-t/\tau)]$$

In these expressions $\hat{V} = 100$ V, $\hat{I} = 5$ A, $\tau = 10$ μs. Estimate the power lost in the switch. It may be assumed that the 1 ms interval between timing pulses is effectively infinite when calculating the energy lost at each change-over.

2.10 In Figure 2.13 is shown a circuit containing two loads carrying the same current and supplied at 50 V DC. The observed voltages are given by the following expressions:

$$v_1(t) = 50[1 - 0.9\cos(\omega t)]$$

$$v_2(t) = 45\cos(\omega t)$$

The current is given by

$$i(t) = 10[1 + \cos(\omega t)]$$

Calculate the power dissipated in each load.

2.11 In the circuit of Figure 2.13 the bottom load, across which $v_1(t)$ is developed, is replaced by a switched load. The voltages remain as in Problem 2.10, but the current now has the waveform shown in Figure 2.14. Calculate the power then dissipated in each load.

2.12 A lightning flash acts as a current source the waveform of which can be approximately described by the superposition of the two waveforms shown in Figure 2.15. The current is carried to earth through a lightning conductor 20 m high and of resistance 150 $\mu\Omega$ m^{-1}. Estimate (i) the total charge passed and (ii) the energy dissipated in the lightning conductor.

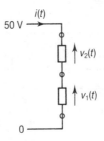

Figure 2.13

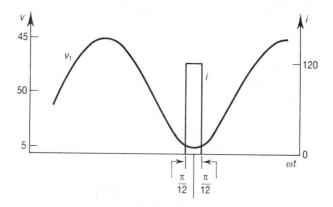

Figure 2.14

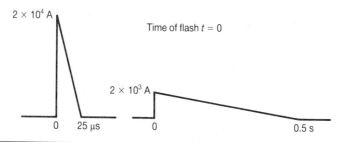

Figure 2.15

2.13 The lightning flash of Problem 2.12 strikes a tree of resistance to ground 100 Ω. Estimate the energy dissipated in the tree.

2.14 The voltage and current for a load supplied from an AC source are given by the following expressions, in which $\omega = 100\pi$:

$$v(t) = 340 \cos (\omega t + \pi/6)$$

$$i(t) = 3 \cos (\omega t) + 7 \sin (\omega t + \pi/4) + \cos (3\omega t)$$

Calculate (i) the RMS current and (ii) the average power supplied.

2.15 The voltage developed across an iron-cored inductor when an AC current of RMS value 0.3 A and angular frequency ω flows through it is assumed to be of the form $A \cos (\omega t) + B \cos (3\omega t)$, with peak voltage occurring at the instant of peak current. The measured RMS voltage is 120 V and the measured power loss is 31 W. Determine the magnitudes of the constants A and B.

3 ✦ Resistance-only Networks

In this chapter the principles of circuit analysis are stated. These principles are then applied to the analysis of a number of networks in which the components are only resistive.

3.1 ✦ Introduction

Although to limit a study to networks containing only resistive-type components might seem unduly restrictive, it is both useful and convenient to do so. It is convenient because the methods of analysing networks containing resistance only and excited by steady, DC, sources can be extended without great conceptual complication to the more general case when inductance and capacitance are present and the sources are time varying. In fact, if only resistance is involved then time variation is irrelevant: it is only when the voltage–current relationship of components depends on rate of change with time that the time variation of the source matters. Apart from demonstrating general techniques, there are some important cases when such simple networks are much used and which repay specific consideration. Some of these cases are considered in the present chapter.

3.2 ✦ Networks

The term 'network' has been used above without a definition, as being a term generally understood. However, a **network** may be defined more specifically as a number of interconnected components excited by one or more sources. The interconnection points, called **nodes**, may be thought of as a number of terminals between which the components are connected. A network is defined completely by identifying the nodes and stating the two nodes to which each component is connected. This information, usually given graphically in the form of a circuit diagram, can also be conveyed in the form of a table. Such a table has the advantage that it can be used as input data for computer-aided network analysis programs. In the present chapter the components are considered to be mainly ideal resistors, although we shall include in some cases the elements defined in Section 1.5.3 as dependent sources. These components are resistive in the sense that time variation does not enter into the defining equations. The importance of these dependent sources lies in the fact that they are used in the modelling of circuits containing such elements as transistors.

3.2.1 ◆ Network analysis and network design

It is very relevant to ask why we should wish to analyse a network. Sometimes indeed we do wish to analyse a given network, in the sense of determining the current through and the voltage across each component. An example in a somewhat generalized sense is afforded by the distribution of electric power by a public utility: components represent factories or urban agglomerations and we need to know the voltage at which power is supplied at each complex. Probably it is more common for us to want to design a network, that is to determine values of components which will produce a desired result, such as a given output voltage. Repeated analysis of the network with values of components guessed at by some means until the desired result is obtained is one way of designing, but usually a very inefficient way. This distinction between analysis and design is mentioned here both because of its importance and because it influences the way we approach problems. Many of the examples given will appear as design problems which use analysis to arrive at an answer.

3.3 ◆ The principles of analysis

The basic relations that are used in network analysis derive from the definitions of voltage and current, considered in Chapter 1. One class of relation concerns voltage. Since voltage represents work done in moving a unit charge, the voltage across two or more components in series is equal to the sum of the separate voltages (this applies whether or not the components are carrying the same current). The voltage difference between two nodes of the network is found by adding, with due regard to sign, the voltage differences across a succession of components or sources forming a path joining the two nodes: it is a basic tenet of network theory that the voltage difference so calculated is the same for all paths which begin and end at the same two points. This result is often put in the form 'the total voltage drop round a closed loop of a network is zero'. In this form it is known as **Kirchhoff's voltage law**.

The other class concerns current: since current represents flow of charge, which is conserved, at any node where several current-carrying connections join the total current leaving the node must be exactly balanced by the total current arriving at it. This result is known as **Kirchhoff's current law**.

Analysing a network means in essence that the voltage and current associated with each component are to be found. To do this, we can set up three groups of equations:

(i) those arising from application of Ohm's law to each component,

(ii) the relations between the component voltages arising from application of the voltage law, and

(iii) the relations between the component currents arising from application of the current law.

The full set of equations will be enough to determine all the unknowns.

The ways in which these various equations are used in formal analysis will be considered in Chapter 6, but in the meantime it can be seen that it is not necessary to introduce explicitly the voltage and current associated with each component. For example, if the voltage of each node with respect to a common reference were known, everything about the network could be calculated. Such a set of voltages may be used as unknowns; the set does in fact automatically ensure that the voltage law equations are satisfied. Alternatively, unknown currents can be chosen so as to satisfy the current law equations. In the examples of the present chapter suitable unknowns will be determined by inspection of the particular circuit.

3.4 ◆ Resistors in series and in parallel

These connections may seem too trivial to regard as circuits, but are nevertheless important.

3.4.1 ◆ Resistors in series

A circuit with two resistors in series is shown in Figure 3.1(a), in which the convention illustrated in Figure 1.1 has been used in defining voltages and

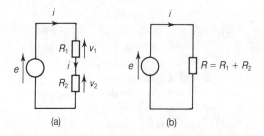

Figure 3.1 Combination of resistors in series: (a) circuit with R_1 and R_2 in series; (b) equivalent circuit with one resistor, $R_1 + R_2$.

currents. Analysis is straightforward. Applying Ohm's law to the resistors, we find

$$v_1 = iR_1$$

$$v_2 = iR_2$$

Since the total voltage across the two resistors must be equal to the applied voltage, we have

$$e = v_1 + v_2$$

$$= i(R_1 + R_2)$$

Thus two resistors in series act as would a single resistor whose resistance is the sum of the separate resistances. The importance of this result lies in the transition from physical circuit to circuit diagram: there may well be two resistors in the physical circuit but in drawing the circuit diagram representing the action of the circuit they may be replaced by one resistor, in so far as calculating the current is concerned. This current can then be used to work out the voltage drops or powers for the actual resistors. In this sense the circuits of Figures 3.1(a) and 3.1(b) may be said to be equivalent.

3.4.2 ◆ Resistors in parallel

A similar result holds for resistors in parallel (Figure 3.2(a)). We have

$$i_1 = \frac{e}{R_1} = eG_1$$

$$i_2 = \frac{e}{R_2} = eG_2$$

Applying the current law we find

$$i = i_1 + i_2 = e(G_1 + G_2)$$

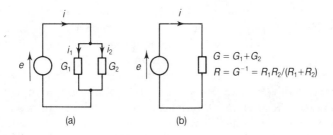

(a)　　　　　　　(b)

Figure 3.2 Combination of resistors in parallel: (a) circuit with R_1 and R_2 in parallel; (b) equivalent circuit with one resistor of conductance $G_1 + G_2$.

Hence

$$e = i\,\frac{R_1 R_2}{R_1 + R_2}$$

Thus the circuit of Figure 3.2(b) is equivalent to that of Figure 3.2(a): the equivalent resistor has a conductance equal to the sum of the individual conductances. As with the series circuit two resistors in parallel in the physical circuit can be represented in the circuit diagram by a single resistor. The voltage across this resistor will be the same as that across each of the individual resistors.

3.5 ◆ The voltage divider

A configuration frequently encountered in electrical circuits is the voltage divider, shown in Figure 3.3(a). The two resistors R_1, R_2 are often configured as the two parts of a single resistance wire or track composed of resistive material to which a third contact is made by an adjustable slider.

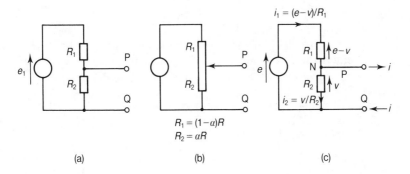

(a)　　　　　　　(b)　　　　　　　(c)

Figure 3.3 Illustrating a voltage divider circuit: (a) using discrete resistors; (b) using resistor with sliding contact; (c) analysis of circuit when supplying a current.

This form is illustrated in Figure 3.3(b), in which the parameter α is related to the slider position (such a device is known as a potentiometer). In this form the circuit may be used to provide a source of adjustable voltage, or as a volume control in an audio circuit. When no connection is made to the terminals P, Q in these figures the same current $e/(R_1 + R_2)$ flows through both R_1 and R_2 so that the voltage between P and Q is simply $eR_2/(R_1 + R_2)$. Hence the name 'voltage divider'. In use a load will be connected between P and Q, and it is important to see the effect on the output voltage. Since we are comparing the terminals P, Q to the terminals of a source, we consider a current to flow from P back to Q, as indicated in Figure 3.3(c), and calculate the voltage between P and Q. If this voltage, v, were known, all the currents could be calculated. We therefore obtain expressions for each current in terms of this unknown and then apply the current law at the node N. The voltage across R_1 is $e - v$, so the currents are as shown in the diagram. Balancing currents at the node N, we have

$$(e - v)\frac{1}{R_1} = i + \frac{v}{R_2}$$

Rearranging, we find

$$v = e\frac{R_2}{R_1 + R_2} - i\frac{R_1 R_2}{R_1 + R_2} \tag{3.1}$$

By comparison with Equation 1.9 we see that the terminals P, Q behave as though they were connected to a constant voltage source of strength $eR_2/(R_1 + R_2)$ in series with a resistance equal to that of R_1 and R_2 in parallel. In the case of the variable divider (Figure 3.3(b)) we have

$$R_1 = (1 - \alpha)R, \qquad R_2 = \alpha r$$

Hence

$$\frac{R_1 R_2}{R_1 + R_2} = \alpha(1 - \alpha)R \tag{3.2}$$

Thus the internal resistance of the equivalent source is zero when the slider is at either end, rising to a maximum of $R/4$ in the centre. Consider the source to be connected to a load of resistance R_L. Using Equations 3.1 and 3.2 we have

$$v = e\frac{\alpha R_L}{R\alpha(1 - \alpha) + R_L}$$

or

$$\frac{v}{e} = \frac{\alpha}{1 + (R/R_L)\alpha(1 - \alpha)} \tag{3.3}$$

This equation is illustrated graphically in Figure 3.4 for several values of

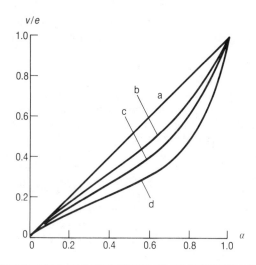

Figure 3.4 Graph showing variation of output voltage with slider position for a divider of resistance R supplying a load of resistance R_L: curve a, $R/R_L = 0$; curve b, $R/R_L = 1$; curve c, $R/R_L = 2$; curve d, $R/R_L = 4$.

R/R_L. The ordinate can be regarded as output and, in the case of the linear potentiometer, will be proportional to the slider movement. With the load connected the output voltage no longer varies linearly with slider position. The effect of the non-linearity is to make adjustment of the slider position for a given voltage more sensitive, an undesirable feature in an adjustable source or a volume control. To avoid this effect becoming too significant, the total potentiometer resistance, R, must be kept low compared with the load resistance, R_L, at the expense of a greater steady current flowing even with no load connected. It will be observed that the worst case shown in Figure 3.4, $R/R_L = 4$, corresponds to the maximum internal resistance, $R/4$, equal to the load resistance.

Another way of avoiding the bunching at the upper end is to use a track which is graded in resistance, so that α is no longer proportional to slider position. An approximately logarithmic grading is often used.

EXAMPLE 3.1 A voltage divider of the form of Figure 3.3(a) in which the source e is a 9 V cell is required to deliver 3 V across a load drawing 0.1 mA. The value of 3 V is to be achieved to within 5%. The designer guesses that this requirement will be met if with no load connected the divider chain takes 1 mA and delivers 3 V. Calculate the values of R_1 and R_2 that this design process yields and check whether or not the requirement is satisfied.

With no load connected, 1 mA is taken from the 9 V cell when the total resistance $R_1 + R_2 = 9$ kΩ; to give 3 V out requires that

$$9R_2/(R_1 + R_2) = 3$$

Hence $R_2 = 3$ kΩ, $R_1 = 6$ kΩ.

With these values, the output resistance $R_1R_2/(R_1 + R_2) = 2$ kΩ, giving a drop of 0.2 V when a current of 0.1 mA is taken. The percentage error is therefore equal to $(0.2/3) \times 100 = 7\%$, outside the limit.

The suggested design process could easily be improved as follows: considering the obvious choice of $R_2 = 3$ kΩ, $R_1 = 6$ kΩ as above and realizing that it gives a 0.2 V drop, recalculate to give 3.2 V off-load. This requires $R_2 = 3.2$ kΩ, $R_1 = 5.8$ kΩ, when the internal resistance is 2.06 kΩ. The actual drop is then reduced to 0.2%.

[It may be noted that like many design problems there is no unique solution: why not use a 60 kΩ resistor in series rather than a divider? The answer probably lies in the fact that it is the 3 V which is important and the 0.1 mA may be somewhat uncertain.]

3.6 ◆ The theorems of Thévenin and Norton

Equation 3.1 is an example of a very general result: it concerns a network with two free terminals which can be connected to an external circuit (in this case the divider and cell, with terminals P, Q). The equation shows that the voltage between those free terminals is related to the current drawn through them by a linear equation. This result is a consequence of the components of the network being linear, in the sense defined in Section 1.4, rather than of the exact nature of the network: any network whatsoever made up of linear components and sources and having two free terminals will likewise show a linear relationship. A formal proof of this will be given in a later chapter, but examples in the present chapter will show its validity. The consequence will be seen to be of great importance, as an aid both to analysis and to the understanding of circuit operation. The general situation is illustrated in Figure 3.5(a). It is easily conceived that experiment would show a relationship between v and i of the type shown in Figure 3.5(b), described by an equation of the form

$$v = v_0 - Ri_i \tag{3.4}$$

The constants v_0, R_0 can be interpreted in terms of measurable quantities. If the free terminals in Figure 3.5(a) are left on open circuit, $i = 0$, and the voltage between them will be v_0. Thus v_0 may be interpreted as the open-circuit voltage. If the terminals are short circuited, $v = 0$, and from Equation 3.4 the current, i_0, will be equal to v_0/R_0. Thus R_0 is the ratio of open-circuit voltage to short-circuit current. In principle, two measurements or two calculations are adequate to find the two constants in Equation 3.4.

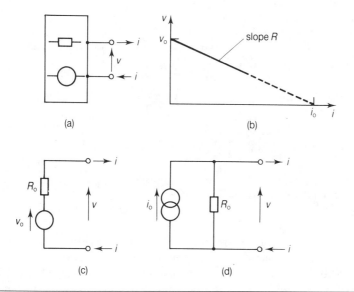

Figure 3.5 The theorems of Thévenin and Norton (Section 3.6): (a) schematic showing an arbitrary network of resistors and sources with two free terminals delivering current *i* at voltage *v*; (b) showing the linear variation of *v* with *i*; (c) the Thévenin equivalent circuit; (d) the Norton equivalent circuit.

Equation 3.4 is precisely that which would be found for the source–resistance circuit of Figure 3.5(c), which is said to be the Thévenin equivalent circuit for the network of Figure 3.5(a). The equivalence is known as Thévenin's theorem. It is important to note that the equivalence only refers to the external connections, and not to anything that might be happening inside the network, although of course if Figure 3.5(c) is used to calculate *v* in particular circumstances that value of *v* is applicable to Figure 3.5(a) with the same external connections.

Equation 3.4 can be rearranged into the form

$$i = v_0 \frac{1}{R_0} - v \frac{1}{R_0}$$

$$= i_0 - v \frac{1}{R_0} \tag{3.5}$$

In this form the equation represents the circuit of Figure 3.5(d), showing a current source of strength equal to the short-circuit current in parallel with the resistance R_0. The equivalence between the configurations of Figure 3.5(d) and Figure 3.5(a) is known as Norton's theorem and the circuit of Figure 3.5(d) as the Norton equivalent circuit for the network of Figure 3.5(a). The equivalence between the two source configurations has already been mentioned in Section 1.5.2.

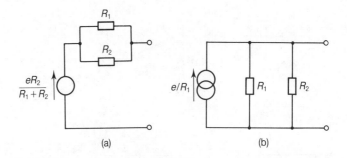

Figure 3.6 Equivalent circuits for the voltage divider of Figure 3.3(a): (a) Thévenin circuit: (b) Norton circuit.

Another way of determining R_0 is often useful. Because of the linearity of the components a reduction in strength of all the sources would result in a reduction of the open-circuit voltage by the same factor. If they are all reduced to zero then the open-circuit voltage likewise becomes zero and according to Equation 3.4 $R_0 = v/(-i)$. Bearing in mind the conventional direction for i, this is saying that R_0 is the resistance which would be measured between the terminals with the network sources reduced to zero in this way. (It must be appreciated that 'reducing the sources to zero' means replacing fixed voltage sources by short circuits and fixed current sources by open circuits.)

3.6.1 ◆ Application to the voltage divider

Consider the application of these theorems to the voltage divider circuit of Figure 3.3(a). For this circuit the open-circuit condition gives

$$v_0 = eR_2/(R_1 + R_2)$$

For the short-circuit condition

$$i_0 = e/R_1$$

Hence

$$R_0 = v_0/i_0 = R_1R_2/(R_1 + R_2)$$

This result also follows from reducing the voltage e to zero, when it is apparent that the equivalent Thévenin resistance is R_1 in parallel with R_2. The Thévenin equivalent circuit is shown in Figure 3.6(a), displaying agreement with Equation 3.1. The Norton equivalent circuit is shown in Figure 3.6(b). Both of these equivalent circuits could have been written down by inspection, the calculation being minimal.

Perhaps the most important consequence of these theorems is that if we are dealing with a network in which external connections are only made

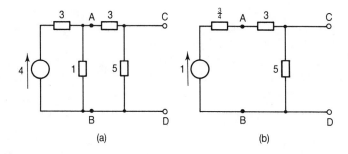

Figure 3.7 Example 3.2: (a) circuit; (b) reduction by use of Thévenin's theorem.

to two free terminals then, however complicated that network may be, we can correctly represent its behaviour to the external circuit by a source–resistance combination. Actually determining the two constants for this equivalent circuit may not be straightforward, but once done that part of the network will not have to be analysed again.

The following two examples illustrate how the theorems may be used.

EXAMPLE 3.2 For the circuit of Figure 3.7(a), determine the voltage between terminals C and D.

The circuit to the left of AB can be replaced by its Thévenin equivalent. The open-circuit voltage is

$$v_{OC} = 4 \times 1/(1 + 3) = 1 \text{ V}$$

The short-circuit current is

$$i_{SC} = 4/3 \text{ A}$$

Hence

$$R_0 = 3/4 \ \Omega$$

This is also the value obtained by replacing the 4 V source by a short circuit. Using these values we arrive at the equivalent circuit shown in Figure 3.7(b), from which

$$v_{CD} = 1 \times 5/(0.75 + 3 + 5) = 4/7 \text{ V}$$

We note that the result of changing the value of the 5 Ω resistor can be predicted without further analysis.

EXAMPLE 3.3 For the circuit of Example 3.2, find the power dissipated in each resistor.

The result already obtained enables us to say that the power dissipated in the 5 Ω resistor is $(4/7)^2/5 = 0.065$ W. The current through the

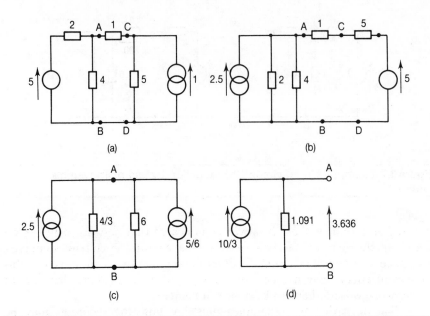

Figure 3.8 Example 3.4: (a) circuit; (b), (c), (d) stages of reduction by use of the theorems of Thévenin and Norton.

3 Ω resistor joining A and C is (4/35) A, so that the power in that resistor is $3 \times (4/35)^2 = 0.039$ W. To find the powers in the remaining resistors we need to know the voltage between A and B. For this purpose we can correctly use the circuit of Figure 3.7(b), whence $v_{AB} = 1 \times 8/(8 + 0.75) = 32/35$ V. Therefore the power in the 1 Ω resistor is $(32/35)^2/1 = 0.836$ W. Finally, returning to the original circuit of Figure 3.7(a), the voltage drop across the 3 Ω resistor is $4 - 32/35$, whence the current is 36/35 A and the power dissipation 3.174 W. The total power consumed is thus 4.114 W. This should equal the power provided from the source: the source current is 36/35 A, correctly giving the source power as $4 \times 36/35 = 4.114$ W.

EXAMPLE 3.4 For the circuit of Figure 3.8(a), determine the voltages between A and B and between C and D. Determine also the power dissipated in each resistor.

A sequence of transformations is shown in Figures 3.8(b)–3.8(d), in the process of which nodes C and D are eliminated but A and B remain. From the final circuit

$$v_{AB} = 3.636 \text{ V}$$

Using this value in the circuit of Figure 3.8(b), the current through the 1 Ω resistor joining A and C is given by

$$(5 - 3.636)/6 = 0.227 \text{ A}$$

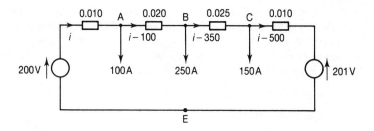

Figure 3.9 Circuit for Example 3.5.

Hence

$$v_{CD} = 5 - 5 \times 0.227 = 3.865 \text{ V}$$

Transferring these values to the original circuit, we can draw up a table showing voltage drop and power for each resistor:

Resistor (Ω)	Drop (V)	Power (W)
2	1.364	0.930
4	3.636	3.305
1	0.229	0.052
5	3.865	2.988

The total power dissipated is 7.275 W. The 5 V source delivers a current of 0.682 A, and the voltage across the current source is 3.865 V. Therefore the total source power is

$$5 \times 0.682 + 3.865 \times 1 = 7.275 \text{ W}$$

This agrees, as it should, with the previous figure.

3.7 ◆ Application of the principles of circuit analysis

The following sections contain examples of circuit problems found in the various branches of electrical and electronic engineering. These are chosen to illustrate how the principles of analysis laid down in Section 3.3 can be applied to commonly encountered circuits in a straightforward manner.

EXAMPLE 3.5 This example illustrates the type of problem encountered in the distribution of mains power. The circuit shown in Figure 3.9 models a simple distribution system. The various loads, each specified by its current drain, are separated by resistors modelling the resistance of the connecting cable. It is required to find the voltage at each load.

Since the total current drain is known, knowledge of the current supplied by one of the generators would enable each voltage to be calculated. Taking, therefore, the current from the 200 V generator as the unknown, i, we can express the current in each resistor as in the diagram. We then draw up a list of voltages as follows:

$$v_{AE} = 200 - 0.010i$$

$$v_{BE} = v_{AE} - 0.020(i - 100)$$

$$v_{CE} = v_{BE} - 0.025(i - 350)$$

$$201 = v_{CE} - 0.010(i - 500)$$

Adding we find

$$201 = 200 + 0.020 \times 100 + 0.025 \times 350 + 0.01 \times 500 - 0.065i$$

Hence

$$i = 14.75/0.065 = 226.9 \text{ A}$$

and

$$v_{AE} = 197.7 \text{ V}$$

$$v_{BE} = 195.2 \text{ V}$$

$$v_{CE} = 198.3 \text{ V}$$

The object of a calculation of this sort is to obtain estimates of the voltage drops in the system when loaded to capacity, in order to check that they will not be excessive. The circuit of Figure 3.9 will already contain approximations. It might be thought, for example, more accurate to represent loads by resistors rather than current sinks. Such a problem could be solved by the methods used in Examples 3.2 and 3.4, but at the cost of much greater complexity. The specification of the present example is adequate because the voltage drops are small, as they should be in a well-designed system, so that estimation of the current in each load can be made by assuming a supply voltage of 200 V.

EXAMPLE 3.6 This example shows the simple application of Ohm's law to a purely DC problem encountered in transistor circuits, that of providing the correct biasing conditions. In Figure 3.10 is shown the symbol for an n–p–n transistor together with a set of voltages and currents which have to be realized when the transistor is placed in circuit. (It is assumed in this problem that these values have been derived according to the manufacturer's specification for the transistor.)

The circuit to be studied is shown in Figure 3.11, and the first point to be realized is that nothing need be known about the transistor except the information given in Figure 3.10. We have to obtain suitable values for R_1 and R_2. The process is straightforward: we have

$$R_1 = (9 - 0.7)/(4 \times 10^{-5}) = 2.08 \times 10^5 \ \Omega$$

$$R_2 = (9 - 4)/(5 \times 10^{-3}) = 1000 \ \Omega$$

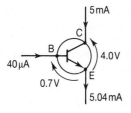

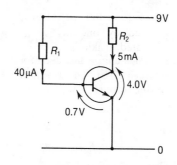

Figure 3.10 Example 3.6: currents and voltages associated with an n–p–n transistor.

Figure 3.11 Circuit for Example 3.6.

The power dissipated in the various components is of interest. For R_1 we have

$$(2.08 \times 10^5)(40 \times 10^{-6})^2 = 0.3 \text{ mW}$$

for R_2,

$$1000(5 \times 10^{-3}) = 25 \text{ mW}$$

for the transistor,

$$4.0 \times 5 \times 10^{-3} + 0.7 \times 40 \times 10^{-6} = 20 \text{ mW}$$

EXAMPLE 3.7 Figure 3.12(a) shows a type of circuit frequently encountered in transistor circuit design. The conditions for the transistor itself are the same as the last example. In order to have a uniquely soluble problem two further pieces of information must be specified by the designer: the voltage across R_4 is to be 1.5 V, and the output (Thévenin) resistance of the voltage divider formed by the resistors R_1, R_2 must be 5 kΩ.

Values for R_3 and R_4 are straightforward to obtain:

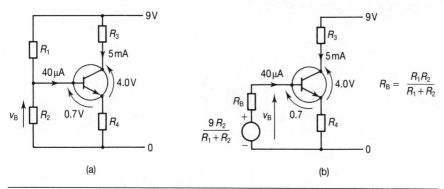

Figure 3.12 Example 3.7: (a) circuit; (b) reduced circuit.

$$R_3 = (9 - 4 - 1.5)/(5 \times 10^{-3}) = 700\ \Omega$$

$$R_4 = 1.5/(5.04 \times 10^{-3}) = 298\ \Omega$$

The Thévenin equivalent for the voltage divider has been used to obtain Figure 3.12(b). The resistance R_B is required to be 5 kΩ and with 40 µA flowing through it will cause a voltage drop of 0.2 V. Hence

$$9 \times R_2/(R_1 + R_2) - 0.2 = 1.5 + 0.7$$

or

$$9 \times R_2/(R_1 + R_2) = 2.4$$

Also

$$R_1 R_2/(R_1 + R_2) = 5 \times 10^3$$

Solving, we find

$$R_1 = 5 \times 10^3 \times 9/2.4 = 18.8\ \text{k}\Omega$$

$$R_2 = 6.8\ \text{k}\Omega$$

3.8 ◆ Circuits involving dependent sources

Dependent sources were introduced in Section 1.5.3. These sources are used in the modelling of physical devices such as transistors. Any source represents an input of energy, controlled in some way according to the nature of the source. While a fixed source keeps either output voltage or output current constant, in a dependent source control is effected by a third variable. This adds a slight complication in the analysis of a circuit, since the value of the control variable has to be found as part of the analysis. In examples of this section four circuits are considered, three of which arise from the small-signal analysis of transistor circuits, and a fourth is derived from an operational amplifier circuit. The examples are chosen to illustrate the process of analysis with circuits containing dependent sources. The actual transistor circuits from which the first three are derived will be found in Section 13.7.2.

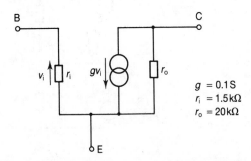

Figure 3.13 Circuit containing dependent source, Section 3.8.

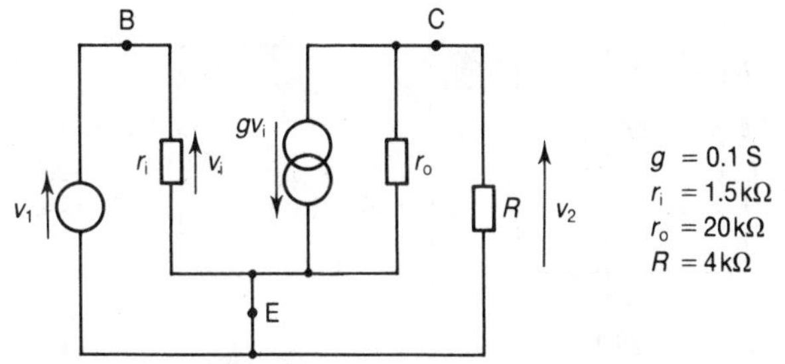

Figure 3.14 Circuit for Example 3.8.

The configuration in which the dependent source appears in the transistor problems is shown in Figure 3.13. The values to be given to the transconductance g and the resistances r_i, r_o depend on the type of device being modelled and on the conditions set by the designer. When numbers are used, the values taken will be $g = 0.1$ S, $r_o = 20$ kΩ, $r_i = 1.5$ kΩ. These are values which might be found with bipolar transistors. Although in the end numerical results may be wanted, it is usually better to treat this type of problem algebraically and finally to substitute the relevant values.

EXAMPLE 3.8 The circuit is shown in Figure 3.14. In this case there is no problem about expressing the control voltage, v_i, in terms of the externally applied voltage, v_1, since they are the same. Thus the strength of the current source is gv_1 and is directed through r_o and R in parallel. Hence

$$v_2 = -gv_1 \frac{Rr_o}{R + r_o}$$

We also have

$$i_1 = v_1 \frac{1}{r_i}$$

Thus the source supplying the input voltage 'sees' a resistance of r_i; the load resistance R 'looks back' at a source of internal resistance r_o. This is often large compared with R (as with the numerical values given for the problem). In the present context these two resistances would be called the **input resistance** and the **output resistance** respectively.

Using the value of 4 kΩ for R and for the other parameters the values shown in Figure 3.13, we find

$$v_2/v_1 = -333$$

The negative sign indicates that a change in v_1 produces a change in v_2 which is in the opposite sense.

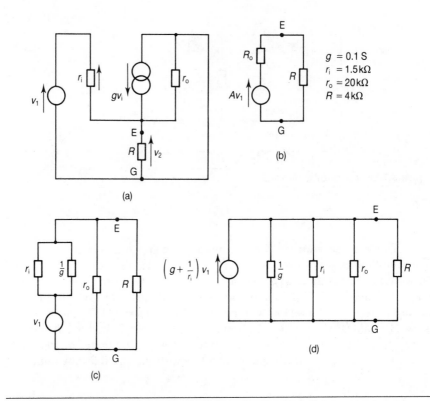

Figure 3.15 Example 3.9: (a) circuit; (b) equivalent circuit for output; (c), (d) alternative equivalent circuits.

EXAMPLE 3.9 The circuit shown in Figure 3.15(a) is a re-arrangement of the same components, in which the control voltage is no longer equal to the input voltage: some analysis is required in order to express v_i in terms of v_1. Taking a single nodal voltage v_2 as the unknown, we have

$$v_1 = v_i + v_2$$

Balancing currents at the node

$$v_i \frac{1}{r_i} + g v_i - v_2 \frac{1}{r_o} = v_2 \frac{1}{R}$$

Hence

$$v_1 \left(g + \frac{1}{r_i} \right) = v_2 \left(\frac{1}{R} + \frac{1}{r_o} + g + \frac{1}{r_i} \right)$$

We notice that the expression for v_2 can be put in the form

$$v_2 = A v_1 \frac{R}{R + R_0}$$

where

$$R_0 = \left(g + \frac{1}{r_i} + \frac{1}{r_o}\right)^{-1}$$

and

$$A = \left(g + \frac{1}{r_i}\right)R_0$$

With the numerical values for the problem, the dominating quantity is g, larger than the other two terms by a factor of over 100. In this case, therefore, $A \approx 1$ and $R_0 = 10\ \Omega$ (a more exact figure for A is 0.9995). It will be noticed that, when this approximation is valid, it is equivalent to ignoring the current through r_i and taking r_o to be infinite.

The input current i_1 is of interest. We have

$$i_1 = (v_1 - v_2)\frac{1}{r_i}$$

Substituting for v_2 we find

$$i_1 = \frac{v_1}{r_i} \frac{(1 - A)R + R_0}{R + R_0}$$

By putting in the full expressions for A and R_0 an exact but complicated expression for v_1/i_1 can be obtained. When, as in the present case, $A \approx 1$ and $R_0 \approx 1/g$ a much simpler expression is obtained, giving

$$\frac{v_1}{i_1} = r_i(gR + 1)$$

This value is much greater than r_i. With the numerical values given, $gR = 400$, so that the input resistance is 600 kΩ. It is to be noticed that the input resistance is in this instance dependent on the value of R. In the same way, if the source which supplies v_1 had an internal resistance it would be found that the output resistance R_0 would depend on the value of the resistance.

The form of the expression for v_2 shows that the terminals E, G can be regarded as the terminals of a source of voltage Av_1 and internal resistance R_0, as shown in Figure 3.15(b). This is as it should be, being an example of Thévenin's theorem. It should be possible to find the Thévenin equivalent directly by considering the terminals E, G as a source connected to the resistance R. Leaving an open circuit between E and G is equivalent to making R infinite, and rather than repeating the analysis we can put this condition in the results obtained above. We then find

$$v_2 = v_{OC} = Av_1$$

With a short circuit between E and G

$$v_i = v_1$$

Hence

$$i_{SC} = \left(g + \frac{1}{r_i}\right)v_1$$

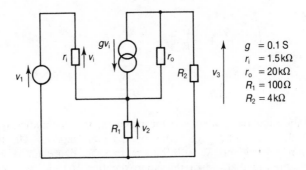

Figure 3.16 Circuit for Example 3.10.

The output resistance of the source equivalent to terminals E, G is therefore $A/(g + 1/r_i)$, which is the same as the expression for R_0 found above. Examination of the form of A and R_0 in the Thévenin equivalent circuit of Figure 3.15(b) shows that it can be rearranged as in Figure 3.15(c). This could have been obtained directly by regarding the load applied to terminals E, G as a single resistance equal to r_o and R in parallel. The Norton equivalent circuit then appears as shown in Figure 3.15(d).

If the second method given in Section 3.6 for determining R_0 is used it must be realized that it is v_1 in Figure 3.15(a) that is reduced to zero: measuring the resistance looking back from R will apply a voltage between E and G which prevents the control voltage from being zero. Its value must be found as in any other analysis problem.

EXAMPLE 3.10 The circuit of Figure 3.16 embraces the circuits of both Figures 3.14 and 3.15(a), in the sense that both can be derived from it by taking suitable value for R_1 and R_2.

In this circuit there are two nodes with the unknown nodal voltages, v_2 and v_3. In terms of these variables we have

$$v_1 = v_i + v_2$$

$$v_3 = -R_2\left(gv_i + \frac{v_3 - v_2}{r_o}\right)$$

$$\frac{v_i}{r_i} + gv_i + \frac{v_3 - v_2}{r_o} = \frac{v_2}{R_1}$$

Clearly these three equations are enough to eliminate v_i and then to solve for v_2 and v_3 in terms of v_1, but the expressions will be complicated, and in practice no useful purpose is served by obtaining them. When all the quantities are specified numerically the simultaneous equations can be solved as accurately as desired, but otherwise use of suitable approximations may lead to better understanding. If, as in the last problem, it is permissible to assume that $r_o \gg r_i \gg 1/g$, much simpler expressions result: neglecting terms in $1/r_o$ and $1/r_i$, the equations become

$$v_1 = v_i + v_2$$

$$v_3 = -R_2 g v_i$$

$$g v_i = \frac{v_2}{R_1}$$

giving

$$v_2 = v_1 \frac{gR_1}{1 + gR_1}$$

$$v_3 = -\frac{R_2}{R_1} v_2$$

$$= -v_1 \frac{gR_2}{1 + gR_1}$$

With the values shown in the Figure 3.16 and taking $v_1 = 0.1$ V, we find that with approximations

$$v_2 = 0.091 \text{ V}, \qquad v_3 = -3.64 \text{ V}$$

The values obtained without making any approximation are

$$v_2 = 0.089 \text{ V}, \qquad v_3 = -3.54 \text{ V}$$

The errors are of the order of 2% and answers to this degree of accuracy are usually quite acceptable in the context of transistor circuits. It is to be observed that the approximate solution would also be obtained directly from the circuit if r_o were to be made infinite and the current v_i/r_i ignored in comparison with $g v_i$.

One way of obtaining a numerical solution such as that which has just been given is by setting up the relevant equations and solving; another way is by using a computer program of the type discussed in Chapters 6 and 15 which accepts the circuit configuration and component values. For the purpose of obtaining actual values the latter is very useful, but does not directly lead to an understanding of the way in which a particular component affects circuit operation. It is in fact an easy way of doing an experiment. Better understanding can be obtained through an approximate algebraic analysis: very simple expressions form a useful basis for initial designs.

EXAMPLE 3.11 In the circuit of Figure 3.17 is shown a circuit containing a voltage-controlled voltage source. The elements within the dotted line form an equivalent circuit for the integrated circuit device called the operational amplifier, remarkable in that the input nodes P and Q are very largely isolated from the output node R and that the gain A is very large. The overall circuit can be regarded as an amplifier, with v_1 as the input and v_2 as the output.

In this circuit, the voltage v is really the only unknown since, if it is known, the current leaving node R can be calculated and hence v_2. It is,

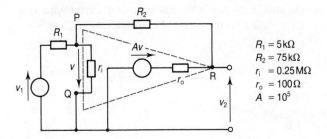

Figure 3.17 Circuit for Example 3.11. The elements within the dashed triangle form a simple equivalent circuit for an operational amplifier.

however, convenient to take v_2 as an additional unknown from the beginning. With the symbols shown in the diagram, a current balance at node P gives

$$\frac{v_1 - (-v)}{R_1} = \frac{(-v) - v_2}{R_2} + \frac{-v}{r_i}$$

or

$$\frac{v_1}{R_1} + \frac{v_2}{R_2} = -v\left(\frac{1}{R_1} + \frac{1}{R_2} + \frac{1}{r_i}\right)$$

At node R, a current balance gives

$$\frac{Av - v_2}{r_o} = \frac{v_2 - (-v)}{R_2}$$

or

$$v\left(\frac{A}{r_o} - \frac{1}{R_2}\right) = v_2\left(\frac{1}{R_2} + \frac{1}{r_o}\right)$$

The variable v can be eliminated from these two equations, leading to an expression for v_2 in terms of v_1. However, before doing this it is useful to consider the implications of the numerical values given for the problem. To a very good approximation, in the first equation the term in r_i can be neglected in comparison with those in R_1 and R_2, giving

$$\frac{v_1}{R_1} + \frac{v_2}{R_2} \approx -v\left(\frac{1}{R_1} + \frac{1}{R_2}\right)$$

In the second equation the terms involving R_2 can be neglected in comparison with those involving r_o, giving

$$Av \approx v_2$$

With these approximations we therefore find

$$\frac{v_1}{R_1} = -\frac{v_2}{R_2}\left[1 + \frac{1}{A}\left(1 + \frac{R_2}{R_1}\right)\right]$$

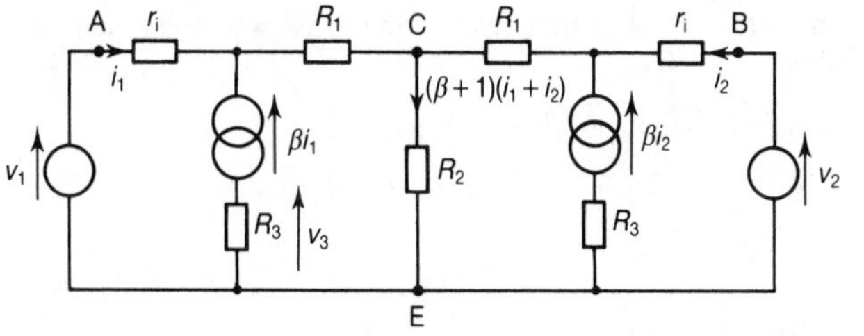

Figure 3.18 Circuit for Example 3.12, derived from the 'longtailed pair' transistor circuit.

Since $AR_1/R_2 = 7 \times 10^3$, to a very close approximation

$$\frac{v_1}{R_1} = \frac{v_2}{R_2}$$

This result could have been obtained directly by realizing that with the great value of A, the voltage v would be very much smaller than either v_1 or v_2, and could be placed equal to zero. Direct elimination of v, with no approximations, leads to

$$\frac{v_1}{R_1} + \frac{v_2}{R_2} = -v_2\frac{K}{R_2}$$

where

$$K = \frac{r_o + R_2}{AR_2 - r_o}\left[1 + R_2\left(\frac{1}{R_1} + \frac{1}{r_i}\right)\right]$$

Hence

$$v_2 = -v_1\frac{R_2}{R_1}\frac{1}{1 + K}$$

For the numerical values given, $K = 1.63 \times 10^{-4}$.

EXAMPLE 3.12 The circuit of Figure 3.18 is derived from the 'longtailed pair' transistor circuit. The circuit takes two input voltages, v_1 and v_2, and gives an output voltage v_3. The equivalent circuit contains two current-controlled dependent current sources. We shall derive an expression for v_3 in terms of v_1 and v_2.

Referring to Figure 3.18, we have

$$v_1 = r_i i_1 + R_1(\beta + 1)i_1 + R_2(\beta + 1)(i_1 + i_2)$$

$$v_2 = R_2(\beta + 1)(i_1 + i_2) + r_i i_2 + R_1(\beta + 1)i_2$$

$$v_3 = -R_3\beta i_1$$

The first two of these equations can be solved for i_1 and i_2. Substitution in the third equation will then yield the required expression. To make use of the symmetry of the first two equations we form sum and difference equations:

$$v_1 - v_2 = [r_i + (\beta + 1)R_1](i_1 - i_2)$$

$$v_1 + v_2 = [r_i + (\beta + 1)(R_1 + 2R_2)](i_1 + i_2)$$

Hence

$$2i_1 = (v_1 - v_2)\frac{1}{r_i + (\beta + 1)R_1}$$

$$+ (v_1 + v_2)\frac{1}{r_i + (\beta + 1)(R_1 + 2R_2)}$$

We can thus express

$$v_3 = -A_{\text{diff}}(v_1 - v_2) - A_{\text{com}}\tfrac{1}{2}(v_1 + v_2)$$

where

$$A_{\text{diff}} = \frac{1}{2}\frac{\beta R_3}{r_i + (\beta + 1)R_1}$$

$$A_{\text{com}} = \frac{\beta R_3}{r_i + (\beta + 1)(R_1 + 2R_2)}$$

A_{diff} is termed the **differential** gain, applying when $v_1 = -v_2$; A_{com} is the **common mode** gain, applying when $v_1 = v_2$. Possible values in a circuit might be

$$r_i = 50 \text{ k}\Omega, \quad \beta = 200, \quad R_1 = 1 \text{ k}\Omega, \quad R_2 = 75 \text{ k}\Omega, \quad R_3 = 75 \text{ k}\Omega$$

We find

$$r_i + (\beta + 1)R_1 \approx 250 \text{ k}\Omega$$

$$r_i + (\beta + 1)(R_1 + 2R_2) \approx 30 \text{ M}\Omega$$

so that

$$A_{\text{diff}} \approx 30, \quad A_{\text{com}} \approx 0.5$$

It can be seen that the output from the circuit will arise mainly from the difference between v_1 and v_2. The circuit is said to form a differential amplifier.

3.9 ◆ Attenuators

A circuit configuration which involves only resistance is that used in attenuators, designed to reduce the power from a source in a controllable way. It is often required in electronic systems to reduce the output from a source, very likely of a high frequency, in a known way. It is possible that the reduction ratio may be very large, as much as 10^6:1. The accepted solution to this problem in systems is to use a configuration of the type shown in Figure 3.19. To ensure interchangeability between systems all

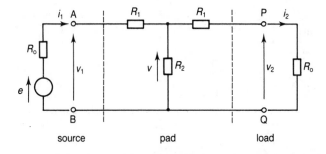

Figure 3.19 Circuit for a T-pad attenuator of characteristic resistance R_0.

sources and all loads are engineered to have a standard resistance, R_0 in the figure. This value is known as the characteristic resistance for the system (common values are 600 Ω for audio equipment, 50 Ω or 75 Ω for high-frequency equipment). We first show that it is possible to choose the values of R_1 and R_2 in the circuit of Figure 3.19 so that the resistance presented to the source by terminals A, B is equal to the characteristic resistance. This requires

$$R_1 + \frac{R_2(R_1 + R_0)}{R_2 + R_1 + R_0} = R_0$$

or

$$(R_0 - R_1)(R_2 + R_1 + R_0) = R_2(R_1 + R_0)$$

Hence

$$R_2 = \frac{1}{2R_1}(R_0^2 - R_1^2) \tag{3.6}$$

Assuming that R_2 has been so chosen for a given value of R_1 then $i_1 = v_1/R_0$, so that

$$v = v_1 - R_1 i_1 = v_1\left(1 - \frac{R_1}{R_0}\right)$$

Also

$$v_2 = v\frac{R_0}{R_1 + R_0}$$

Hence

$$\frac{v_1}{v_2} = n = \frac{R_0 + R_1}{R_0 - R_1} \tag{3.7}$$

The ratio n is termed the attenuation of the attenuator pad. (The particular pad shown in Figure 3.19 is known from its configuration as a **T-pad**.)

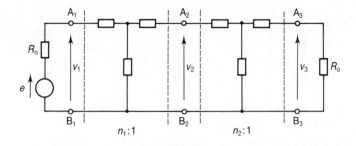

Figure 3.20 Circuit showing two T-pad attenuator sections in cascade.

Example 3.13 Determine the correct values of resistance for a 600 Ω T-pad attenuator of attenuation ratio 10:1.

Using Equation 3.7,

$$R_1/R_0 = (10 - 1)/(10 + 1) = 9/11$$

Hence

$$R_1 = 491 \ \Omega$$

From Equation 3.6,

$$R_2/R_0 = (11/9 - 9/11)/2 = 0.202$$

Hence

$$R_2 = 121 \ \Omega$$

3.9.1 ◆ Cascading attenuator sections

It is customary to limit an individual pad to less than 10:1: higher values are made by cascading sections as shown in Figure 3.20. Assuming that both pads have been designed to the same value of characteristic resistance, the resistance presented at AB by the circuit to the right will be R_0. Thus the first pad is correctly terminated in its characteristic resistance, so that $v_1/v_2 = n_1$.

Hence

$$v_1/v_3 = n_1 n_2 \tag{3.8}$$

By this means the use of single circuits of large attenuation ratio is avoided. In an arrangement such as that of Figure 3.20, source, load and attenuator pads are said to be matched to the characteristic resistance R_0.

3.9.2 ◆ Decibels

It is usual to express attenuation ratios in terms of a logarithmic measure of power ratios. This is convenient because of the very large power ratios regularly encountered in electronic systems. The measure adopted is based on the logarithm to base 10 of the ratio. The plain logarithm $\log(P_2/P_1)$ is said to be the measure of the power ratio in bels, 1 bel being defined as a ratio of 10:1. This unit is too large, and in practice power ratios are measured in decibels (dB). In decibels the ratio is given by

$$N = 10 \log(P_2/P_1) \tag{3.9}$$

Thus a ratio of 10:1 corresponds to 10 dB. Conversely, 1 dB corresponds to a power ratio of 1.259:1, and 0.1 dB to 1.023:1.

It is sometimes convenient to express an absolute power level by a decibel measure. Two units are used, decibels above 1 W (dBW) and decibels above 1 mW (dBm). Thus a power of P watts can be expressed as

$$10 \log(P) \, \text{dBW}$$

or

$$10 \log(P/10^{-3}) \, \text{dBm}$$

For example, a power of 2 mW may be expressed as 3 dBm.

The attenuation ratios n_1, n_2 used in Section 3.9.1 were voltage ratios. In a matched arrangement, such as that shown in Figure 3.20, the power entering the first pad is $P_1 = v_1^2/R_0$; the power leaving that pad and entering the second is $P_2 = v_2^2/R_0$. Thus

$$P_2/P_1 = (v_2/v_1)^2$$

The attenuation expressed in decibels is therefore given by

$$\begin{aligned} N_1 &= 10 \log(P_1/P_2) \\ &= 20 \log(v_1/v_2) \\ &= 20 \log n_1 \end{aligned} \tag{3.10}$$

In a similar way the attenuation of the second pad, N_2, and the total attenuation, N, are calculated. Equation 3.8 then leads to the expression

$$N = N_1 + N_2 \tag{3.11}$$

It is to be noted that the expression $20 \log(v_1/v_2)$ only applies when the two voltages are developed across equal resistances. If v_1 appears across R_1 and v_2 across R_2, then the decibel measure is correctly given by

$$10 \log[(v_1^2/R_1)/(v_2^2/R_2)] \tag{3.12}$$

The simpler expression of Equation 3.10 is sometimes incorrectly used as a logarithmic measure of voltage ratio: this practice leads to confusion and is best avoided.

One particular value is usefully committed to memory: a power ratio of 3 dB corresponds to 2:1 and, in the equal resistance case, to a voltage ratio of $\sqrt{2}$:1.

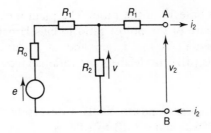

Figure 3.21 Source and T-pad attenuator regarded as an equivalent source supplying current i_2 at voltage v_2.

3.9.3 ◆ Output resistance

In the diagram of Figure 3.19 the source is shown with an internal resistance of R_0, a fact which has not hitherto been used. The final load resistance R_0 is effectively excited by a source comprising all the circuit to its left: this part of the circuit considered as a source is shown in Figure 3.21. In this circuit

$$v = v_2 + R_1 i_2$$

$$i_1 = \frac{e - v_2 - R_1 i_2}{R_0 + R_1}$$

$$i_1 = i_2 + \frac{v_2 + R_1 i_2}{R_2}$$

Equating the two expressions for i_1 we find

$$\frac{e}{R_0 + R_1} - i_2\left(1 + \frac{R_1}{R_1 + R_0} + \frac{R_1}{R_2}\right) = v_2\left(\frac{1}{R_0 + R_1} + \frac{1}{R_2}\right)$$

Using Equation 3.6 we find

$$\frac{1}{R_0 + R_1} + \frac{1}{R_2} = \frac{1}{R_0 + R_1} + \frac{2R_1}{R_0^2 - R_1^2} = \frac{1}{R_0 - R_1}$$

Hence

$$\frac{e}{R_0 + R_1} - i_2\left(1 + \frac{R_1}{R_0 - R_1}\right) = v_2\frac{1}{R_0 - R_1}$$

$$v_2 = e\frac{R_0 - R_1}{R_0 + R_1} - R_0 i_2 \tag{3.13}$$

Thus the terminals A, B in Figure 3.21 appear as the terminals of a source of voltage e/n and internal resistance R_0: the effect of the attenuator pad has been to produce a source identical except for a reduced voltage. This result is only true if the original source has the correct, matched, value of internal resistance.

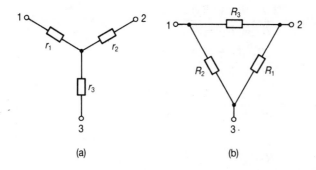

Figure 3.22 The star–delta transform: (a) star-connected resistors; (b) delta-connected resistors.

This result can be much more simply obtained by use of Thévenin's theorem. The output impedance can be found by reducing the source e to zero, that is a short circuit, when the resistance between A and B is seen to be R_0 by reason of the original choice of R_1 and R_2. The open-circuit voltage is given by

$$v_{OC} = e\frac{R_2}{R_0 + R_1 + R_2}$$

Substituting for R_2 from Equation 3.6 we find

$$v_{OC} = e\frac{R_0 - R_1}{R_0 + R_1}$$

$$= e\frac{1}{n}$$

These are the results incorporated in Equation 3.13.

3.10 ♦ The star–delta transform

The theorems of Thévenin and Norton provide a way of simplifying networks. Another useful theorem concerns the transformation of a star connection of three resistors into a delta connection, or vice versa. These two networks are shown in Figure 3.22. To determine the values of the equivalent resistors, we make the networks behave in the same way under three different sets of conditions. We firstly choose to make the same the resistances measured between any two terminals with the third left unconnected. The resulting equalities give the equations

$$r_1 + r_2 = \frac{R_3(R_1 + R_2)}{R_1 + R_2 + R_3}$$

$$r_2 + r_3 = \frac{R_1(R_2 + R_3)}{R_1 + R_2 + R_3}$$

$$r_3 + r_1 = \frac{R_2(R_3 + R_1)}{R_1 + R_2 + R_3}$$

Hence

$$r_1 = \frac{R_2 R_3}{R_1 + R_2 + R_3}$$

$$r_2 = \frac{R_3 R_1}{R_1 + R_2 + R_3} \qquad\qquad \textbf{(3.14)}$$

$$r_3 = \frac{R_1 R_2}{R_1 + R_2 + R_3}$$

These formulae provide values for the resistors in the star connection equivalent to a given delta connection. In order to find values for the resistors in the delta connection to be equivalent to a given star connection it is more convenient to deal with the resistance between one terminal and the other two shorted together. In this way we find

$$\frac{1}{R_2} + \frac{1}{R_3} = (r_2 + r_3)\,\frac{1}{S}$$

$$\frac{1}{R_3} + \frac{1}{R_1} = (r_3 + r_1)\,\frac{1}{S}$$

$$\frac{1}{R_1} + \frac{1}{R_2} = (r_1 + r_2)\,\frac{1}{S}$$

where

$$S = r_1 r_2 + r_2 r_3 + r_3 r_1$$

Hence

$$R_1 = \frac{1}{r_1}\,(r_1 r_2 + r_2 r_3 + r_3 r_1)$$

$$R_2 = \frac{1}{r_2}\,(r_1 r_2 + r_2 r_3 + r_3 r_1) \qquad\qquad \textbf{(3.15)}$$

$$R_3 = \frac{1}{r_3}\,(r_1 r_2 + r_2 r_3 + r_3 r_1)$$

If in either configuration the resistors are equal we have

$$R = 3r$$

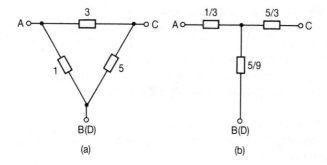

Figure 3.23 Circuits for Example 3.14: (a) delta circuit; (b) star equivalent

EXAMPLE 3.14   We can use the star–delta transform to solve the problem of Example 3.2, Figure 3.7(a), in a different way. The delta connection of resistors joining terminals A, B, C, shown in Figure 3.23(a), may be replaced by the star connection shown in Figure 3.23(b) in which, using Equations 3.14,

$$r_A = 1 \times 3/(1 + 5 + 3) = 1/3\ \Omega$$

$$r_B = 3 \times 5/9 = 5/3\ \Omega$$

$$r_C = 1 \times 5/9 = 5/9\ \Omega$$

Connecting this star in place of the delta connection in Figure 3.7(a) we have

$$v_{CD} = (5/9) \times 4/(3 + 1/3 + 5/9)$$

$$= 4/7\ \text{V}$$

This is the answer previously obtained.

3.11 ◆ Summary

Kirchhoff's Voltage Law
The total voltage drop round any closed path in a network is zero. Alternatively, any path which joins the same two nodes has the same cumulative voltage drop.

Kirchhoff's Current Law
The net current arriving at a node is zero.

Combinations of resistors
Resistors in series: $R = R_1 + R_2$.
Resistors in parallel: $G = G_1 + G_2$, or $R = R_1 R_2/(R_1 + R_2)$.

The theorems of Thévenin and Norton

These theorems concern a network with two free terminals which can be connected to a load (Figure 3.5). If for this network the voltage between the terminals on open circuit is v_{OC}, and the current on short circuit i_{SC}, with due regard to the sign convention in the figure, and the resistance R_0 is equal to v_{OC}/i_{SC}, then the following theorems apply.

> *Thévenin's theorem.* The network may be replaced, as far as a load is concerned, by an ideal voltage source of strength v_{OC} in series with a resistance R_0.

> *Norton's theorem.* The network may be replaced, as far as a load is concerned, by an ideal constant current source of strength i_{SC} in parallel with a resistance R_0.

Decibels

The measure of a power ratio in decibels is

$$N = 10\log\left(\frac{P_2}{P_1}\right)$$

$$= 10\log\left(\frac{v_2^2/R_2}{v_1^2/R_1}\right)$$

Star–delta transform

With notation as in Figure 3.22 the networks are equivalent if

$$r_1 = \frac{R_2R_3}{R_1 + R_2 + R_3}$$

$$r_2 = \frac{R_3R_1}{R_1 + R_2 + R_3}$$

$$r_3 = \frac{R_1R_2}{R_1 + R_2 + R_3}$$

or if

$$R_1 = \frac{1}{r_1}(r_1r_2 + r_2r_3 + r_3r_1)$$

$$R_2 = \frac{1}{r_2}(r_1r_2 + r_2r_3 + r_3r_1)$$

$$R_3 = \frac{1}{r_3}(r_1r_2 + r_2r_3 + r_3r_1)$$

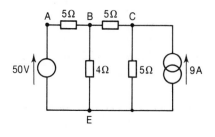

Figure 3.24

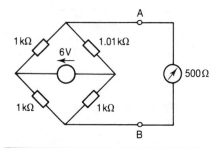

Figure 3.25

PROBLEMS

3.1 Three batteries have the following values of open-circuit voltage and internal resistance:

      6.3 V   0.3 Ω

      6.0 V   0.2 Ω

      6.6 V   0.4 Ω

They are connected in parallel with the same polarity. Determine the voltage appearing between the terminals.

3.2 The three batteries of Problem 3.1 connected in parallel supply a load drawing 3.5 A. Calculate the terminal voltage and the current supplied by each battery.

3.3 A voltage divider has the form shown in Figure 3.3(c), with $e = 15$ V. It is required that (i) when the current drawn from P is 1 mA, the voltage between P and Q shall be 5 V, and (ii) when the current is 2mA the voltage shall not be less than 4.8 V. Determine suitable values for R_1 and R_2.

3.4 For the circuit of Figure 3.24, determine the current in each resistor. Determine also the power provided by each source.

3.5 For the circuit of Figure 3.24, determine (i) the voltage between nodes A and C, and (ii) the current which would flow in a resistanceless link joining these two nodes. Hence determine the current which would flow in a link of resistance 3 Ω joining A and C.

3.6 In Figure 3.25 is shown the circuit of a Wheatstone bridge, in which the 6 V source has negligible internal resistance. Derive a Thévenin equivalent circuit for the circuit to the left of the terminals A, B. Hence estimate the current flowing in a galvanometer of resistance 500 Ω connected between A and B.

3.7 In the transistor circuit of Figure 3.26 it is required to set up the voltages and currents at the levels shown. Determine the values of the various resistors in the

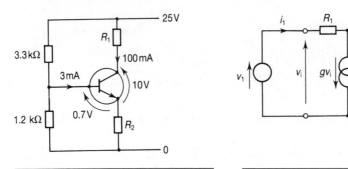

Figure 3.26   **Figure 3.27**

circuit. It may be presumed that the specified voltage and current levels are consistent with the properties of the transistor.

3.8   The circuit of Figure 3.27 contains a voltage-controlled dependent current source. Obtain an expression for the ratio v_2/v_1 and for the input resistance v_1/i_1. Show that if $R_1 \gg R_2$ these expressions approximate to

$$\frac{v_2}{v_1} \approx -gR_2, \qquad \frac{v_1}{i_1} \approx \frac{R_1}{gR_2}$$

The values in a particular circuit are

$$R_1 = 100 \text{ k}\Omega, \qquad R_2 = 1 \text{ k}\Omega, \qquad g = 0.08 \text{ S}$$

The voltage v_1 is provided from a source of 0.1 V open-circuit voltage and internal resistance 1 kΩ. Determine the value of v_2.

3.9   In Figure 3.28 is shown a circuit containing a current-controlled dependent current source. Determine the current through the 4.7 kΩ resistor and the voltage between nodes A and E.

3.10   In Figure 3.29 is shown a circuit containing a voltage-controlled dependent voltage source in which the current meter A has negligible resistance. Obtain an expression for the ratio i_2/v_1.

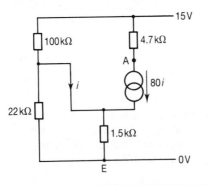

Figure 3.28

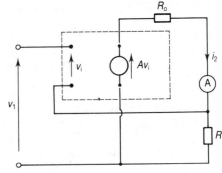

Figure 3.29

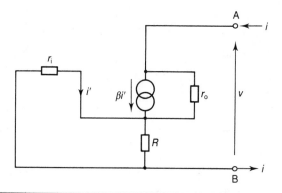

Figure 3.30

The circuit is used as an electronic voltmeter by calibrating the current meter reading i_2 on the basis that $i_2 = v_1/R$. Determine the percentage error when $R = 1 \text{ k}\Omega$, $R_0 = 100 \, \Omega$, $A = 10^4$.

3.11 The circuit of Figure 3.30, containing a current-controlled dependent current source, is connected to an external circuit at the terminals A, B. Show that the load it presents to such an external circuit is equivalent to a resistance of value

$$\frac{R[r_i + (\beta + 1)r_o] + r_i r_o}{R + r_i}$$

Determine the value when $r_o = 20 \text{ k}\Omega$, $r_i = 1.5 \text{ k}\Omega$, $\beta = 150$, $R = 2 \text{ k}\Omega$.

3.12 Determine the values of the resistors in a T-pad attenuator of 50 Ω characteristic resistance and 3 dB attenuation.

Such an attenuator is supplied by a source of 2 V open-circuit voltage and internal resistance 50 Ω. The output is connected to a mismatched load of resistance 55 Ω. Determine the voltage appearing across the load and the power dissipated in it. Express the latter in dBm.

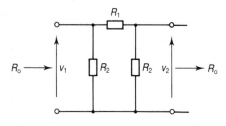

Figure 3.31

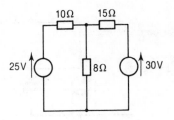

Figure 3.32

3.13 In Figure 3.31 is shown the circuit of a Π-pad attenuator. Derive the design formulae

$$R_1 = 2R_0 \frac{R_0 R_2}{R_2^2 - R_0^2}$$

$$n = \frac{v_1}{v_2} = \frac{R_2 + R_0}{R_2 - R_0}$$

in which R_0 is the characteristic resistance.

3.14 The final part of Problem 3.4 was to find the current flowing through a 3 Ω resistor connected between nodes A and C in Figure 3.24. Determine this current by use of a star–delta transformation.

3.15

(a) By using a star–delta transformation determine the current supplied by each source in the circuit of Figure 3.22.

(b) By use of a star–delta transformation derive the design formulae of Problem 3.13 from the design formulae for the T-pad attenuator, Section 3.9.

4 ◆ *Alternating Current*

This chapter considers the excitation of networks containing inductance and capacitance as well as resistance by sinusoidally varying waveforms. The device of phasor representation for sinusoidally varying quantities is introduced, leading to the characterization of each component by a complex impedance.

4.1 ◆ Introduction

In the previous chapters the principles of analysis of circuits containing only resistances have been presented. With such circuits the shape of the exciting waveform is irrelevant, since the voltage–current relationship for resistance does not involve rate of change. It is quite otherwise when the circuit contains inductance and/or capacitance: as put forward in Sections 1.3.2 and 1.3.3, the voltage–current relationships for these components are respectively

$$v = L\frac{\mathrm{d}i}{\mathrm{d}t} \tag{4.1}$$

$$i = C\frac{\mathrm{d}v}{\mathrm{d}t} \tag{4.2}$$

Thus in general the waveshape of the voltage across either of these components will be different from the waveshape of the current. The one case when the waveshapes are similar is when the excitation is sine wave AC, as described in Section 2.2.2. The trigonometrical functions $\sin(\omega t)$ and $\cos(\omega t)$ have the same shape, differing only in position on the time axis. Further, summing two such trigonometrical functions leads to another similar trigonometrical function. Thus in a linear network excited with a sinusoidal waveform, all currents and voltages will have similar waveforms, differing only in amplitude and phase. This fact simplifies analysis. It may be commented that in a strict sense steady-state AC excitation in the form of a sinusoidal waveform present throughout time is an impossibility: in practice it has to be applied at a definite instant. It will be shown in a later chapter that the network response in such a case consists of the steady-state AC response together with a switching transient. This transient in all practical cases dies away to zero in a relatively few cycles. It is thus profitable to discuss the steady-state AC response of networks.

4.2 ◆ Inductance and capacitance

When an AC waveform

$$i(t) = \hat{I}\cos(\omega t) \tag{4.3}$$

passes through an inductance application of Equation 4.1 shows that the voltage developed across it is given by

$$\begin{aligned}
v(t) &= L\hat{I}(-\omega)\sin(\omega t) \\
&= \omega L\hat{I}\cos(\omega t + \pi/2) \\
&= \hat{V}\cos(\omega t + \pi/2)
\end{aligned} \tag{4.4}$$

In accordance with the definitions given in Section 2.2.2, the voltage leads the current by $\pi/2$ (or the current lags the voltage by $\pi/2$). The coefficient ωL, equal to $\hat{V}/\hat{I}$, has the dimensions of ohms.

A similar result holds for capacitance. When a current given by Equation 4.3 flows through the capacitance we have, using Equation 4.3,

$$C\frac{\mathrm{d}v}{\mathrm{d}t} = \hat{I}\cos(\omega t)$$

Integrating, we find

$$\begin{aligned}
v(t) &= \hat{I}\,\frac{1}{\omega C}\sin(\omega t) \\
&= \hat{I}\,\frac{1}{\omega C}\cos(\omega t - \pi/2) \\
&= \hat{V}\cos(\omega t - \pi/2)
\end{aligned} \tag{4.5}$$

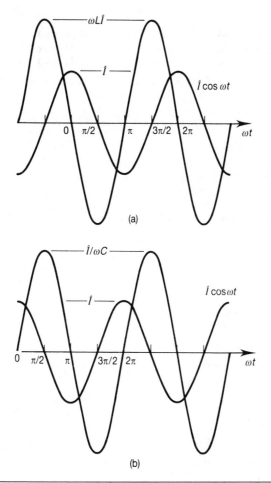

Figure 4.1 Current and voltage waveforms relating to (a) inductance and (b) capacitance.

Thus in the case of a capacitance the voltage lags the current (or the current leads the voltage) by $\pi/2$. The ratio $\hat{V}/\hat{I}$ is equal to $1/\omega C$, which also has dimensions of ohms.

These waveforms are illustrated in Figure 4.1.

4.3 ◆ Impedance

The results of the last section have shown that for both inductance and capacitance the voltage developed in response to a sinusoidal current flow is a sine wave of the same frequency, related to the current waveform by a

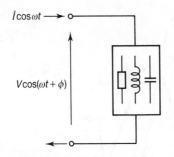

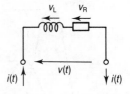

Figure 4.3 The series resistance–inductance circuit.

Figure 4.2 Schematic diagram illustrating excitation of an arbitrary circuit containing inductance, capacitance and resistance by steady-state AC.

magnitude and a phase angle. For comparison, when a current given by Equation 4.3 flows through a resistance, R, the voltage developed is given by

$$v(t) = R\hat{I} \cos(\omega t)$$
$$= \hat{V} \cos(\omega t) \tag{4.6}$$

Thus resistance is characterised by a magnitude, equal to the resistance R, with a phase angle of zero between voltage and current.

We next consider a generalized component with two terminals, composed of any sequence of interconnections between those terminals of resistance, inductance and capacitance, as represented schematically in Figure 4.2. As with the other cases, analysis shows that the voltage resulting from the current of Equation 4.3 can be written in the form

$$v(t) = Z\hat{I} \cos(\omega t + \phi)$$
$$= \hat{V} \cos(\omega t + \phi) \tag{4.7}$$

The action of this component can also be characterized by a magnitude, Z, giving the ratio of peak voltage to peak current, together with a phase angle ϕ. As with ωL and $1/\omega C$, Z is measured in ohms. It is referred to as the **impedance** of the component.

4.3.1 ◆ Series *R–L* circuit

As an example consider the circuit consisting of resistance in series with inductance (Figure 4.3). Taking the expression of Equation 4.3 for $i(t)$, we have from Equations 4.4 and 4.6

$$v_R = R\hat{I} \cos(\omega t)$$
$$v_L = \omega L\hat{I} \cos(\omega t + \pi/2) = -\hat{I}\omega L \sin(\omega t)$$

Hence

$$v(t) = v_R + v_L = \hat{I}[R\cos(\omega t) - \omega L \sin(\omega t)$$
$$= \hat{I}\,(R^2 + \omega^2 L^2)^{1/2}\cos(\omega t + \phi) \tag{4.8}$$

in which $\phi = \tan^{-1}(\omega L/R)$. Thus for this circuit

$$Z = (R^2 + \omega^2 L^2)^{1/2}$$

EXAMPLE 4.1 An inductor is equivalent to a resistance of 15 Ω in series with an inductance of 80 mH. Determine the impedance and phase angle at (i) 50 Hz and (ii) 1 kHz.

(i) $Z = [15^2 + (100\pi \times 0.08)^2]^{1/2} = 29.3 \,\Omega$
   $\phi = \tan^{-1}(100\pi \times 0.08/15) = 59°$

(ii) $Z = 503 \,\Omega$, $\phi = 88°$

4.4 ◆ Representation of sinusoidal waveforms

The analysis of an AC network is basically similar in form to the analysis encountered in Chapter 3, the main difference being that all the voltages and currents are not just numbers but sine waves of the same frequency but each with its own amplitude and phase. There is thus a need for a simple way of adding or subtracting a number of such expressions. An example of this process occurred in the derivation of Equation 4.8 in the last section. The fact that the result must also be a sine wave of the same frequency makes a simple technique possible.

Consider the sum

$$v(t) = v_1(t) + v_2(t) + v_3(t) \ldots \tag{4.9}$$

in which

$$v_1(t) = \hat{V}_1\cos(\omega t + \phi_1)$$
$$v_2(t) = \hat{V}_2\cos(\omega t + \phi_2)$$
$$v_3(t) = \hat{V}_3\cos(\omega t + \phi_3)$$

and similarly for further terms. The sum can be expressed in the form

$$v(t) = \hat{V}\cos(\omega t + \phi)$$

A graphical method of performing the addition is illustrated in Figure 4.4(a). In this figure, the first term is represented by the line OP_1, of length $\hat{V}_1$ and making an angle with the reference axis of $\omega t + \phi_1$; the second term is represented by the line P_1P_2 of length $\hat{V}_2$ at an angle of

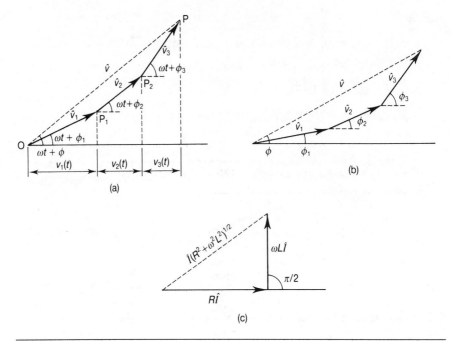

Figure 4.4 Illustrating a graphical method for the addition of several sinusoids of the same frequency: (a) $\sum \hat{v}_i \cos(\omega t + \phi_i)$; (b) reduced form; (c) applied to the series R–L circuit, Equation 4.8.

$\omega t + \phi_2$; the other terms can be represented by similar lines. The actual value of each term in Equation 4.9 is given by the projection of its representative line on the reference axis. In the same representation the line OP has the correct projection on the reference axis, so that $\hat{V}$ and ϕ can be determined. Clearly these values can equally well be determined with the aid of the diagram of Figure 4.4(b): there is no need to have the actual voltage represented on the diagram since it can easily be calculated from $\hat{V} \cos(\omega t + \phi)$ once $\hat{V}$ and ϕ are known. Using this method, the result expressed by Equation 4.8 can be derived from the diagram of Figure 4.4(c).

Another way of regarding the diagram of Figure 4.4(b) is as an Argand diagram for complex numbers: in this light the diagram shows the addition of the complex numbers $\hat{V}_1 \exp(j\phi_1)$, $\hat{V}_2 \exp(j\phi_2)$, $\hat{V}_3 \exp(j\phi_3)$, etc. to give a resultant $\hat{V} \exp(j\phi)$. In this representation the process of forming the resultant of the three terms can be expressed as the addition

$$\hat{V} \exp(j\phi) = \hat{V}_1 \exp(j\phi_1) + \hat{V}_2 \exp(j\phi_2) + \hat{V}_3 \exp(j\phi_3) \qquad (4.10)$$

Granted that sine waves of the same angular frequency are being referred to, the connection between an individual term and its complex representation is simple to write down in algebraic form. Consider $v_2(t)$. We have

$$\hat{V}_2 \exp(j\phi_2) = \hat{V}_2 \cos(\phi_2) + j\hat{V}_2 \sin(\phi_2) \tag{4.11}$$

We regard the real part of this expression as the amplitude of a term $\cos(\omega t)$, and the imaginary part as the amplitude of a term $\cos(\omega t + \pi/2)$. This process interprets the imaginary j as signifying a phase angle of $\pi/2$ leading. With this interpretation, Equation 4.11 would represent a physical voltage given by

$$
\begin{aligned}
v_2(t) &= \hat{V}_2 \cos(\phi_2) \cos(\omega t) + \hat{V}_2 \sin(\phi_2) \cos(\omega t + \pi/2) \\
&= \hat{V}_2 [\cos(\phi_2) \cos(\omega t) - \sin(\phi_2) \sin(\omega t)] \\
&= \hat{V}_2 \cos(\omega t + \phi_2)
\end{aligned}
\tag{4.12}
$$

This is the correct expression for $v_2(t)$.

An even simpler way of deriving the expression for $v_2(t)$, easily verified, is

$$v_2(t) = \mathrm{Re}[\hat{V}_2 \exp(j\phi_2) \exp(j\omega t)] \tag{4.13}$$

We see therefore that the resultant of several sinusoidal voltages can be found both by graphical means and by complex representation.

4.5 ♦ Complex impedance

It has been seen in Section 4.3 that the relation between voltage and current for a two-terminal component can be characterized by a magnitude and a phase angle, there referred to as Z and ϕ respectively. These can be combined to form a complex number

$$\mathbf{Z} = Z \exp(j\phi) \tag{4.14}$$

A frequently used alternative is to write this quantity in the form

$$\mathbf{Z} = Z \angle \phi$$

The complex number $\mathbf{Z}$ is known as the **complex impedance** of the component: it will be shown to play a part in AC analysis analogous to that played by resistance in DC analysis. In AC analysis it is nearly always the complex impedance which is being dealt with, so this quantity often becomes abbreviated to 'impedance'. This term was used in Equation 4.7 for the magnitude Z. It is now seen to be the modulus of the complex impedance $\mathbf{Z}$. Which of the two is being referred to is usually clear from the context.

Equation 4.7 was derived assuming a current waveform given by $i(t) = \hat{I} \cos(\omega t)$, for which the complex representation formed in the last section will be

$$\mathbf{I} = \hat{I} \exp{(\mathrm{j}0)}$$

The representation for the voltage $v(t)$ in Equation 4.7 will be

$$\mathbf{V} = Z\hat{I} \exp{(\mathrm{j}\phi)}$$

This latter expression may be obtained directly by multiplying the representation for $i(t)$ by the complex impedance defined in Equation 4.14:

$$\hat{V} \exp{(\mathrm{j}\phi)} = \mathbf{Z}[\hat{I} \exp{(\mathrm{j}0)}]$$

or

$$\mathbf{V} = \mathbf{Z}\mathbf{I}$$

This is strictly comparable with the relation

$$v = Ri$$

as used for a resistance. For an inductance

$$\mathbf{Z} = \omega L \exp{(\mathrm{j}\pi/2)} = \mathrm{j}\omega L \tag{4.15}$$

For a capacitance

$$\mathbf{Z} = \frac{1}{\omega C} \exp{(-\mathrm{j}\pi/2)} = -\mathrm{j}\frac{1}{\omega C}$$

$$= \frac{1}{\mathrm{j}\omega C} \tag{4.16}$$

For a resistance, since there is no phase difference, the impedance will be

$$\mathbf{Z} = R \exp{(\mathrm{j}0)}$$

This is a real number equal to the resistance. Conversely, a purely real impedance is equivalent to a resistance. A purely imaginary impedance, such as that of an inductance or capacitance, is called a **reactance**. The numerical value, which may be positive or negative, is usually denoted by the symbol X. For inductance and capacitance, the reactances are respectively

$$X_{\mathrm{L}} = \omega L$$
$$X_{\mathrm{C}} = -\frac{1}{\omega C} \tag{4.17}$$

Like resistance, values of reactance are given in ohms.

In general, a complex impedance will have both real and imaginary parts, and is usually expressed in the form

$$\mathbf{Z} = R + \mathrm{j}X \tag{4.18}$$

In such an expression R is referred to as the **resistive** component of the impedance and X as the **reactive** component.

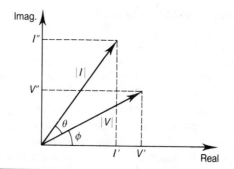

Figure 4.5 Representation of a phasor on an Argand diagram, Section 4.6.

4.6 ◆ **Phasors**

The use of a complex representation for voltage and current is systematized in phasors. The only difference between phasors and the complex representation just used lies in the use of RMS values rather than peak values. The reason for this is, as shown in Section 2.5.3, that with sinusoidal waveforms power is most simply expressed in terms of RMS values. A typical phasor

$$\mathbf{V} = V' + jV'' = |\mathbf{V}|\exp(j\phi) \tag{4.19}$$

represents a physical voltage of RMS value $|\mathbf{V}|$ given by the expression

$$v(t) = \sqrt{2}[V'\cos(\omega t) + V''\cos(\omega t + \pi/2)]$$
$$= \sqrt{2}|\mathbf{V}|\cos(\omega t + \phi) \tag{4.20}$$

An alternative expression from which both forms can be derived is

$$v(t) = \sqrt{2}\,\mathrm{Re}\,[\mathbf{V}\exp(j\phi)] \tag{4.21}$$

The phasor of Equation 4.19 can be represented on an Argand diagram as shown in Figure 4.5. In a similar way, the current phasor

$$\mathbf{I} = I' + jI'' = |\mathbf{I}|\exp(j\theta) \tag{4.22}$$

represents the physical current

$$i(t) = \sqrt{2}\,\mathrm{Re}\,[\mathbf{I}\exp(j\theta)]$$
$$= \sqrt{2}\,[I'\cos(\omega t) + I''\cos(\omega t + \pi/2)]$$
$$= \sqrt{2}|\mathbf{I}|\cos(\omega t + \theta) \tag{4.23}$$

This phasor representation also appears in Figure 4.5.

It is difficult to be entirely consistent in the use of symbols for voltage and current, but in general in this book physical quantities will be denoted by lower-case letters and their phasor equivalents by the corresponding bold capital letter.

4.7 ◆ AC analysis using phasors

The principles used in the analysis of resistance-only networks in Chapter 3 also hold good whatever the nature of the components: the difference in application is that all voltages and currents are functions of time, and equalities must be satisfied at all instants. In the case of sinusoidal AC excitation, the introduction of phasors allows us to express the phasor representation of a number of voltages or currents as the sum of the individual phasor representations. Thus, if currents $i_1(t)$, $i_2(t)$, $i_3(t)$ are each sine waves of the form $\hat{I}\cos(\omega t + \theta)$, then the typical equation

$$i_1(t) + i_2(t) + i_3(t) = 0$$

is satisfied if

$$\mathbf{I}_1 + \mathbf{I}_2 + \mathbf{I}_3 = 0$$

in which $\mathbf{I}_1 = (1/\sqrt{2})\hat{I}_1\exp(j(\theta)$ and similarly for $\mathbf{I}_2$ and $\mathbf{I}_3$. In the same way, the equality of two voltages $v_1(t)$, $v_2(t)$ is ensured by the equality of the phasor representations, $\mathbf{V}_1$, $\mathbf{V}_2$. It follows that the equations derived from network connections by the application of Kirchhoff's laws are exactly the same in phasor form as in real form. It is also the case that the voltage–current relationships for the components have, when expressed in phasor form, the same algebraic form as Ohm's law for resistance. These similarities mean that analysis of networks with sinusoidal AC excitation can be carried out using phasor representations in a manner algebraically identical to that used for resistance-only networks, the only difference being that the quantities involved are complex rather than real. This similarity leads to a convenient but potentially confusing terminology in which phasor equivalents are often referred to as 'voltage' or 'current', quoted in volts and amps. However, it must never be forgotten that phasors are representative of real, physical variables related to the phasors as in Section 4.6.

EXAMPLE 4.2: The series *R–L* circuit Relations for this circuit were derived in Section 4.3.1. Consider now the approach using phasors. The circuit is redrawn in Figure 4.6(a), in which $\mathbf{V}$, $\mathbf{I}$ are phasor equivalents. We now have

$$\mathbf{V}_L = j\omega L\mathbf{I}$$

$$\mathbf{V}_R = R\mathbf{I}$$

As discussed above, we simply add $\mathbf{V}_L$ and $\mathbf{V}_R$ to find $\mathbf{V}$:

$$\mathbf{V} = \mathbf{V}_L + \mathbf{V}_R = j\omega L\mathbf{I} + R\mathbf{I}$$

$$= (j\omega L + R)\mathbf{I} \tag{4.24}$$

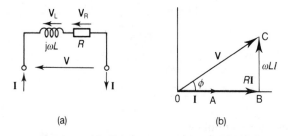

Figure 4.6 Phasor analysis of the *R–L* circuit of Figure 4.3: (a) circuit with phasor notation; (b) phasor diagram.

This shows that the impedance of L and R in series is

$$\mathbf{Z} = j\omega L + R \tag{4.25}$$

Hence impedances add in the same way as resistances. The phasor diagram for this calculation is shown in Figure 4.6(b): the line OA represents the current phasor $\mathbf{I}$, taken as real, the line OB represents $\mathbf{V}_R$ and the line BC represents $\mathbf{V}_L$.

4.8 ◆ Admittance

It has been seen that a resistance can be given a numerical value either by quoting its resistance, R, or its conductance, $G = 1/R$. Similarly, it is useful on occasion to deal with the reciprocal of impedance. This quantity is termed **admittance** and is denoted by the symbol $\mathbf{Y}$:

$$\mathbf{Y} = \frac{1}{\mathbf{Z}} = G + jB \tag{4.26}$$

The real part, G, is termed conductance, and the imaginary part, B, is termed **susceptance**. The conductance and susceptance of a component are related to the resistance and reactance of that component by

$$G + jB = \frac{1}{R + jX}$$

Hence

$$G = \frac{R}{R^2 + X^2}$$

$$B = -\frac{X}{R^2 + X^2} \tag{4.27}$$

Inverse expressions are

$$R = \frac{G}{G^2 + B^2}$$

$$X = -\frac{B}{G^2 + B^2} \tag{4.28}$$

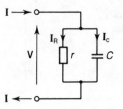

Figure 4.7 Circuit for Example 4.3.

EXAMPLE 4.3 Obtain an expression for the impedance of a resistance r in parallel with a capacitance C.

The circuit is shown in Figure 4.7. We have, using Equation 4.16,

$$\mathbf{V} = \mathbf{I}_R r$$

$$\mathbf{V} = \mathbf{I}_C \frac{1}{j\omega C}$$

Hence

$$\mathbf{I} = \mathbf{I}_R + \mathbf{I}_C = \mathbf{V}\left(\frac{1}{r} + j\omega C\right)$$

The admittance is therefore given by

$$\mathbf{Y} = \frac{1}{r} + j\omega C$$

showing that admittances in parallel add. The impedance is given by

$$\mathbf{Z} = \left(\frac{1}{r} + j\omega C\right)^{-1} = \frac{r}{1 + j\omega Cr}$$

$$= r\frac{1 - j\omega Cr}{1 + (\omega Cr)^2}$$

This is the required expression. The resistive part is given by

$$R = r\frac{1}{1 + (\omega Cr)^2}$$

$$X = r\frac{\omega Cr}{1 + (\omega Cr)^2}$$

4.9 ◆ Power

It is important to remember that phasors have been designed to ease the addition of sinusoidal waveforms, which is a linear operation. Calculation of power, as emphasised in Section 2.5, involves the non-linear operation of multiplication. Straightforward multiplication of phasors as a way of determining power is meaningless. However, it is possible to derive a phasor expression from which power can be calculated. It was shown in Section 2.5.2 that the average power associated with AC excitation could be calculated in the form of Equation 2.16:

$$P = V_{RMS} I_{RMS} \cos(\theta)$$

in which θ is the phase angle between voltage and current. The phasor representation of the voltage will be of the form

$$\mathbf{V} = V_{RMS} \exp(j\alpha)$$

and that of the current

$$\mathbf{I} = I_{RMS} \exp[j(\alpha + \theta)]$$

The inclusion of the angle α allows for the fact that in general neither the voltage nor the current will be in phase with the reference voltage: α is the phase angle of the voltage with respect to the reference voltage. We form the product

$$\mathbf{VI}^* = |\mathbf{V}| |\mathbf{I}| \exp(-j\theta)$$

in which $\mathbf{I}^*$ denotes the complex conjugate of $\mathbf{I}$, formed by changing the sign of j wherever it occurs. Since $|\mathbf{V}|$ and $|\mathbf{I}|$ are RMS values, a correct expression for power is given by

$$P = \text{Re}(\mathbf{VI}^*) \tag{4.29}$$

This is a convenient form from which to calculate the average power directly from phasor expressions.

EXAMPLE 4.4   An impedance consists of a resistance of 100 Ω in parallel with the series combination of a resistance of 10 Ω and an inductive reactance of 50 Ω. The impedance is connected to a supply of (RMS) voltage 240 V. Determine values for

(i) the magnitude and phase of the impedance,
(ii) the magnitude of the phasor current and its phase relative to the supply, and
(iii) the power drawn from the source.

Calculate also the power dissipated in each of the resistors and verify that the total agrees with the value obtained in (iii).

    The circuit is shown in Figure 4.8. Denoting the supply by the (real) phasor $\mathbf{V}$, we have

$$\mathbf{I}_R = \mathbf{V}/100, \quad \mathbf{I}_L = \mathbf{V}/(10 + j50)$$

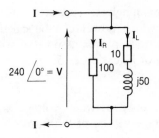

Figure 4.8 Circuit for Example 4.4.

Therefore

$$\mathbf{I} = \mathbf{I}_R + \mathbf{I}_L = \mathbf{V}[1/100 + 1/(10 + j50)]$$

$$= (\mathbf{V}/100)[1 + (10 - j50)/26]$$

$$= (\mathbf{V}/100)[1.385 - j1.923]$$

$$= \mathbf{V} \times 0.023\,70 \angle -54.25°$$

The admittance is therefore

$$\mathbf{Y} = 0.023\,70 \angle -54.25°$$

and the impedance

(i) $\mathbf{Z} = 1/\mathbf{Y} = 42.2 \angle 54.25°$

The magnitude of $\mathbf{Z}$ is 42.2 Ω with phase angle 54.25°. Putting

$$\mathbf{V} = 240 \angle 0°$$

we find

(ii) $\mathbf{I} = 240\mathbf{Y} = 5.69 \angle -54.25°$

The magnitude of the current is therefore 5.69 A, lagging on the supply by 54.3°.

For the power we form

$$P = \mathrm{Re}\,[\mathbf{VI}^*] = 240 \times 5.69 \times \cos(54.25°)$$

(iii) $= 798$ W

The power in the 100 Ω resistor is given by

$$(240)^2/100 = 576 \text{ W}$$

The power in the 10 Ω resistor is given by

$$10|\mathbf{I}_L|^2 = 10|240/(10 + j50)|^2$$

$$= 10 \times 240^2/2600$$

$$= 221.5 \text{ W}$$

The total agrees with the previous figure.

4.10 ◆ Volt–amps reactive

It was shown in the last section that the real part of $\mathbf{VI}^*$ is equal to power. Since energy is conserved, it follows that in a case such as that of the generalized component referred to in Section 4.3 the power delivered from the source must be equal to the total power consumed in the separate resistors. It is not immediately obvious that the imaginary part of $\mathbf{VI}^*$ is conserved in a similar way. A formal proof that this is so will be found in Appendix A, Section A.5. Assuming the correctness of the assertion, we have

$$\mathbf{S} = (\mathbf{VI}^*)_{\text{total}} = P + jQ = \sum_{\text{comp}} \mathbf{VI}^* \qquad (4.30)$$

Although $\mathrm{Im}\,(\mathbf{VI}^*)$ has the same dimensions as power it represents a different physical quantity, and is referred to as **volt–amps reactive**, abbreviated to VAR. It is also sometimes called 'reactive power'. For an inductance

$$\mathbf{V} = j\omega L\mathbf{I}$$

Hence

$$\mathrm{Im}\,(\mathbf{VI}^*) = \omega L|\mathbf{I}|^2 = \frac{1}{\omega L}|\mathbf{V}|^2 \qquad (4.31)$$

For a capacitance

$$\mathbf{V} = \frac{1}{j\omega C}\mathbf{I}$$

and

$$\mathrm{Im}\,(\mathbf{VI}^*) = -\frac{1}{\omega C}|\mathbf{I}|^2 = -\omega C|\mathbf{V}|^2 \qquad (4.32)$$

We shall verify equation 4.30 for some examples.

EXAMPLE 4.5 Verify the conservation of volt–amps reactive in the case of Example 4.4.
 In that example

$$\mathrm{Im}\,(\mathbf{VI}^*) = 240 \times 5.69 \times \sin\,(54.25°)$$

$$= 1108 \text{ VAR}$$

For the inductance $\omega L = 50\ \Omega$, so that

$$\omega L|\mathbf{I}|^2 = 50 \times (240)^2/2600 = 1108 \text{ VAR}$$

EXAMPLE 4.6 An AC source of 100 V RMS supplies the load for which the circuit and values of the component impedances are shown in Figure 4.9. Determine in magnitude and phase the impedance of the load, the

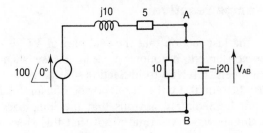

Figure 4.9 Circuit for Example 4.6.

current drawn from the source and the voltage developed between terminals A, B. (Phase is to be expressed relative to the source voltage.) Determine also the power and volt–amps reactive supplied by the source.

In the following solution the units of volt, amp and ohm are implied. Denoting the impedance between A and B by $\mathbf{Z}_{AB}$, we have

$$\frac{1}{\mathbf{Z}_{AB}} = \frac{1}{10} + \frac{1}{-j20}$$
$$= 0.1(1 + j0.5)$$

Hence

$$\mathbf{Z}_{AB} = 10(1 - j0.5)/1.25$$
$$= 8 - j4 = 8.94 \angle -26.6°$$

The total impedance is therefore given by

$$\mathbf{Z} = 5 + j10 + 8 - j4$$
$$= 13 + j6 = 14.3 \angle 24.8°$$

The phasor current **I** is given by

$$\mathbf{I} = 100/\mathbf{Z} = 6.99 \angle -24.8°$$

The phasor representing the voltage between A and B is

$$\mathbf{V}_{AB} = \mathbf{Z}_{AB}\mathbf{I} = 8.94 \times 6.99 \angle(-26.6 - 24.8)°$$
$$= 62.5 \angle -51.4°$$

To find the power and volt–amps reactive we form

$$\mathbf{VI}^* = P + jQ = 100 \times 6.99 \angle 24.8°$$
$$= 634.5 + j293.2$$

Hence $P = 635$ W, $Q = 293$ VAR. Evaluating the amounts for individual components, we have for the power

$$5|\mathbf{I}|^2 + |\mathbf{V}_{AB}|^2/10 = 635$$

For the volt–amps reactive we have

$$10|\mathbf{I}|^2 - |\mathbf{V}_{AB}|^2/20 = 293$$

4.10.1 ◆ Power factor

The expression for average power was derived in Section 2.5.2 and repeated in Section 4.9:

$$P = V_{RMS} I_{RMS} \cos(\theta)$$

It has been pointed out that apart from the factor $\cos(\theta)$ the expression is similar to the expression for power in the case of DC excitation. The factor $\cos(\theta)$ is often referred to as the **power factor**. In the field of AC mains distribution, loads are often specified in terms of power and power factor at a rated voltage rather than in terms of impedance. For example, a load might be specified in the form of 5 kW, 0.8 lagging power factor designed to work at 240 V (RMS). For this load at the rated voltage

$$240 \times I_{RMS} \times 0.8 = 5000$$

$$I_{RMS} = 26.0 \text{ A}$$

The phase angle of the current with respect to the voltage is

$$-\cos^{-1}(0.8) = -36.9°$$

In terms of phasors we may therefore write for the voltage

$$\mathbf{V} = 240 \angle 0°$$

and for the current

$$\mathbf{I} = 26.0 \angle -36.9°$$

The impedance of the load is therefore

$$\mathbf{Z} = \frac{\mathbf{V}}{\mathbf{I}} = \frac{240}{26.0} \angle 36.9°$$

$$= 9.23 \angle 36.9°$$

$$= 9.23(0.8 + j0.6) = 7.38 + j5.54$$

4.10.2 ◆ Complex power

As discussed in the last two sections, both the real and imaginary parts of the complex quantity $\mathbf{S} = P + jQ$ defined in Equation 4.30 have dimensions of power although representing different physical quantities. The name **complex power** or is sometimes given to $\mathbf{S}$. The volt–amps reactive, Q, is often termed **reactive power**.

4.11 ◆ AC measurements and phasors

The closing remarks of Section 4.7 warned against the dangers of confusing phasor representations with the real, physical variable. Nevertheless, for many purposes it is not necessary to make explicit reference to the real variable. The reason for this lies in the use made of AC signals. In most systems, at some point the signal feeds a transducer which responds to a particular feature of the signal, such as power or peak voltage. Both these quantities and many others can be calculated directly from the phasor representation without making other than implicit use of the sinusoidal nature of the signal. The real, numerical outputs from measuring devices are therefore in a sense directly connected with phasor quantities. AC meters are almost always calibrated in RMS values. Some meters, such as those relying on a heating effect, measure true RMS values (and will do so for non-sinusoidal waveforms). Others effectively measure peak values, although the calibration converts to RMS: such meters give correct measure only for sine waves. A useful device is one which accepts two inputs and measures the time average of the product. The output will depend on a product of the form $V_1 V_2 \cos \theta$, in which V_1, V_2 are RMS values and θ the difference in phase between the two inputs. With separate measurement of the amplitudes it is then possible to deduce the phase difference.

Apart from such meter types of measurements, waveform measurements can be made using the oscilloscope. This technique provides the most complete information, although determination of amplitude and phase is limited to an accuracy of a few per cent. Sampling oscilloscopes can be used to obtain a digitized record of signal waveforms, from which other quantities can be derived by computation.

Impedances can be measured directly using AC bridges. Such bridges effectively determine the numerical value of resistive and reactive components at a given frequency.

In these various ways there is a direct link between the mathematical device of phasor representation and the real world of measurement.

4.12 ◆ Impedances in series and parallel

It was shown in Section 3.4 that two resistors in series are equivalent to a single resistor whose resistance is the sum of the individual resistances, and that two resistors in parallel are equivalent to a single resistor whose conductance is the sum of the individual conductances. Similar results hold for impedances.

4.12.1 ◆ Impedances in series

A series circuit is shown in Figure 4.10, in which two impedances $\mathbf{Z}_1$, $\mathbf{Z}_2$ carry the same current. If $\mathbf{I}$ denotes the phasor current, the phasor voltages are

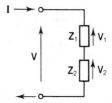

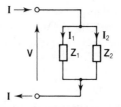

Figure 4.10 Combination of two impedances, $\mathbf{Z}_1$, $\mathbf{Z}_2$ in series, Section 4.12.1.

Figure 4.11 Combination of two impedances, $\mathbf{Z}_1$, $\mathbf{Z}_2$ in parallel, Section 4.12.2.

$$\mathbf{V}_1 = \mathbf{Z}_1\mathbf{I}$$

$$\mathbf{V}_2 = \mathbf{Z}_2\mathbf{I}$$

The phasor equivalent for the voltage across the combination, $\mathbf{V}$, is the sum of the individual phasor voltages, so that

$$\mathbf{V} = \mathbf{V}_1 + \mathbf{V}_2 = \mathbf{Z}_1\mathbf{I} + \mathbf{Z}_2\mathbf{I}$$

$$= (\mathbf{Z}_1 + \mathbf{Z}_2)\mathbf{I}$$

The combination is therefore equivalent to the impedance $\mathbf{Z}_1 + \mathbf{Z}_2$.

4.12.2 ◆ Impedances in parallel

In the parallel combination shown in Figure 4.11, the phasor current $\mathbf{I}$ is the sum of the individual phasor currents $\mathbf{I}_1$, $\mathbf{I}_2$:

$$\mathbf{I} = \mathbf{I}_1 + \mathbf{I}_2 = \mathbf{V}\mathbf{Y}_1 + \mathbf{V}\mathbf{Y}_2$$

$$= (\mathbf{Y}_1 + \mathbf{Y}_2)\mathbf{V}$$

Hence the admittance of the parallel combination is the sum of the individual admittances, or in terms of impedance

$$\mathbf{Z} = \frac{1}{\mathbf{Y}} = \frac{\mathbf{Z}_1\mathbf{Z}_2}{\mathbf{Z}_1 + \mathbf{Z}_2}$$

One important difference between these formulae and those for resistance is to be noted: whereas two resistors in series or in parallel can be replaced physically by single resistors it may not be physically possible to reduce the number of components making up a combination of impedances. This is indicative of the representational, non-physical nature of phasor variables. As an example, consider the two impedances shown in Figure 4.12(a), one incorporating an inductive reactance, the other a capacitative reactance. In the series combination we can amalgamate the resistive components but not the reactive ones, as shown in Figure 4.12(b). In the parallel combination no physical reduction is possible (Figure 4.12(c)). Combination and replacement by different values of the same

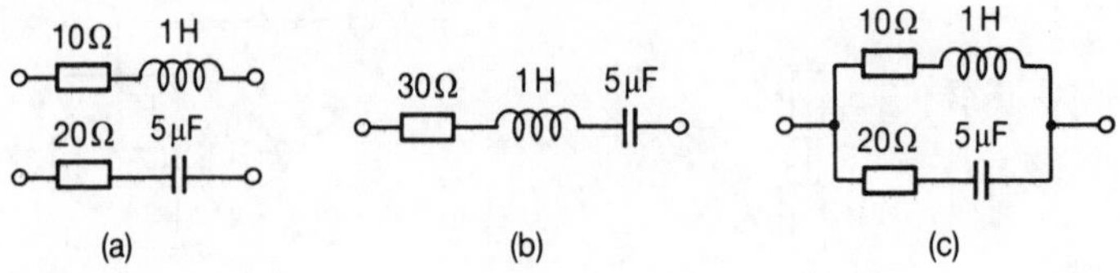

Figure 4.12 Illustrating the combination of sample impedances: (a) circuits; (b) series combination; (c) parallel combination.

component type is possible if one impedance is a real, positive multiple of the other. Thus if $\mathbf{Z}_1 = \lambda\mathbf{Z}_2$, with λ real, then in series

$$\mathbf{Z} = \mathbf{Z}_1 + \mathbf{Z}_2 = (\lambda + 1)\mathbf{Z}_2$$

The parallel combination has an impedance given by

$$\mathbf{Z} = \left(\frac{1}{\mathbf{Z}_1} + \frac{1}{\mathbf{Z}_2}\right)^{-1} = \frac{\lambda}{\lambda + 1}\mathbf{Z}_2$$

These formulae signify the same proportional change in both resistance and reactance, representing a changed value of the type of component. As a consequence, for ideal inductors a series combination gives

$$L = L_1 + L_2$$

For a parallel combination

$$L = \frac{L_1 L_2}{L_1 + L_2}$$

For ideal capacitors in series

$$\frac{1}{C} = \frac{1}{C_1} + \frac{1}{C_2}$$

or

$$C = \frac{C_1 C_2}{C_1 + C_2}$$

In parallel

$$C = C_1 + C_2$$

Combination may also be possible if only one frequency, or perhaps a narrow range of frequencies, is of interest. In this situation the reactive parts can be assigned numerical values which may then be combined and realized by a single component. This process is illustrated in Figure 4.13 for the impedances shown in Figure 4.12(a). At a frequency of 50 Hz the values become those shown in Figure 4.13(a), which when combined become those shown in Figure 4.13(b). This impedance is also given by the circuit of Figure 4.13(c), but at no other frequency is the equivalence correct.

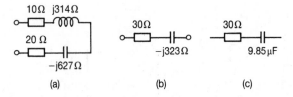

Figure 4.13 Reduction at a single frequency: (a) impedance of Figure 4.12(a) in numerical form at 50 Hz: (b) combined in numerical form; (c) reduced circuit with the same impedance at 50 Hz only.

4.12.3 ◆ Series–parallel equivalents

It was shown in Section 4.8 that an impedance $R + jX$ has an admittance $G + jB$ given by

$$G = \frac{R}{R^2 + X^2}, \ B = -\frac{X}{R^2 + X^2}$$

The admittance $G + jB$ may be represented as the parallel combination of a resistance R' and a reactance X' where

$$R' = 1/G = R\left[1 + \left(\frac{X}{R}\right)^2\right]$$

$$X' = -\frac{1}{B} = X\left[1 + \left(\frac{R}{X}\right)^2\right]$$

(4.33)

Hence we may regard the series impedance $R + jX$ as having the parallel equivalent shown in Figure 4.14. This equivalence can always be realized at a single frequency but not over a range of frequencies. Consider, for example, the case of a resistance R in series with an inductance L, for which $X = \omega L$. It will be seen that R' varies with frequency and so cannot be physically realized by a simple resistor. Likewise, X' could not be realized by a simple inductor.

4.13 ◆ The theorems of Thévenin and Norton

In Section 3.6 these theorems were introduced in connection with resistance-only networks. It was emphasized in that section that the correctness of these theorems was a consequence of the linearity of the network equations. In Section 4.7 it was shown that, when formulated in terms of phasors and complex impedances, the equations for networks with sinusoidal AC excitation are of the same algebraic form as those for resistance-only networks with DC excitation. This being so, a form of Thévenin's theorem must apply. Consider a network containing resistance, inductance and capacitance, excited by a number of sources of precisely the same

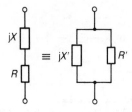

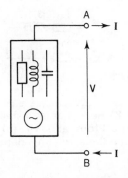

Figure 4.14 Equivalence between series and parallel forms for an impedance at a single frequency, Section 4.12.3.

Figure 4.15 Schematic representation showing phasor variables for a general AC network with two free terminals.

frequency and having two 'free' terminals. This is illustrated schematically in Figure 4.15. Analysis using phasors will give as the equivalent of Equation 3.4 equations of the form

$$V = V_o - Z_o I \tag{4.34}$$

or

$$I = I_o - \frac{V}{Z_o} \tag{4.35}$$

in which $I_o = V_o/Z_o$. A formal proof of this equation will be found in Appendix A, Section A.3 Equation 4.34 can be given the circuit interpretation shown in Figure 4.16(a), which is the Thévenin equivalent; Equation 4.35 can be given the circuit interpretation shown in Figure 4.16(b), the Norton equivalent circuit. The two phasor constants V_o, I_o have respectively the significance of phasor representations of the voltage on

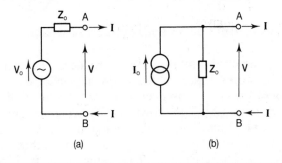

(a)   (b)

Figure 4.16 The theorems of Thévenin and Norton in phasor form for AC networks.

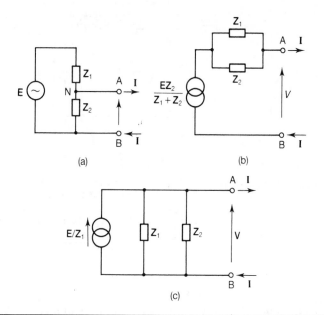

Figure 4.17 An AC voltage divider: (a) the circuit; (b) Thévenin equivalent; (c) Norton equivalent.

open circuit and the current on short circuit. The impedance $\mathbf{Z}_o$ is given by the ratio $\mathbf{V}_o/\mathbf{I}_o$. An alternative way of finding $\mathbf{Z}_o$ is as the impedance between the terminals A and B when all the sources are reduced to zero. As an illustration of these theorems we consider the AC version of the voltage divider studied in Section 3.5.

4.13.1 ◆ The AC voltage divider

Such a divider is shown schematically in Figure 4.17(a), similar to Figure 3.3(a) except that $\mathbf{Z}_1$ and $\mathbf{Z}_2$ are complex impedances. Conservation of current at node N gives

$$\frac{\mathbf{E} - \mathbf{V}}{\mathbf{Z}_1} = \mathbf{I} + \frac{\mathbf{V}}{\mathbf{Z}_2}$$

Hence

$$\mathbf{V} = \mathbf{E}\frac{\mathbf{Z}_2}{\mathbf{Z}_1 + \mathbf{Z}_2} - \mathbf{Z}_o\mathbf{I} \tag{4.36}$$

in which

$$\mathbf{Z}_o = \frac{\mathbf{Z}_1\mathbf{Z}_2}{\mathbf{Z}_1 + \mathbf{Z}_2} \tag{4.37}$$

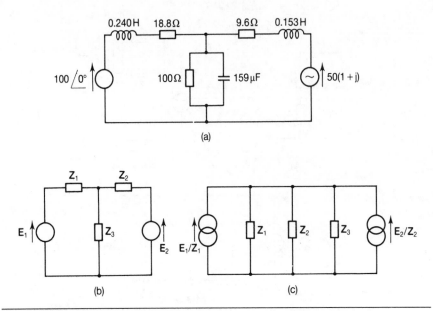

Figure 4.18 Example 4.7: (a) circuit; (b) using short-hand notation for impedances; (c) reduction using theorems of Thévenin and Norton.

Equation 4.36 is to be compared with Equation 4.34, and it is seen that the first term indeed represents the open-circuit voltage. The impedance Z_o is the impedance of Z_1 and Z_2 in parallel, correctly given as the impedance between A and B when the source E is reduced to zero. The Thévenin equivalent circuit is shown in Figure 4.17(b). Rearrangement of Equation 4.36 gives

$$I = \frac{E}{Z_1} - \frac{V}{Z_o} \tag{4.38}$$

In this equation the first term represents the short-circuit current. The equation leads to the Norton equivalent circuit shown in Figure 4.17(c).

The dividing action on open circuit corresponds to the ratio $Z_2/(Z_1 + Z_2)$. This will in general be complex, indicating a shift of phase with respect to the source E. A real ratio requires that Z_1 be a real multiple of Z_2: this condition ensures a real ratio at all frequencies.

EXAMPLE 4.7 The sources in the circuit of Figure 4.18(a) both have a frequency of 50 Hz. Determine the power dissipated in the 100 Ω resistor.

To find the power dissipated in the 100 Ω resistor it suffices to know the voltage developed across it. The use of Norton's theorem to change the voltage sources into current sources will lead straightforwardly to this voltage, on the lines of Examples 3.2 and 3.4. Because the quantities to be manipulated are complex, it is desirable to carry out the calculations in a systematic fashion. This is most easily done if an algebraic approach is adopted. Using the notation shown Figure 4.18(b), the transformation of the voltage sources leads to the circuit of Figure 4.18(c). The phasor voltage between the nodes is given by

$$\mathbf{V} = \mathbf{Z}\left(\frac{\mathbf{E}_1}{\mathbf{Z}_1} + \frac{\mathbf{E}_2}{\mathbf{Z}_2}\right)$$

where

$$\frac{1}{\mathbf{Z}} = \frac{1}{\mathbf{Z}_1} + \frac{1}{\mathbf{Z}_2} + \frac{1}{\mathbf{Z}_3}$$

We have

$$1/\mathbf{Z}_1 = 1/(18.8 + j75.4) = 0.003\,113 - j0.012\,490$$

$$1/\mathbf{Z}_2 = 1/(9.6 + j48.1) = 0.003\,990 - j0.019\,990$$

$$1/\mathbf{Z}_3 = 0.01 + j0.05$$

$$\mathbf{E}_1/\mathbf{Z}_1 = 0.3113 - j1.249$$

$$\mathbf{E}_2/\mathbf{Z}_2 = 50(1 + j) \times (0.003\,990 - j0.019\,990)$$

$$= 1.199 - j0.800$$

$$\mathbf{E}_1/\mathbf{Z}_1 + \mathbf{E}_2/\mathbf{Z}_2 = 1.510 - j2.050$$

$$1/\mathbf{Z}_1 + 1/\mathbf{Z}_2 + 1/\mathbf{Z}_3 = 0.017\,10 + j0.017\,51$$

$$1/(0.017\,10 + j0.017\,51) = 28.55 - j29.23$$

Hence

$$\mathbf{V} = (1.510 - j2.050) \times (28.55 - j29.23)$$

$$= -16.8 - j102.7$$

The required power is therefore

$$|\mathbf{V}|^2/100 = 108\ \text{W}$$

4.14 ◆ The star–delta transformation

This network transformation applied to resistors was introduced in Section 3.10, Equations 3.11 and 3.12. The algebra of the transform still holds good if we replace resistance by impedance, leading to an equivalence between the two networks shown in Figure 4.19. Denoting the impedances in the star configuration by $\mathbf{Z}_1$, $\mathbf{Z}_2$, $\mathbf{Z}_3$ and those in the delta configuration by $\mathbf{P}_1$, $\mathbf{P}_2$, $\mathbf{P}_3$, we find that the equivalence of the star to the delta network requires that

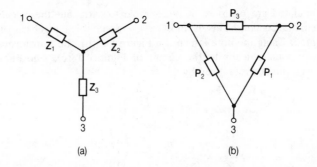

Figure 4.19 The star–delta transform for impedances: (a) star connection; (b) delta connection.

$$\mathbf{Z}_1 = \frac{\mathbf{P}_2\mathbf{P}_3}{\mathbf{P}_1 + \mathbf{P}_2 + \mathbf{P}_3}$$

$$\mathbf{Z}_2 = \frac{\mathbf{P}_3\mathbf{P}_1}{\mathbf{P}_1 + \mathbf{P}_2 + \mathbf{P}_3} \qquad (4.39)$$

$$\mathbf{Z}_3 = \frac{\mathbf{P}_1\mathbf{P}_2}{\mathbf{P}_1 + \mathbf{P}_2 + \mathbf{P}_3}$$

For the equivalence of the delta network to the star network we have

$$\mathbf{P}_1 = \frac{1}{\mathbf{Z}_1}(\mathbf{Z}_1\mathbf{Z}_2 + \mathbf{Z}_2\mathbf{Z}_3 + \mathbf{Z}_3\mathbf{Z}_1)$$

$$\mathbf{P}_2 = \frac{1}{\mathbf{Z}_2}(\mathbf{Z}_1\mathbf{Z}_2 + \mathbf{Z}_2\mathbf{Z}_3 + \mathbf{Z}_3\mathbf{Z}_1) \qquad (4.40)$$

$$\mathbf{P}_3 = \frac{1}{\mathbf{Z}_3}(\mathbf{Z}_1\mathbf{Z}_2 + \mathbf{Z}_2\mathbf{Z}_3 + \mathbf{Z}_3\mathbf{Z}_1)$$

If the star impedances are all of the same type we may write

$$\mathbf{Z}_1 = \lambda_1\mathbf{Z}$$

$$\mathbf{Z}_2 = \lambda_2\mathbf{Z}$$

$$\mathbf{Z}_3 = \lambda_3\mathbf{Z}$$

in which λ_1, λ_2, λ_3 are real positive numbers. The delta impedances will then also be multiples of $\mathbf{Z}$:

$$\mathbf{P}_1 = \mathbf{Z}(\lambda_1\lambda_2 + \lambda_2\lambda_3 + \lambda_3\lambda_1)/\lambda_1$$

with similar expressions for $\mathbf{P}_2$ and $\mathbf{P}_3$. If the impedances are all equal, then

$$\mathbf{P} = 3\mathbf{Z}$$

As with the equivalent impedances derived in other network transforma-

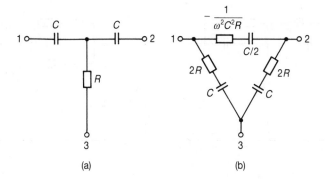

Figure 4.20 Example 4.8: (a) star-connected circuit; (b) delta equivalent.

tions the impedances in the transformed network may not be physically realizable. This is illustrated by the following example.

EXAMPLE 4.8 Derive the delta equivalent for the circuit of Figure 4.20(a).

We have

$$\mathbf{Z}_1 = \mathbf{Z}_2 = \frac{1}{j\omega C}$$

$$\mathbf{Z}_3 = R$$

$$\mathbf{Z}_1\mathbf{Z}_2 + \mathbf{Z}_2\mathbf{Z}_3 + \mathbf{Z}_3\mathbf{Z}_1 = \frac{2R}{j\omega C} + \left(\frac{1}{j\omega C}\right)^2$$

$$= \frac{2R}{j\omega C} - \left(\frac{1}{\omega C}\right)^2$$

Hence

$$\mathbf{P}_1 = \mathbf{P}_2 = 2R + \frac{1}{j\omega C}$$

$$\mathbf{P}_3 = \frac{2}{j\omega C} - \frac{1}{R}\left(\frac{1}{\omega C}\right)^2$$

The resulting network, shown in Figure 4.20(b), contains the unrealizable element of negative resistance, which is also frequency dependent. Although it is not possible to make this delta network it is a perfectly correct equivalent in the mathematical sense. It can also be useful, as shown in the next example.

EXAMPLE 4.9 By using the star–delta transformation show that the component values R and C in the network of Figure 4.21(a) can be chosen so that for one specific frequency no output is obtained.

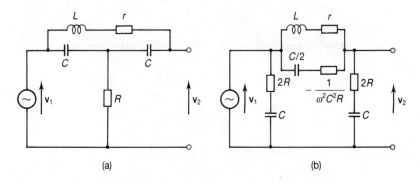

Figure 4.21 Example 4.9: (a) circuit; (b) reduced to delta form.

Since the transformation obtained in Example 4.8 is correct for all terminal connections, the networks of Figures 4.21(a) and 4.21(b) are equivalent. The series arm in Figure 4.2(b) consists of two impedances in parallel: if these impedances are equal and opposite the impedance of the parallel combination becomes infinite, equivalent to no connection between input and output. For this to be the case we require

$$j\omega L + r = -\left[\frac{2}{j\omega C} - \frac{1}{R}\left(\frac{1}{\omega C}\right)^2\right]$$

For this to be satisfied, we must have

$$C = \frac{2}{\omega^2 L}, \ R = \frac{(\omega L)^2}{r}$$

It is possible to choose components with these values.

4.15 ◆ Power matching

It is often required to achieve maximum transmission of power from one part of a system to another. One part of the system acts as a source, and may be characterized as far as its action on a load is concerned by a Thévenin equivalent circuit. The load also has a Thévenin equivalent circuit which, since the load is presumed not to contain any sources, is simply an impedance. This impedance will have a resistive part and will consequently appear to dissipate power. The power thus calculated will be correct in total, although the distribution of that power between the various components in the load can only be determined by deeper analysis: for example, the power might go as mechanical power in an electric motor, or as radio waves from a transmitting antenna. The equivalent circuit of the source–load arrangement is shown in Figure 4.22. The source imped-ance will in general have both resistive and reactive parts. We therefore

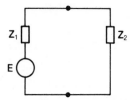

Figure 4.22 Equivalent circuit for a source–load configuration.

write

$$\mathbf{Z}_1 = R_1 + jX_1$$

Similarly, the equivalent load impedance must be written

$$\mathbf{Z}_2 = R_2 + jX_2$$

Analysing this circuit we have

$$\mathbf{I} = \mathbf{E} \frac{1}{\mathbf{Z}_1 + \mathbf{Z}_2}$$

$$= \mathbf{E} \frac{1}{R_1 + R_2 + j(X_1 + X_2)}$$

The power transmitted to the load is that dissipated in the resistance R_2, and is given by

$$P = R_2|\mathbf{I}|^2$$

$$= |\mathbf{E}|^2 \frac{R_2}{(R_1 + R_2)^2 + (X_1 + X_2)^2}$$

We can always modify the load reactance X_2 without loss of power by adding another reactance, which may be positive or negative, in series: adding reactance to make $X_1 + X_2$ zero increases the power transmitted to R_2 to the value

$$P = |\mathbf{E}|^2 \frac{R_2}{(R_1 + R_2)^2}$$

A graph of this expression as a function of R_2/R_1 appears as Figure 4.23. It exhibits a maximum when $R_2 = R_1$, when the value is

$$P = |\mathbf{E}|^2 \frac{1}{4R_1} \tag{4.41}$$

This maximum value is termed the **available power**, and is characteristic of the source. The condition for maximum power transfer can be written

$$\mathbf{Z}_2 = \mathbf{Z}_1^*$$

When the condition for maximum power transfer is satisfied, the load is said to be **matched** to the source.

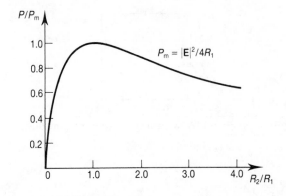

Figure 4.23 Showing variation of power in a resistive load R_2 from a source of resistive internal impedance R_1.

4.15.1 ◆ Matching circuits

A network containing only pure reactances interposed between source and load does not introduce additional power loss. Such a network can be used to power match when source and load have different resistances. One such circuit is shown in Figure 4.24, in which source and load are assumed to have purely resistive impedances. If the reactances can be chosen so that the impedance presented between the terminals A, B is equal to the source resistance R_1, the available power will be drawn from the source and must be delivered to the load R_2. Letting the reactance of the series component be X, the susceptance of the shunt component be B, and writing $G = 1/R$, this condition requires that

$$R_1 = \mathrm{j}X + \frac{1}{G_2 + \mathrm{j}B}$$
$$= \mathrm{j}X + \frac{G_2 - \mathrm{j}B}{G_2^2 + B^2}$$

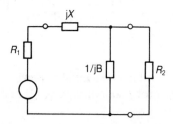

Figure 4.24 A network for power matching, Section 4.16.1.

Hence we must have

$$X = \frac{B}{G_2^2 + B^2}$$

and

$$R_1 = \frac{G_2}{G_2^2 + B^2}$$

From the second of these equations we find

$$B = \left(\frac{G_2}{R_1} - G_2^2\right)^{1/2}$$

$$= \frac{1}{R_2}\left(\frac{R_2}{R_1} - 1\right)^{1/2} \tag{4.42}$$

Division of the two equations gives

$$\frac{X}{R_1} = \frac{B}{G_2}$$

and hence

$$X = [R_1(R_2 - R_1)]^{1/2} \tag{4.43}$$

In order for B to be real, it is necessary that $R_2 > R_1$. The circuit can therefore be used to match when the load resistance is greater than the source resistance. The susceptance B can be chosen to be either positive, when it will be a capacitance, or negative, when it will be an inductance. The corresponding reactance X is in the first case an inductance, in the second case a capacitance.

EXAMPLE 4.10   A microwave detector operating at a frequency of 1 GHz has an impedance equivalent to 95 Ω in parallel with a capacitance of 0.5 pF. Design a circuit which will match it to a source of 50 Ω resistive impedance.

Using the values $R_1 = 50$, $R_2 = 95$ in Equation 4.42, we find $B = 9.986$ mS. Since the load has a built-in shunt capacitance it is sensible to try making B a capacitance. A susceptance of 9.986 mS corresponds at 1 GHz to a capacitance of 1.59 pF, so that the 0.5 pF already present can be regarded as part of this total. An extra shunt capacitance of 1.09 pF is required. With this choice for B, the series reactance will be inductive, and from Equation 4.43 is found to be of magnitude 47.4 Ω, corresponding at 1 GHz to an inductance of 7.5 nH.

4.16 ♦ Excitation at more than one frequency

The previous sections have dealt specifically with sinusoidal AC excitation, which by definition is single frequency. However, practical waveforms often contain several sine waves of different frequencies. For example, the voltage of a supply might be represented by a form such as

$$v(t) = \hat{V}_1 \cos(\omega t) + \hat{V}_3 \cos(3\omega t + \phi) \tag{4.44}$$

In the case of linear circuits, to which the phasor representation developed in the previous sections applies, a phasor-related method exists for dealing simply with waveforms such as that of Equation 4.44. This method makes use of the principle of superposition, a formal proof of which will be found in Appendix A, Section A.2. This principle, which applies only to linear networks, can be stated as follows: suppose that an excitation $v_1(t)$ results in a current $i_1(t)$ through a particular component, and an excitation $v_2(t)$ results in a current $i_2(t)$ through the same component, then the application of the excitation $v_1(t) + v_2(t)$ will result in the current through that component being $i_1(t) + i_2(t)$. We may thus consider the separate application of each of the terms in an excitation such as that of Equation 4.44 and perform the addition as a final step. (It is important to note that in this formulation the voltages and currents are the real, physical voltages.)

Consider the application of the waveform of Equation 4.44 to the series L–R circuit of Figure 4.3. The current resulting from an excitation $\hat{V}_1 \cos(\omega t)$ can be evaluated using phasors: the phasor representation of this waveform is

$$\mathbf{V}_1 = \frac{1}{\sqrt{2}} \hat{V}_1 \angle 0$$

The complex impedance at this frequency is $j\omega L + R$, so that the phasor current is

$$\mathbf{I}_1 = \frac{\mathbf{V}_1}{j\omega L + R}$$

$$= \frac{1}{\sqrt{2}} \hat{V}_1 [(\omega L)^2 + R^2]^{-1/2} \exp(-j\theta_1)$$

where

$$\theta_1 = \tan^{-1}(\omega L/R)$$

The physical current represented by this phasor is, using Equation 4.23, given by

$$i_1(t) = \hat{V}_1 [(\omega L)^2 + R^2]^{-1/2} \cos(\omega t - \theta_1) \tag{4.45}$$

In a similar way we can evaluate the current resulting from an excitation $\hat{V}_3 \cos(3\omega t + \phi)$. It is sine wave of angular frequency 3ω, and can be given a phasor representation:

$$\mathbf{V}_3 = \frac{1}{\sqrt{2}} V_3 \angle \phi$$

The complex impedance is now $R + j3\omega L$, giving a phasor current

$$\mathbf{I}_3 = \frac{\mathbf{V}_3}{j3\omega L + R}$$

$$= \frac{1}{\sqrt{2}}\hat{V}_3[(3\omega L)^2 + R^2]^{-1/2}\exp[j(\phi - \theta_3)]$$

where

$$\theta_3 = \tan^{-1}(3\omega L/R)$$

Equation 4.23 is now used to give the corresponding physical current, in the form

$$i_3(t) = \mathrm{Re}[\sqrt{2}\mathbf{I}_3\exp(j3\omega t)]$$

$$= \hat{V}_3[(3\omega L)^2 + R^2]^{-1/2}\cos(3\omega t + \phi - \theta_3) \qquad \textbf{(4.46)}$$

The principle of superposition now allows us to add together expressions of Equations 4.45 and 4.46 in order to obtain the current resulting from the application of the voltage waveform of Equation 4.44:

$$i(t) = i_1(t) + i_3(t)$$

$$= \hat{V}_1[(\omega L)^2 + R^2]^{-1/2}\cos(\omega t - \theta_1)$$

$$+ \hat{V}_3[(3\omega L)^2 + R^2]^{-1/2}\cos(3\omega t + \phi - \theta_3) \qquad \textbf{(4.47)}$$

Following Section 2.3.4, it may be observed that the mean-square current is given by

$$\langle i^2 \rangle = \tfrac{1}{2}\{\hat{V}_1^2[(\omega L)^2 + R^2]^{-1} + \hat{V}_3^2[(3\omega L)^2 + R^2]^{-1}\}$$

This expression is also given by

$$\langle i^2 \rangle = |\mathbf{I}_1|^2 + |\mathbf{I}_3|^2$$

EXAMPLE 4.11 In the circuit of Figure 4.25 the source voltage $e(t)$ is given by

$$e(t) = 10\sqrt{2}\cos(1000\pi t) + 2\sqrt{2}\sin(3000\pi t)$$

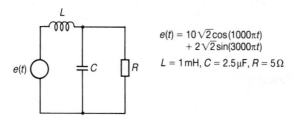

$$e(t) = 10\sqrt{2}\cos(1000\pi t)$$
$$+ 2\sqrt{2}\sin(3000\pi t)$$
$$L = 1\,\mathrm{mH}, C = 2.5\,\mu\mathrm{F}, R = 5\,\Omega$$

Figure 4.25 Circuit for Example 4.11.

Obtain an expression for the voltage across the 5 Ω resistor. Determine the RMS value of this voltage and the power dissipated in the resistor.

Analysing the circuit in phasor form for an arbitrary frequency we have

$$\frac{\mathbf{E} - \mathbf{V}}{j\omega L} = \frac{\mathbf{V}}{\mathbf{R}} + \mathbf{V}j\omega C$$

Hence

$$\mathbf{E} = \mathbf{V}\left(\frac{j\omega L}{R} + 1 - \omega^2 LC\right)$$

The computation is best presented in tabular form:

f (Hz)	500	1500
ωL (Ω)	3.141	9.425
$\omega^2 LC$	0.0247	0.222
$1 - \omega^2 LC$	0.975	0.778
$\omega L/R$	0.628	1.885
$\mathbf{E}/\mathbf{V}$	0.975 + j0.628	0.778 + j1.885
$\mathbf{E}/\mathbf{V}$	1.160 ∠32.7°	2.039 ∠67.6°
$\mathbf{E}$	10 ∠0°	2 ∠−90°
$\mathbf{V}$	8.62 ∠−32.7°	0.981 ∠−157.6°

The required expression is therefore

$$v(t) = 8.62 \sqrt{2}\cos\left(1000\pi t - 32.7°\right)$$
$$+ 0.981 \sqrt{2}\cos\left(3000\pi t - 157.6°\right)$$

The mean-square value of $v(t)$ is given in this case by the sum of the individual mean squares (Section 2.4, Equations 2.9):

$$\langle v^2 \rangle = (8.62)^2 + (0.981)^2$$

The RMS value is therefore 8.68 V, and the power dissipated in the resistor $(8.68)^2/5 = 15.1$ W.

4.17 ◆ Summary

Phasors
The representation of sinusoidally varying quantities by phasors is summarized in the following table:

Time-variation

Phasor

$v(t) = \hat{V} \cos(\omega t + \phi)$
$\mathbf{V} = \dfrac{1}{\sqrt{2}} \hat{V} \exp(j\phi)$

$\sqrt{2}\,\mathrm{Re}\,[\mathbf{V} \exp(j\omega t)]$
$\mathbf{V}$

$\sqrt{2}[V' \cos(\omega t) + V'' \cos(\omega t + \pi/2)]$
$\mathbf{V} = V' + jV''$

$\sqrt{2}\,\mathrm{Re}[(\mathbf{V}_1 + \mathbf{V}_2) \exp(j\omega t)]$
$\mathbf{V} = \mathbf{V}_1 + \mathbf{V}_2$

$$\text{impedance} = \frac{\text{phasor voltage}}{\text{phasor current}} = \text{resistance} + j \times \text{reactance}$$

$$\mathbf{Z} = \frac{\mathbf{V}}{\mathbf{I}} = R + jX$$

$$\text{admittance} = \frac{\text{phasor current}}{\text{phasor voltage}} = \text{conductance} + j \times \text{susceptance}$$

$$\mathbf{Y} = \frac{\mathbf{I}}{\mathbf{V}} = G + jB$$

Series–parallel equivalence:

$$G = \frac{R}{R^2 + X^2}, \quad B = -\frac{X}{R^2 + X^2}$$

$$R = \frac{G}{G^2 + B^2}, \quad X = -\frac{B}{G^2 + B^2}$$

Power:

$$P = \mathrm{Re}\,(\mathbf{VI}^*) \text{ (W)}$$

Volt–amps reactive:

$$Q = \mathrm{Im}\,(\mathbf{VI}^*) \text{ (VAR)}$$

Both power and volt–amps reactive are conserved.

Theorems of Thévenin and Norton

These theorems relate to a network with two free terminals which can be connected to a load. As far as the load connected to these terminals is concerned, the network can be replaced by the Thévenin equivalent of Figure 4.16(a) or the Norton equivalent of Figure 4.16(b). In these figures the phasor $\mathbf{V}_o$ is the representation of the voltage between the terminals when they are open-circuited the phasor $\mathbf{I}_o$ the representation of the current when the terminals are short-circuited and the impedance $\mathbf{Z}_o = \mathbf{V}_o/\mathbf{I}_o$.

Star–delta equivalence (Figure 4.19)

Delta to star:

$$\mathbf{Z}_1 = \frac{\mathbf{P}_2 \mathbf{P}_3}{\mathbf{P}_1 + \mathbf{P}_2 + \mathbf{P}_3}$$

Star to delta:

$$P_1 = \frac{1}{Z_1}(Z_1Z_2 + Z_2Z_3 + Z_3Z_1)$$

Formulae for the other two arms can be obtained by cycling the indices in the order 1–2–3.

PROBLEMS

4.1 Calculate the reactance of a 100 mH inductance at frequencies of 150 Hz and 1.2 kHz. Calculate also the reactance of a 2.2 μF capacitance at these frequencies. At what frequency will the components have equal but opposite reactances?

4.2 Determine the complex impedance, in the form $R + jX$, of the combination of a 0.5 H inductor in series with a 1 kΩ resistor at a frequency of 970 Hz. Express this impedance as a magnitude and a phase angle. At what frequency will the phase angle be 45° and what is then the magnitude?

4.3 For the circuit of Problem 4.2, at the same frequency, calculate the admittance in the form $G + jB$. Hence deduce the values of resistance and inductance which when connected in parallel will have the same impedance.

4.4 Calculate the complex impedance and admittance of the circuit consisting of a 1 μF capacitor in series with a 100 Ω resistor at a frequency of 1 kHz. Express both quantities in the form of magnitude and phase. Calculate also the values of resistance and capacitance which when placed in parallel will have the same impedance.

4.5 A 2 H inductor and a 2 μF capacitor are connected (i) in series and (ii) in parallel. Determine in both cases the reactance of the combination at a frequency of 100 Hz. In case (ii) a resistor of 100 Ω is connected in series with the inductor. Calculate the impedance of the parallel combination.

4.6 A circuit consists of a 2 H inductor, a 2 μF capacitor and an 800 Ω resistor connected in series. The circuit carries an AC current of 0.25 A (RMS), 50 Hz. Calculate the voltage developed across each component in magnitude and phase, measuring phase with respect to the current. Calculate also the magnitude and phase angle of the total voltage across the circuit. Represent all these quantities on a phasor diagram.

4.7 For the circuit of Figure 4.26 determine the power and volt–amps reactive supplied by the source. Verify the conservation of both these quantities.

4.8 An industrial load can be represented by an impedance consisting of a resistance of 6 Ω in series with an inductance of 20 mH. The voltage across this load is 240 V (RMS) at a frequency of 50 Hz. The generator is connected to the load by a cable that can be represented by a resistance of 1 Ω in series with an inductance of 10 mH.

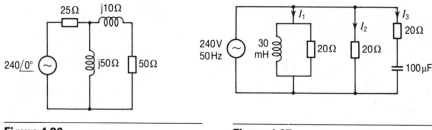

Figure 4.26   **Figure 4.27**

(a) Calculate values for the load current, power, volt–amps reactive and power factor.

(b) Calculate the value of the capacitance which when placed in parallel with the load will give a combined power factor of 1.

(c) When the power factor has been so corrected, calculate the generator voltage necessary to maintain the load voltage at 240 V.

4.9 A 240 V AC mains at 50 Hz supplies three loads in parallel, as shown in Figure 4.27. Calculate the magnitude of the current in each load and its phase angle with respect to the supply voltage. Represent these currents on a phasor diagram.

Calculate also the current supplied by the generator and the total power delivered.

4.10 In the circuit of Figure 4.28 the source is of angular frequency ω. Show that the value of C may be chosen so that the reading of the current meter M is independent of the position of the switch S, and that this value depends only on the value of the inductance and the frequency. Determine this value when $L = 1$ H and the source frequency is 50 Hz.

4.11 In the circuit shown in Figure 4.29 the two AC sources are identical, and the resistance R is variable. Show that the voltage of node A with respect to node B as R is varied remains constant in magnitude but varies in phase. In a particular case the source frequency is 1 kHz, the capacitance 0.22 μF and R can be varied between 10 Ω and 5 kΩ. Determine the range of phase difference which can be obtained.

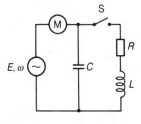

Figure 4.28   **Figure 4.29**

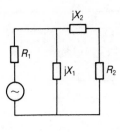

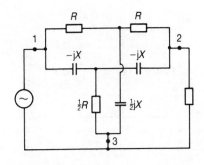

Figure 4.30
Figure 4.31

4.12 In the circuit of Problem 4.11, the sources are of frequency 1 kHz and of magnitude 10 V (RMS), and $C = 0.22\ \mu$F. The value of R is set so that the phase of the voltage between A and B is 90° with respect to the source voltage. Give a Thévenin equivalent circuit for the source represented by terminals A, B. Hence determine the magnitude and phase of the voltage between A and B when a resistance of 10 kΩ is connected between them.

4.13 Show that the reactances X_1 and X_2 in the circuit of Figure 4.30 can be chosen so that the source and load are matched when the source resistance R_1 is greater than the load resistance R_2. Determine suitable components for matching at a frequency of 10 kHz when $R_1 = 600\ \Omega$ and $R_2 = 400\ \Omega$.

4.14
 (a) A network consists of three inductances of values $L - M$, $L - M$, M connected in star. Derive an equivalent delta network.
 (b) Three components of impedances jX, $-jX$, X (X real) are connected in the delta configuration. Derive an equivalent star network.

4.15 The network shown in Figure 4.31 connects a source to a load. Show, by transforming each star network into the delta configuration, that when the condition $R = X$ is satisfied there will be no output to the load.

5 ◆ *Frequency Response*

This chapter is concerned with the process of determining the behaviour of networks as the frequency of the exciting source is varied. Results are given for some commonly encountered circuits. The concept of Bode plots is introduced.

5.1 ◆ Introduction

Chapter 4 was concerned with determining the response of a circuit to excitation by a sinusoidal AC waveform, showing that with the phasor method the effect of inductance and capacitance could be described by the reactance of the component, as given in Equation 4.17. In that chapter the frequency was considered as fixed, thereby assigning a numerical value to reactance, on par with resistance. In some practical situations only one frequency is indeed of interest, so that a numerical specification of reactance is satisfactory. AC power distribution is such a case. Some systems, such as a communications channel, are required to handle a wide range of frequencies. In such a situation frequency has to be regarded as variable, and it is necessary to know how the response changes with frequency. The response must therefore be formulated algebraically.

In the present chapter the frequency response of a number of circuits likely to be encountered in practice is investigated. In considering the significance of these results it is desirable to keep in mind two facts:

(i) as mentioned in Section 2.2.2, any waveform can be regarded as made up of a number of sinusoidal waveforms of different frequencies, and

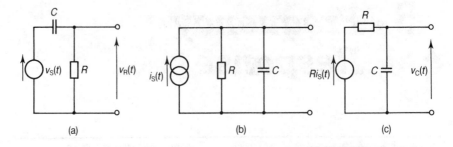

Figure 5.1 Resistance–capacitance circuits: (a) the *R–C* coupling circuit; (b) parallel *R–C* circuit current fed; (c) reduction of (b) using Thévenin's theorem.

(ii) the principle of superposition for linear circuits stated in Section 4.16.

The first of these means that a voltage source producing a non-sinusoidal waveform can be replaced by a number of AC sources in series, each of the magnitude and phase appropriate to the frequency. The second then implies that we may consider each of these AC sources acting separately and subsequently sum the responses to these individual sources. Taken together, therefore, these two facts show that the response of a linear circuit to any waveform can be found if the response to sinusoidal AC excitation is known for all frequencies: the frequency response becomes a basic description of the action of the circuit.

5.2 ◆ The series resistance–capacitance circuit

The association of a single resistor with a single capacitor is very frequently found in electronic circuits, in both of the forms illustrated in Figures 5.1(a) and 5.1(b). In the first of these, the capacitor has the effect of isolating the load (represented by the resistor R) from any steady voltage component present in the source $v_s(t)$. The circuit is frequently termed the **R–C coupling** circuit. In the second configuration the current source causes an output voltage to be developed across the resistor R, which inevitably has some capacitance in parallel. Using Thévenin's theorem this circuit can be transformed into that of Figure 5.1(c), which is also in the series configuration. Thus an analysis of the series configuration covers both circuits. In terms of phasor variables, the circuit is shown in Figure 5.2. We then have

$$\mathbf{E} = \mathbf{I}R + \mathbf{I}\,\frac{1}{j\omega C}$$

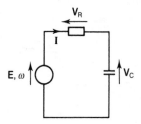

Figure 5.2 Circuit with phasor notation.

Hence

$$V_C = I \frac{1}{j\omega C} = E \frac{1}{1 + j\omega CR} \tag{5.1}$$

$$V_R = IR = E \frac{1}{1 + 1/j\omega CR} \tag{5.2}$$

If we write

$$\theta = \tan^{-1}(\omega CR)$$

then these expressions may be written in the form

$$\frac{V_C}{E} = \frac{1}{[1 + (\omega CR)^2]^{1/2}} \exp(-j\theta) \tag{5.3}$$

and

$$\frac{V_R}{E} = \frac{j\omega CR}{1 + j\omega CR} \tag{5.4}$$

$$= \frac{\omega CR}{[1 + (\omega CR)^2]^{1/2}} \exp[j(\pi/2 - \theta)] \tag{5.5}$$

5.2.1 ◆ The *R–C* coupling circuit

For the circuit of Figure 5.1(a) Equations 5.2 and 5.4 apply. The ratio V_R/E is regarded as the frequency response for the circuit. It is described by magnitude and phase:

$$|V_R/E| = \frac{\omega CR}{[1 + (\omega CR)^2]^{1/2}}$$

$$\angle(V_R/E) = \pi/2 - \theta = \tan^{-1}(1/\omega CR) \tag{5.6}$$

The way in which these quantities vary with frequency is illustrated in Figure 5.3. Figure 5.3(a) shows that as expected the magnitude is zero at DC, for which ω is zero, and steadily increases to unity as the frequency

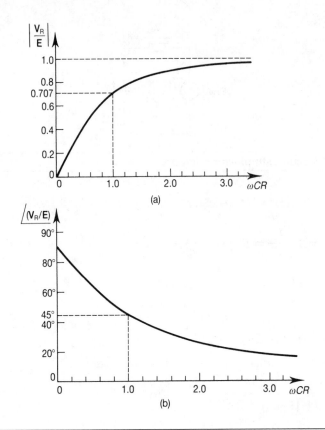

Figure 5.3 Frequency response with the output across the resistance in the circuit of Figure 5.2: (a) gain, $|V_R/E|$; (b) phase, $\angle(V_R/E)$.

increases. The rate at which the response approaches unity depends on the value of the product RC. This quantity has dimensions of time (as may be verified by comparing Equations 1.1 and 1.4) and is referred to as the **time constant** of the circuit. A useful marker value is obtained by noting that when ωCR is unity the magnitude of the response is $1/\sqrt{2}$, or 70.7%, of the maximum. The variation of phase with frequency is illustrated in Figure 5.3(b): at very low frequencies V_R leads E by $\pi/2$ and becomes more nearly in phase as the frequency increases. At the marker point $\omega CR = 1$ the phase is $\pi/4$. In general terms the behaviour may be regarded as a result of the reactance of the capacitor varying from very high values at low frequencies to very small values at high frequencies.

EXAMPLE 5.1 The load R in the circuit of Figure 5.1(a) has a value of $1\ M\Omega$ and the source $v_s(t)$ contains frequencies in the range 50 Hz–20 kHz.

Estimate a value for the capacitor C such that any frequency component of the output voltage $v_L(t)$ shall be not less than 80% of its value in the source.

The lowest frequency component is reduced most, so we need to ensure that $|\mathbf{V_R}/\mathbf{E}| \geqslant 0.8$ at 50 Hz. For this condition to be satisfied, ωCR must have the value determined from

$$\frac{\omega CR}{[1 + (\omega CR)^2]^{1/2}} \geqslant 0.8$$

or

$$1 + (1/\omega CR)^2 \leqslant (1/0.8)^2$$

Hence

$$\omega CR \geqslant 1.33$$

Putting $\omega = 2\pi \times 50$ and $R = 10^6$ we find

$$C \geqslant 3.2\ \text{nF}$$

5.2.2 ◆ Capacitance in parallel with a resistive load

This is the configuration of Figure 5.1(b), or its transform in Figure 5.1(c). The frequency response in this case is the ratio $\mathbf{V_C}/\mathbf{E}$ (Equation 5.3). We have

$$|\mathbf{V_C}/\mathbf{E}| = \frac{1}{[1 + (\omega CR)^2]^{1/2}}$$

$$\angle(\mathbf{V_C}/\mathbf{E}) = -\theta = -\tan^{-1}(\omega CR)$$

The behaviour of this ratio in magnitude and phase as the frequency varies is illustrated in Figure 5.4. With this circuit the output $|\mathbf{V_C}|$ falls off as the frequency increases, corresponding to the reduction of capacitor reactance. It lags on the input $\mathbf{E}$ by an angle small at low frequencies but increasing to $\pi/2$ at high frequencies. Once again the marker $\omega CR = 1$ is useful in describing the action of the circuit: at this point $|\mathbf{V_C}/\mathbf{E}|$ has the value $1/\sqrt{2}$, and the phase lag is $\pi/4$.

EXAMPLE 5.2 The capacitance in the circuit of Figure 5.1(c) is 20 pF and the source $Ri_s(t)$ contains frequencies up to 1 MHz. Determine the maximum value that R may have if the frequency components in the output $v_C(t)$ are to be no less than 80% of the corresponding value in the source. With this value of R determine the phase shift at 0.5 MHz.

In this case we require $|\mathbf{V_C}/\mathbf{E}| \geqslant 0.8$, or

$$1 + (\omega CR)^2 \leqslant (1/0.8)^2$$

Hence

$$\omega CR \leqslant 0.75$$

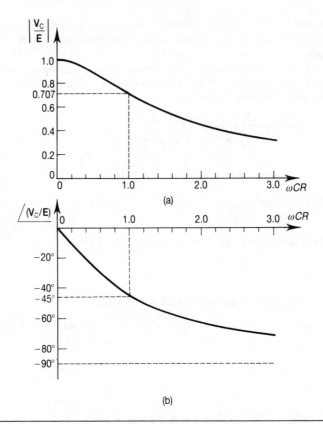

Figure 5.4 Frequency response with the output across the capacitance in the circuit of Figure 5.2: (a) gain $|V_C/E|$; (b) phase, $\angle(V_C/E)$.

At 1 MHz with $C = 20$ pF we find

$$R \leqslant 5.968 \text{ k}\Omega$$

The time constant is given by

$$RC = 5.968 \times 20 \times 10^{-12}$$

$$= 0.12 \ \mu s$$

At 0.5 MHz

$$\theta = -\tan^{-1}(2\pi \times 0.5 \times 0.12)$$

$$= -21°$$

Hence at this frequency the output lags the input by 21°.

5.2.3 ◆ Some more *R–C* circuits

Since any neighbouring pair of conductors constitutes a capacitor, all circuits have built-in capacitance. Whether or not these are important depends on the frequency of operation and the magnitude of the resistive

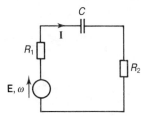

Figure 5.5 Series R–C circuit with source impedance.

impedances associated with them: a practical situation is illustrated by Example 5.2, showing that for an 80% response at 1 MHz the stray capacitance of 20 pF limits the resistance to some 6 kΩ. This circuit is typical of those encountered in amplifier design. Some of the configurations encountered in practical circuits are analysed in the following examples.

EXAMPLE 5.3   The circuit shown in Figure 5.5 is a modification of that shown in Figure 5.1(a), allowing the source to have internal resistance, as a real source must have. The output voltage appears across R_2.

In this case

$$\mathbf{I} = \frac{\mathbf{E}}{R_1 + R_2 + 1/j\omega C}$$

$$\mathbf{V} = R_2 \mathbf{I}$$

$$= \frac{\mathbf{E}R_2}{R_1 + R_2} \frac{1}{1 + 1/j\omega T}$$

where $T = (R_1 + R_2)C$.

By comparison with Equation 5.2 it is seen that the frequency response has the same form as that for the circuit of Figure 5.1(a) save for a dividing action in the ratio $R_2:(R_1 + R_2)$.

EXAMPLE 5.4   It is often required to reduce the magnitude of a signal before applying it to another part of the circuit. A voltage divider can be used, as discussed in Section 3.5, but whereas a purely resistive divider affects all frequency components equally the presence of a capacitor across the output will reduce response at high frequencies. The size of the effect can be assessed by analysis of the circuit of Figure 5.6(a). Applying Thévenin's theorem to that part of the circuit to the left of terminals A, B results in the circuit of Figure 5.6(b). This is essentially the circuit of Figure 5.2. Hence

$$\mathbf{V} = \frac{\mathbf{E}R_2}{R_1 + R_2} \frac{1}{1 + j\omega T} \tag{5.7}$$

where $T = CR_1R_2/(R_1 + R_2)$. Thus the effective time constant governing the frequency response is that of C and R_1 and R_2 in parallel.

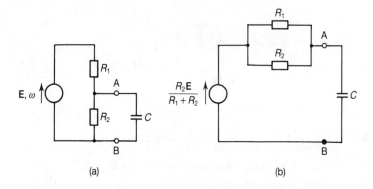

Figure 5.6 Example 5.4: (a) circuit; (b) reduction using Thévenin's theorem.

EXAMPLE 5.5   When a signal is displayed on an oscilloscope the input impedance of the oscilloscope, typically $1\,\text{M}\Omega$ in parallel with $20\,\text{pF}$, is necessarily placed in parallel with the signal source and may well affect its output. The capacitance of the connecting cable further increases the shunt capacitance. This difficulty is often alleviated by using an attenuating probe. Such a probe has basically the circuit of Figure 5.6(a), but with a capacitor C_1 in parallel with R_1 as shown in Figure 5.7. Denoting by $\mathbf{Z}_1$ the impedance of R_1 in parallel with C_1 and by $\mathbf{Z}_2$ that of R_2 in parallel with C_2, the circuit constitutes the voltage divider considered in Section 4.14.1. Hence

$$\mathbf{V} = \mathbf{E}\,\frac{\mathbf{Z}_2}{\mathbf{Z}_1 + \mathbf{Z}_2} = \frac{E}{1 + \mathbf{Z}_1/\mathbf{Z}_2}$$

We have

$$\frac{1}{\mathbf{Z}_1} = \frac{1}{R_1} + j\omega C_1$$

$$\frac{1}{\mathbf{Z}_2} = \frac{1}{R_2} + j\omega C_2$$

(5.8)

Hence

$$\frac{\mathbf{Z}_1}{\mathbf{Z}_2} = \frac{R_1}{R_2}\,\frac{1 + j\omega C_2 R_2}{1 + j\omega C_1 R_1}$$

If C_1 is varied so that $C_1 R_1 = C_2 R_2$ then

$$\frac{\mathbf{Z}_1}{\mathbf{Z}_2} = \frac{R_1}{R_2}$$

and

$$\frac{V}{E} = \frac{R_2}{R_1 + R_2}$$

(5.9)

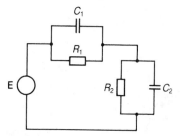

Figure 5.7 Circuit for Example 5.5.

Thus the attenuation ratio $\mathbf{E}/\mathbf{V}$ is independent of frequency and is given by

$$A = \frac{R_1 + R_2}{R_2} \tag{5.10}$$

The impedance seen by the source $\mathbf{E}$ is $\mathbf{Z}_1 + \mathbf{Z}_2$. Using the expressions of Equation 5.8 and the fact that the time constants have been equalized we have

$$\mathbf{Z}_1 + \mathbf{Z}_2 = \frac{R_1 + R_2}{1 + j\omega C_2 R_2}$$

$$= \frac{AR_2}{1 + j\omega C_2 R_2}$$

This corresponds to a resistance AR_2 in parallel with a capacitance C_2/A.

A common value for the attenuation ratio is 10:1. Allowing for the capacitance of the connecting cable a value for C_2 of 100 pF is typical. With $R_2 = 1\,\text{M}\Omega$, Equation 5.10 gives $R_1 = 9\,\text{M}\Omega$ so that C_1 has a value of $100/9$ pF. The impedance seen by the source is 10 MΩ in parallel with 10 pF.

EXAMPLE 5.6 Figure 5.8(a) shows a circuit containing a dependent generator in a configuration encountered in transistor circuitry. The quantities of interest are the ratio $\mathbf{V}_2/\mathbf{E}$ and the impedance seen by the source $\mathbf{E}$.

Since no current is taken by the control terminal of the dependent source we have

$$\mathbf{I} = j\omega C(\mathbf{E} - \mathbf{V}_2) \tag{5.11}$$

Balancing currents at node B, noting that $\mathbf{V}_1 = \mathbf{E}$,

$$-\mathbf{I} + G\mathbf{V}_2 + g\mathbf{E} = 0$$

where $G = 1/R$. Substituting for $\mathbf{I}$ from Equation 5.11 we find

$$\frac{\mathbf{V}_2}{\mathbf{E}} = -\frac{g - j\omega C}{G + j\omega C}$$

$$= -gR\frac{1 - j\omega C/g}{1 + j\omega CR} \tag{5.12}$$

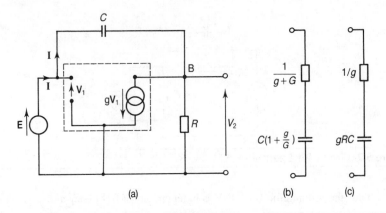

Figure 5.8 Example 5.5: (a) circuit; (b) equivalent input circuit; (c) input circuit when $g \gg G$.

To find **I** we substitute for V_2 in Equation 5.11:

$$\mathbf{I} = j\omega C\left(\mathbf{E} + \mathbf{E}\,\frac{g - j\omega C}{G + j\omega C}\right)$$

$$= \mathbf{E}j\omega C\,\frac{g + G}{G + j\omega C}$$

Hence

$$\frac{\mathbf{E}}{\mathbf{I}} = \frac{1}{g + G}\left(1 + \frac{G}{j\omega C}\right)$$

The circuit for which this is the impedance is shown in Figure 5.8(b). Commonly in applications $g \gg G$ in which case this circuit becomes approximately that of Figure 5.8(c), showing that the source **E** becomes loaded by a capacitance equal to gRC. It may be noticed that gR is the magnitude of $V_2/\mathbf{E}$ at low frequencies.

EXAMPLE 5.7 The configuration shown in Figure 5.9 is commonly encountered as part of the 'van der Pol' oscillator circuit. It provides a phase

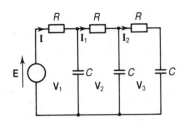

Figure 5.9 Example 5.7, the van der Pol phase shifting network.

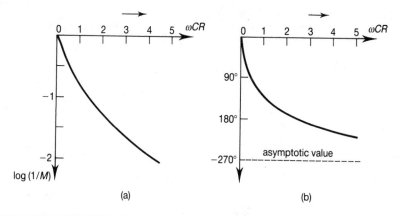

Figure 5.10 Frequency response for the van der Pol network: (a) gain, $|V_3/E|$; (b) phase, $\angle(V_3/E)$.

delay of 180° of output over input at one particular frequency. It is included both because of its importance and in order to show the need for a systematic approach to circuit calculations.

Ladder networks are often best analysed by starting from the output end. Using the notation in the diagram, and denoting the product $j\omega CR$ by x, we have

$$I_2 = j\omega C V_3$$
$$RI_2 = x V_3$$
$$V_2 = RI_2 + V_3 = (1 + x)V_3$$
$$RI_1 = R(I_2 + j\omega C V_2) = (2x + x^2)V_3$$
$$V_1 = RI_1 + V_2 = (1 + 3x + x^2)V_3$$
$$RI = R(I_1 + j\omega C V_1) = (3x + 4x^2 + x^3)V_3$$
$$E = RI + V_1 = (1 + 6x + 5x^2 + x^3)V_3$$

Hence

$$E/V_3 = 1 + 6(j\omega CR) + 5(j\omega CR)^2 + (j\omega CR)^3$$
$$= 1 - 5(\omega CR)^2 + (j\omega CR)[6 - (\omega CR)^2]$$
$$= M \exp(j\phi)$$

We find

$$M^2 = 1 + 26(\omega CR)^2 + 13(\omega CR)^4 + (\omega CR)^6 \tag{5.13}$$

$$\cos\phi = [1 - 5(\omega CR)^2]/M$$
$$\sin\phi = \omega CR[6 - (\omega CR)^2]/M \tag{5.14}$$

Both these latter equations are needed to find ϕ unambiguously in the range $0° < \phi < 360°$. Figure 5.10 shows the form of frequency variation of the quantities

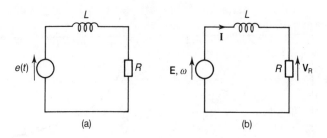

Figure 5.11 Resistance and inductance in series: (a) circuit; (b) phasor notation.

$$|\mathbf{V}_3/\mathbf{E}| = 1/M$$
$$\angle(\mathbf{V}_3/\mathbf{E}) = -\phi$$

The frequency for which the phase is 180° is given by

$$f = \frac{\sqrt{6}}{2\pi CR}$$

At this frequency the attenuation M is equal to 29.

5.3 ◆ The series inductance–resistance circuit

The circuit shown in Figure 5.11(a) has previously been used as an example of the application of the phasor method (Example 4.2). Whereas many variants of circuits containing only resistance and capacitance are encountered in practice, circuits involving inductance usually involve capacitance as well. Such circuits will be considered in the following sections. The simple $L-R$ configuration is found in 'smoothing' circuits in power supplies which produce a DC output from an AC input. In such cases the source $e(t)$ in Figure 5.11(a) would have a steady, DC, component associated with smaller amplitudes of AC of various frequencies. The load for the power supply is represented by the resistor R. The function of the inductor is, in virtue of a reactance increasing with increase of frequency, to reduce the amplitude of the AC components in the voltage across the load relative to the DC. We are thus concerned with the frequency response $\mathbf{V}_R/\mathbf{E}$ (Figure 5.11(b)). We have

$$\mathbf{I} = \mathbf{E}\,\frac{1}{\mathrm{j}\omega L + R}$$
$$\frac{\mathbf{V}_R}{\mathbf{E}} = \frac{1}{1 + \mathrm{j}\omega L/R}$$

Hence

$$|\mathbf{V_R/E}| = \frac{1}{[1 + (\omega L/R)^2]^{1/2}}$$

$$\angle(\mathbf{V_R/E}) = -\tan^{-1}(\omega L/R)$$

These equations have precisely the same form as Equation 5.3, but with the time constant CR replaced by L/R, the time constant of the $L–R$ circuit. The variation of magnitude and phase with frequency is that illustrated in Figure 5.4 but with the abscissa representing $\omega L/R$ rather than ωCR.

> **EXAMPLE 5.8** The output from a power supply contains a component of 300 Hz frequency and magnitude 2 V (RMS). The supply feeds a load of resistance 200 Ω. The largest inductor that can be obtained has inductance 1.2 H. What will be the voltage level of the component of this frequency appearing across the load?
>
> The reduction factor is $[1 + (\omega L/R)^2]^{1/2}$. Here
>
> $$\omega L/R = 2\pi \times 300 \times 1.2/200 \approx 11$$
>
> so that
>
> $$[1 + (\omega L/R)^2]^{1/2} \approx 11$$
>
> The 300 Hz component in the voltage will be $2/11 \approx 0.2$ V RMS.

5.4 ◆ The series *R–L–C* circuit

The simplicity of behaviour of circuits containing resistance and either inductance or capacitance, but not both, is due to the fact that in both cases energy is stored in one form only: electrostatic energy in the case of capacitors and magnetic energy in the case of inductors. This energy is dissipated in the resistive part of the circuit. When both inductance and capacitance are present a new possibility arises, that energy can be exchanged between the two types of stores, much as kinetic and potential energy are exchanged in mechanical systems. This possibility gives rise to the important phenomenon of resonance, most simply illustrated by analysis of the series $R–L–C$ circuit.

The circuit is shown in Figure 5.12. Using the notation in that figure we have

$$\mathbf{E} = \left(j\omega L + R + \frac{1}{j\omega C}\right)\mathbf{I} \qquad (5.15)$$

The magnitude of the current flowing is given by

$$|\mathbf{I}| = \frac{E}{[R^2 + (\omega L - 1/\omega C)^2]^{1/2}}$$

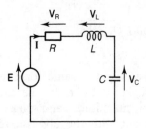

Figure 5.12 Circuit with inductance, capacitance and resistance in series, with phasor notation.

This clearly has a maximum at the frequency for which

$$\omega L = \frac{1}{\omega C}$$

or

$$\omega = \omega_0 = \frac{1}{(LC)^{1/2}}$$

The frequency

$$f_0 = \frac{\omega_0}{2\pi} = \frac{1}{2\pi(LC)^{1/2}} \tag{5.16}$$

is termed the **resonant** frequency of the circuit. Using this definition we can express Equation 5.15 in the form

$$\mathbf{E} = \mathbf{I}\left[R + j\left(\frac{L}{C}\right)^{1/2}\left(\frac{\omega}{\omega_0} - \frac{\omega_0}{\omega}\right)\right]$$

$$= IR\left[1 + jQ\left(\frac{f}{f_0} - \frac{f_0}{f}\right)\right] \tag{5.17}$$

where

$$Q = \frac{1}{R}\left(\frac{L}{C}\right)^{1/2} \tag{5.18}$$

This parameter Q is, for reasons to be discussed, termed the **quality factor** for the circuit. We will consider first the way in which the current varies as the frequency is altered, holding the source voltage $\mathbf{E}$ constant. From Equation 5.17 we have

$$|\mathbf{I}| = \frac{|\mathbf{E}|/R}{[1 + Q^2(f/f_0 - f_0/f)^2]^{1/2}} \tag{5.19}$$

$$\angle(\mathbf{I/E}) = -\tan^{-1}\left[Q\left(\frac{f}{f_0} - \frac{f_0}{f}\right)\right] \tag{5.20}$$

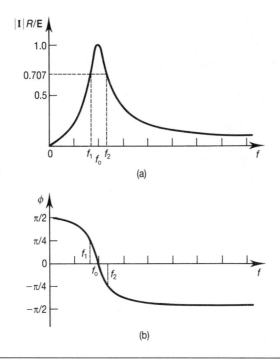

Figure 5.13 Frequency response for the current in the series $L–C–R$ circuit: (a) magnitude, $|\mathbf{I}|$; (b) phase, $\angle\mathbf{I}$.

From Equation 5.19 we see that $|\mathbf{I}|$ has a maximum value of $|\mathbf{E}|/R$ which occurs when $f = f_0$. This is the resonant frequency defined in Equation 5.16, at which the reactances of the inductance and the capacitance are equal and opposite. The value of $|\mathbf{I}|$ decreases steadily to zero as the frequency either increases to higher values or decreases to zero. This behaviour is illustrated in Figure 5.13(a). Equation 5.20 shows that when $f = f_0$ the current is in phase with the source voltage. As the frequency tends to higher values so the current increasingly lags on the voltage, the phase becoming $-\pi/2$ at very high frequencies. This corresponds to the high reactance of the inductance dominating the response. At frequencies lower than resonant the current leads the voltage, the phase tending towards $+\pi/2$ as the reactance of the capacitance becomes the dominating factor. This behaviour is illustrated in Figure 5.13(b).

From this discussion it is seen that the circuit is frequency selective: it responds more to some frequencies than to others. One way of quantizing the degree of selectivity is in terms of the frequency deviation required to reduce the response to a given level. This will be shown to depend on the parameter Q. From Equation 5.19 it is seen that the current is reduced to $1/\sqrt{2}$ of its maximum value (E/R) when

$$Q\left(\frac{f}{f_0} - \frac{f_0}{f}\right) = \pm 1 \tag{5.21}$$

Each sign leads to a quadratic equation in f/f_0. With the positive sign we have

$$\left(\frac{f}{f_0}\right)^2 - \frac{1}{Q}\left(\frac{f}{f_0}\right) - 1 = 0$$

Hence

$$\frac{f}{f_0} = \frac{1}{2Q} \pm \left(1 + \frac{1}{4Q^2}\right)^{1/2}$$

For the negative sign the equation becomes

$$\left(\frac{f}{f_0}\right)^2 + \frac{1}{Q}\left(\frac{f}{f_0}\right) - 1 = 0$$

giving

$$\frac{f}{f_0} = -\frac{1}{2Q} \pm \left(1 + \frac{1}{4Q^2}\right)^{1/2}$$

Choosing the two positive frequencies from among these roots, the lower frequency is found to be

$$f_1 = f_0\left[\left(1 + \frac{1}{4Q^2}\right)^{1/2} - \frac{1}{2Q}\right] \tag{5.22}$$

for which the expression in Equation 5.21 is -1. Hence, from Equation 5.20, the current leads the voltage, the phase lead at this frequency being $\pi/4$. The upper frequency, for which the expression is $+1$, is given by

$$f_2 = f_0\left[\left(1 + \frac{1}{4Q^2}\right)^{1/2} + \frac{1}{2Q}\right] \tag{5.23}$$

At this frequency the current lags the voltage, the phase being $-\pi/4$. The width of the curve at this level of response is

$$\Delta f = f_2 - f_1 = \frac{f_0}{Q} \tag{5.24}$$

This width is usually termed the **bandwidth**. To be quite specific it is the 'bandwidth to $1/\sqrt{2}$ down'. Equation 5.24 shows in what sense the Q factor determines the selectivity: the fractional bandwidth defined as

$$\frac{\Delta f}{f_0} = \frac{1}{Q} \tag{5.25}$$

depends only on the value of Q.

EXAMPLE 5.9   In a series L–R–C circuit

$$L = 5\text{ mH}, \; C = 4.7\text{ nF}, \; R = 500\text{ }\Omega$$

Determine the resonant frequency, the value of the Q factor and the bandwidth to $1/\sqrt{2}$.

The circuit is connected to a source adjusted to the resonant frequency and delivering 10 V (RMS) between its terminals. Calculate the magnitude and phase (with respect to the source) of the voltage across each component.

$$f_0 = 1/2\pi(5 \times 10^{-3} \times 4.7 \times 10^{-9})^{1/2} = 32.8 \text{ kHz}$$

$$Q = 2\pi f_0 L/R = 2.06$$

$$\Delta f = f_0/Q = 15.9 \text{ kHz}$$

At f_0, $\mathbf{I} = \mathbf{E}/R = (10/500)\angle 0 = 0.02\angle 0$. Hence

$$\mathbf{V}_R = 10\angle 0$$

$$\mathbf{V}_L = Q \times 10\angle 90 = 20.6\angle 90$$

$$\mathbf{V}_C = Q \times 10\angle -90 = 20.6\angle -90$$

EXAMPLE 5.10   In the circuit of Example 5.9, the frequency of the source is adjusted so that the voltage across the resistor is (a) 5 V and (b) 8 V. Calculate in each case the possible values of the frequency and the difference in phase between the resistor voltage and the source.

We can write the phasor voltage in the form

$$\mathbf{V}_R = \mathbf{E}/(1 + jx)$$

in which

$$x = \frac{1}{Q}\left(\frac{f}{f_0} - \frac{f_0}{f}\right)$$

Hence

$$|\mathbf{V}_R/\mathbf{E}| = (1 + x^2)^{-1/2}$$

$$\angle(\mathbf{V}_R/\mathbf{E}) = -\tan^{-1} x$$

To find the frequency we have to solve

$$x = Q\left(\frac{f}{f_0} - \frac{f_0}{f}\right)$$

This leads to the quadratic equation

$$\left(\frac{f}{f_0}\right)^{1/2} - \frac{x}{Q}\frac{f}{f_0} - 1 = 0$$

Solving for the positive root, we have

$$\frac{f}{f_0} = \frac{x}{2Q} + \left[1 + \left(\frac{x}{2Q}\right)^2\right]^{1/2}$$

(a) $(1 + x^2)^{-1/2} = 0.5$

$$x = \pm\sqrt{3}$$

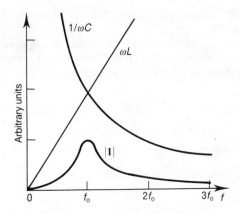

Figure 5.14 The series $L-C-R$ circuit, showing variation with frequency of the magnitude of the current, $|\mathbf{I}|$, and the component reactances ωL and $1/\omega C$.

The corresponding values for phase are therefore

$$x = +\sqrt{3} \qquad \angle V_R = -60°$$
$$x = -\sqrt{3} \qquad \angle V_R = +60°$$

For the frequency we have

$$x = +\sqrt{3} \qquad f = 49.4 \text{ kHz}$$
$$x = -\sqrt{3} \qquad f = 21.8 \text{ kHz}$$

(b) $(1 + x^2)^{-1/2} = 0.8$

$$x = \pm 0.75$$

For these values we find

$$x = +0.75 \qquad f = 39.3 \text{ kHz} \qquad \angle(V_R/E) = -36.9°$$
$$x = -0.75 \qquad f = 27.4 \text{ kHz} \qquad \angle(V_R/E) = +36.9°$$

5.4.1 ◆ Voltage maxima

It has been shown that the current in the series $R-L-C$ circuit (and hence the voltage across the resistor) reaches a maximum at the resonant frequency, f_0. The voltage appearing across the capacitor and that across the inductor also attain maximum values at some frequency. Consider the voltage across the capacitor:

$$|\mathbf{V}_C| = \frac{1}{\omega C} |\mathbf{I}| \tag{5.26}$$

The quantities appearing in this equation are illustrated in Figure 5.14, from which it will be realised that the maximum of the product of $1/\omega C$

and $|\mathbf{I}|$ will be at a somewhat lower frequency than f_0. Substituting for $\mathbf{I}$ from Equation 5.15 we have

$$\mathbf{V}_{\mathrm{C}} = \frac{\mathbf{E}}{1 - \omega^2 LC + j\omega CR} \tag{5.27}$$

Using the expressions for ω_0 and Q this becomes

$$\mathbf{V}_{\mathrm{C}} = \frac{\mathbf{E}}{1 - (f/f_0)^2 + j(1/Q)f/f_0}$$

Hence

$$|\mathbf{V}_{\mathrm{C}}| = \frac{|\mathbf{E}|}{\{[1 - (f/f_0)^2]^2 + (1/Q^2)(f/f_0)^2\}^{1/2}} \tag{5.28}$$

By differentiating with respect to f it may be shown that the maximum of $|\mathbf{V}_{\mathrm{C}}|$ occurs when

$$\left(\frac{f}{f_0}\right)^2 = 1 - \frac{1}{2Q^2} \tag{5.29}$$

Substitution in Equation 5.28 then shows that the value of the maximum is

$$\frac{|\mathbf{E}|Q}{(1 - 1/4Q^2)^{1/2}} \tag{5.30}$$

Thus when Q is large the voltage maximum occurs at a frequency very near to f_0; at this frequency the voltage reaches a value of the source voltage multiplied by Q, and may thus be very much greater than the source voltage.

The voltage across the inductor cam be investigated in a similar fashion:

$$|\mathbf{V}_{\mathrm{L}}| = \left| \frac{\mathbf{E}}{R + j\omega L + 1/j\omega C} \right|$$

$$= \frac{|\mathbf{E}|}{\{[1 - (f_0/f)^2]^2 + (1/Q^2)(f_0/f)^2\}^{1/2}}$$

The maximum of this expression is found to have the value given by Equation 5.30. The frequency at which it occurs is given by

$$\left(\frac{f_0}{f}\right)^2 = 1 - \frac{1}{2Q^2} \tag{5.31}$$

This gives a frequency slightly higher than f_0, as to be expected from inspection of Figure 5.14.

We note that the voltages across the inductor and the capacitor are in antiphase, and therefore cancel when added together.

EXAMPLE 5.11 For the circuit of Example 5.10 determine the frequency at which the voltage across (a) the capacitor and (b) the inductor has its maximum value. Determine also the magnitudes of those voltages.

The frequency for which the voltage across the capacitor is a maximum is given by Equation 5.29. Using the more precise value $f_0 = 32.831$ kHz, we find

$$f = 30.842 \text{ kHz}$$

For the voltage across the inductor to be maximized, we use Equation 5.31 to find

$$f = 34.949 \text{ kHz}$$

The maxima both have the value given by Equation 5.30: 21.26 V.

5.4.2 ◆ The Q factor

The Q factor is an important parameter of the circuit: it has been shown to determine the bandwidth, Δf, and together with the source voltage to determine the maximum value of the voltage appearing across the reactances. When Q is large, the expression of Equation 5.30 becomes closely equal to Q: if $Q > 5$, $|V_C/E| = Q$ to within 0.5%. For this reason Q is alternatively known as the magnification factor.

Equation 5.18 can be expressed in alternative forms:

$$Q = \frac{\omega_0 L}{R} = \frac{1}{\omega_0 CR} \tag{5.32}$$

In practice it is frequently the case that the bulk of the resistance is associated with the inductor: the ratio $\omega L/R$ is referred to as the Q factor of the inductor and is taken as a measure of its goodness. At radio frequencies the Q factor of an inductor may well be as high as 100.

For large values of Q certain approximations may be made. Thus

$$\left(1 - \frac{1}{2Q^2}\right)^{1/2} \approx 1 - \frac{1}{4Q^2}$$

$$\left(1 - \frac{1}{2Q^2}\right)^{-1/2} \approx 1 + \frac{1}{4Q^2}$$

Using these approximations in Equations 5.29 and 5.31 we see that the maxima of the voltages across the inductor and the capacitance occur at frequencies equally spaced on either side of f_0, by an amount $f_0/4Q^2$. This is well within the bandwidth f_0/Q. It also follows in this case of high Q that the two frequencies f_1, f_2 defined in Equations 5.22 and 5.23 are symmetrically placed on either side of f_0.

A useful approximation can be obtained for the expression

$$1 + jQ\left(\frac{f}{f_0} - \frac{f_0}{f}\right)$$

Since the frequency range of interest is of the order of $f_0/Q \ll f_0$ we write

$$f \approx f_0 + \delta f$$

Assuming $\delta f \ll f_0$ and using the binomial theorem we have

$$(f_0 + \delta f)^{-1} = \frac{1}{f_0}\left(1 + \frac{\delta f}{f_0}\right)^{-1} \approx \frac{1}{f_0}\left(1 - \frac{\delta f}{f_0} + \ldots\right)$$

Hence

$$\left(\frac{f}{f_0} - \frac{f_0}{f}\right) \approx \frac{2\,\delta f}{f_0}$$

and Equation 5.17, for example, becomes

$$\mathbf{E} = \mathbf{I}R\left(1 + \mathrm{j}2Q\,\frac{\delta f}{f_0}\right)$$

This form illustrates the symmetry about the resonant frequency when the Q factor is high.

EXAMPLE 5.12 Compare the answers of Example 5.11 with those obtained by assuming Q is large enough for the results of Section 5.4.2 to apply.

Since Q is only about 2, some errors are to be expected. For the frequency we have

$$f = f_0(1 \pm 1/4Q^2)$$

$$= 34.760, 30.902 \text{ kHz}$$

These are to be compared with the earlier results

$$34.949, 30.842$$

The maximum error is in the upper value, which is about 0.2 kHz. The important parameter is the bandwidth, determined in Example 5.9 to be some 16 kHz. Relative to this, the error is only about 3%, even with the low value of Q.

EXAMPLE 5.13 In a series $R–L–C$ circuit the capacitor is variable; the source is of fixed frequency, f, and of value E. The voltage across the capacitor is monitored as C is varied. Obtain, in terms of R, L and f, an expression for the value of C at which the capacitor voltage is a maximum.

From Equation 5.27 $|\mathbf{V}_C|$ may be expressed in the form

$$|\mathbf{V}_C| = E/S$$

where

$$S^2 = (1 - \omega^2 LC)^2 + (\omega CR)^2$$

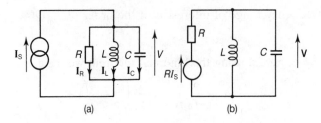

Figure 5.15 Parallel connection of inductance, capacitance and resistance: (a) current-fed circuit; (b) reduction using Thévenin's theorem.

In order to find the maximum we have to form the equation $d|V_C|/dC = 0$. This condition is satisfied by

$$dS/dC = -2\omega^2 L(1 - \omega^2 LC) + 2C(\omega R)^2 = 0$$

Solving for C we find

$$C = L/[R^2 + (\omega L)^2]$$

$$= \frac{1}{\omega^2 L} \left(1 + \frac{1}{Q^2}\right)^{-1}$$

where $Q = \omega L/R$. Thus for high values of Q the value of C is nearly that which makes the circuit resonant at the frequency of the source.

5.5 ◆ The parallel *L–C–R* circuit

A configuration perhaps more commonly encountered than the series form is that shown in Figure 5.15(a), or in an equivalent form in Figure 5.15(b). For this circuit we have

$$I_S = V \left(\frac{1}{R} + \frac{1}{j\omega L} + j\omega C\right) \tag{5.33}$$

In this case we introduce parameters

$$f_0 = \frac{\omega_0}{2\pi} = \frac{1}{2\pi(LC)^{1/2}} \tag{5.34}$$

$$Q = R \left(\frac{C}{L}\right)^{1/2} \tag{5.35}$$

The first of these is identical with the resonant frequency defined in Equation 5.16; the second appears to be the inverse of the Q factor defined in Equation 5.18, but will be shown to have the same significance in the parallel configuration as the previous definition does for the series case. Using these definitions Equation 5.33 can be rewritten in the form

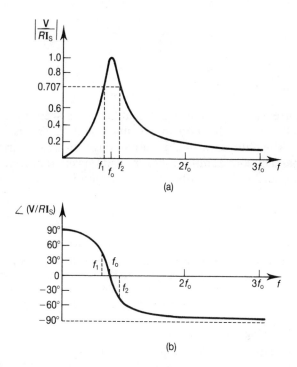

Figure 5.16 Frequency response for circuit of Figure 5.15(b): (a) gain, $|V/RI_s|$; (b) phase, $\angle(V/RI_s)$.

$$I_S = \frac{V}{R}\left[1 + jQ\left(\frac{f}{f_0} - \frac{f_0}{f}\right)\right] \tag{5.36}$$

This is identical to Equation 5.17 in so far as variation with frequency is concerned: the variation of $|V|$ and $\angle V$ with frequency is similar to that expressed in Equations 5.19 and 5.20 and is illustrated in Figure 5.16.

Whereas in the series configuration it was the voltages across the reactive elements which were magnified, in the case of the parallel configuration it is the currents through those elements. We have, using the notation of Figure 5.15(a),

$$I_L = \frac{V}{j\omega L}$$

$$I_C = j\omega CV$$

Exact analysis follows that used to derive Equations 5.29, 5.30 and 5.31. As before, provided that the Q factor is large, the frequencies at which the maxima occur are very close to the resonant frequency. At this frequency we then have to a good approximation

$$V = \mathbf{I}_S R$$

$$\mathbf{I}_L \approx \mathbf{I}_S \frac{R}{j\omega_0 L} = -jQ\mathbf{I}_S \qquad (5.37)$$

$$\mathbf{I}_C \approx \mathbf{I}_S j\omega_0 CR = jQ\mathbf{I}_S$$

These two currents are in antiphase and may be seen to constitute a current circulating round the loop comprising L and C. This current can be large. The 'tank' circuit used in the output stage of a radio transmitter is often in the parallel configuration, and in high-power transmitters the very large circulating current has to be taken into account when designing the components.

EXAMPLE 5.14 The output circuit of a radio transmitter designed to operate at 10 MHz is equivalent to a current source, a 2.5 μH inductance, a capacitor and the antenna all in parallel. The antenna can be assumed to be purely resistive. The overall Q factor of the circuit is 10. When the power going into the circuit at the resonant frequency is 1 kW, calculate

   (i) the voltage across the circuit,

  (ii) the current provided by the source, and

 (iii) the current through the inductance.

The circuit is that of Figure 5.15(a). By eliminating C between Equations 5.34, 5.35 we have

$$Q = \frac{R}{\omega_0 L}$$

Hence in this case

$$R = 10 \times 2\pi \times 10^7 \times 2.5 \times 10^{-6} = 1.571 \text{ k}\Omega$$

The power can only be dissipated in the resistance, so that the voltage V across the circuit can be found from

$$V^2/R = 10^3$$

Hence

$$V = 1.253 \text{ kV}$$

   (i) At the resonant frequency the current from the source is the same as that through the resistance. Hence

  (ii) $I_S = 1.25/1.57 = 0.80$ A

      From Equation 5.37 the current through the inductor is

$$I_L = QI_S = 8.0 \text{ A}$$

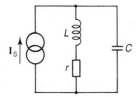

Figure 5.17 Equivalent circuit for a real inductor in parallel with a capacitor.

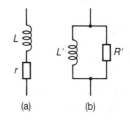

(a) (b)

Figure 5.18 Series–parallel equivalence, circuits for inductor (Equations 5.38, Section 5.5.1): (a) series; (b) parallel.

5.5.1 ♦ The series–parallel transformation

It will have been seen that the different definitions adopted for Q factor in the two configurations lead to the parameter playing the same part in both cases. As mentioned in Section 5.4.1 the resistive element is commonly associated with the inductor, and is usually regarded as a series element arising from the wire used to wind the inductor. This matter is more fully discussed in Chapter 14. If this is accepted it means that the circuit for an inductor in parallel with a capacitor should be as in Figure 5.17 rather than that of Figure 5.15(a). In the case of high Q the response of these two circuits is very similar, as may be seen by using the series–parallel equivalents discussed in Section 4.12.3. Applied to the impedance of L in series with r (Figure 5.18(a)), Equations 4.33 lead to the equivalent parallel circuit shown in Figure 5.18(b), in which

$$L' = L\left[1 + \left(\frac{r}{\omega L}\right)^2\right]$$
$$R' = r\left[1 + \left(\frac{\omega L}{r}\right)^2\right]$$

(5.38)

As has been pointed out in Section 4.12.3, L' and R' are not in general simply realizable in a physical sense since they both vary with frequency. In the case of high Q, however, we are interested only in a narrow range of frequencies, a few times f_0/Q in the vicinity of f_0. Within this narrow frequency range therefore, we may to a good approximation replace ω by ω_0 in Equations 5.38. Then, using the definition of Q given in Equation 5.32, we have

$$L' = L\left(1 + \frac{1}{Q^2}\right)$$
$$R' = r(1 + Q^2)$$

If $Q \gg 1$ these expressions become

$$L' \approx L$$

$$R' \approx rQ^2$$

Thus over the range of frequencies concerned, the circuit of Figure 5.17 may be expected to behave in very similar fashion to that of Figure 5.15(a) provided that we take $R = rQ^2$. When this is done, we can see that the two definitions of Q, Equations 5.18 and 5.34, are identical: denoting

$$R = r \left(\frac{\omega_0 L}{r} \right)^2$$

then

$$\frac{R}{\omega_0 L} = \frac{\omega_0 L}{r} = Q \tag{5.39}$$

An inductor is often specified in terms of its inductance and its Q factor. A circuit to represent the inductor may take the form of either an inductance L in series with a resistance $\omega L/Q$ or an inductance L in parallel with resistance ωLQ.

EXAMPLE 5.15 An AC bridge operates at a frequency of 1 kHz, and when used to measure an inductor displays at choice values of inductance and resistance for either the series equivalent circuit or the parallel circuit. For a particular inductor the series form reads 120 μH in series with 1.5 Ω. Calculate the readings to be expected from the parallel form.

For the series combination $Q = \omega L/R$. In this case, therefore,

$$Q = 2\pi \times 10^3 \times 120 \times 10^{-6}/1.5 = 0.503$$

Since the Q factor is not large, we have to use the full form of Equations 5.38, giving

$$L' = 120(1 + 3.958) = 595 \ \mu\text{H}$$

$$R' = 1.5(1 + 0.253) = 1.88 \ \Omega$$

A similar analysis can be applied to capacitors. A real capacitor can be represented by either of the circuits shown in Figure 5.19. If these are to have the same impedance then, using Equations 4.33,

$$r_P = r_S \left[1 + \left(\frac{1}{\omega C_S r_S} \right)^2 \right]$$

$$C_P = C_S \left[1 + (\omega C_S r_S)^2 \right] \tag{5.40}$$

By comparison with the case of the inductor, we see that the quantity $(\omega C_S r_S)^{-1}$ plays the part of a Q factor and if this is large, or $\omega C_S r_S$ is small, then

$$C_P \approx C_S$$

$$r_P \approx \frac{r_S}{(\omega C_S r_S)^2}$$

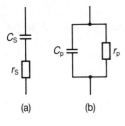

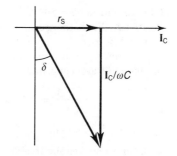

Figure 5.19 Series parallel equivalence for capacitor (Equation 5.40, Section 5.5.1).

Figure 5.20 Phasor diagram for series model of capacitor, illustrating the significance of the loss angle δ.

The 'goodness' of a capacitor is usually specified in terms of a **loss angle**, nearly always denoted by δ. The significance of this angle is illustrated by the phasor diagram of Figure 5.20 which applies to the series circuit of Figure 5.19(a). In this phasor diagram $\mathbf{V_C}$ and $\mathbf{I_C}$ are the phasor representations respectively of the voltage across the capacitor and the current through it: if the capacitor were perfect (that is, lossless) $\mathbf{V_C}$ would lag on $\mathbf{I_C}$ by exactly $\pi/2$. With a real capacitor the angle will be less than $\pi/2$ by δ, termed the loss angle. From the diagram

$$\tan \delta = \omega C_S r_S \tag{5.41}$$

Alternatively

$$\tan \delta = 1/\omega C_P r_P \tag{5.42}$$

Thus, provided that $\tan \delta$ is small and the frequency range narrow, the real capacitor of capacitance C and loss angle δ can be represented either by a capacitance C in series with a resistance $\tan \delta / \omega C$ or by a capacitance C in parallel with a resistance $1/\omega C \tan \delta$. These equivalences are very useful in the analysis and design of resonant circuits.

EXAMPLE 5.16 The dielectric used in a capacitor has a loss angle of 2×10^{-3} rad. Estimate for a 100 pF capacitor the effective loss resistance at a frequency of 50 MHz in (i) the series form and (ii) the parallel form.
Using Equations 5.41 and 5.42, we find

$$r_S = 2 \times 10^{-3}/2\pi \times 50 \times 10^6 \times 100 \times 10^{-12} = 0.064 \ \Omega$$

$$r_P = r_S/(2 \times 10^{-3})^2 = 15.9 \ \text{k}\Omega$$

EXAMPLE 5.17 A circuit takes the form shown in Figure 5.15(a). The inductor is of inductance 1 mH, Q factor 20; the capacitor is of capacitance

0.01 μF and $\tan \delta = 5 \times 10^{-3}$; the parallel resistance provided by the source is 10 kΩ. Find the resonant frequency, the overall Q factor and the bandwidth.

Since this is a parallel circuit we use parallel equivalent circuits for the inductor and capacitor. We have

$$\omega_0 = \frac{1}{(LC)^{1/2}} = 3.16 \times 10^5$$

$$f_0 = 50.3 \text{ kHz}$$

The equivalent parallel resistance for the inductor is

$$\omega_0 L Q = 6.32 \text{ k}\Omega$$

For the capacitor the equivalent parallel resistance is

$$1/\omega_0 C \tan \delta = 63.3 \text{ k}\Omega$$

To find the resultant parallel resistance we take these two in parallel with the source resistance of 10 kΩ, giving 3.65 kΩ.

The overall Q factor is $R/\omega_0 L$, giving

$$Q = 11.6$$

It may be noted that

$$\frac{1}{Q_T} = \frac{\omega_0 L}{R_S} + \frac{1}{Q_L} + \tan \delta$$

in which R_S is the source resistance.

The bandwidth to $1/\sqrt{2}$ is 4.3 kHz.

EXAMPLE 5.18 In the circuit of Example 5.14, the intrinsic Q factor of the inductor is 80, contributing to the overall Q factor of 10. When the source is providing 1 kW at the resonant frequency estimate the power radiated by the antenna and the power dissipated in the inductor.

Since the Q factor of the inductor is high, we may use the parallel form of equivalent circuit. For the effective parallel resistance at 10 MHz we have (Equation 5.39)

$$R_L = Q\omega_0 L = 80 \times 2\pi \times 10^7 \times 2.5 \times 10^{-6} = 12.7 \text{ k}\Omega$$

The effective parallel resistance resulting from the combination of the inductor resistance and the antenna resistance was calculated in Example 5.11 to be

$$R = 1.571 \text{ k}\Omega$$

The resistance of the antenna can then be calculated from

$$\frac{1}{R} = \frac{1}{R_A} + \frac{1}{R_L}$$

Hence

$$R_A = 1.793 \text{ k}\Omega$$

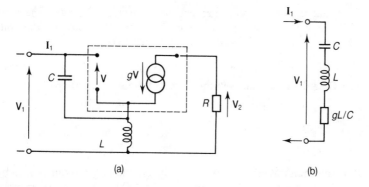

Figure 5.21 Example 5.19: (a) circuit; (b) equivalent input circuit.

It was calculated in Example 5.14 that the voltage across the circuit is 1.253 kV. Hence

$$P_A = (1253)^2/1793 = 876 \text{ W}$$

$$P_L = (1253)^2/12700 = 124 \text{ W}$$

As a check, the total power is correctly 1 kW.

EXAMPLE 5.19 In Figure 5.21(a) is shown an $L-C-R$ circuit involving a dependent source. It models a situation encountered in high-frequency amplifiers when inductance associated with a connecting lead results in unexpected effects. In this circuit the inductance L should ideally be zero: its presence leads to an undesirable drop in input impedance.

We have

$$\mathbf{V}_1 = \mathbf{V} + \mathbf{V}_L$$

$$\mathbf{I}_1 = j\omega C\mathbf{V}$$

$$\mathbf{V}_L = (\mathbf{I}_1 + g\mathbf{V})j\omega L$$

$$\mathbf{V}_2 = -gR\mathbf{V}$$

Eliminating **V**, we find

$$\mathbf{V}_L = \mathbf{I}_1 \left(j\omega L + \frac{gL}{C} \right)$$

and

$$\mathbf{V}_1 = \mathbf{I}_1 \left(\frac{1}{j\omega C} + j\omega L + \frac{gL}{C} \right)$$

The expression in parentheses is the impedance 'seen' by the source providing the voltage $\mathbf{V}_1$, and is equivalent to the circuit shown in Figure 5.21(b). In the situation on which the example is based, the inductance L is small and usually its reactance can be neglected. However, the resistance gL/C to which it gives rise is often important. Consider the following values, typical of a transistor circuit designed to operate at 1 GHz:

$$g = 12 \text{ mS}, \ C = 2 \text{ pF}, \ L = 1 \text{ nH}$$

At 1 GHz the impedances in the circuit of Figure 5.21(b) are

1 nH	6.3 Ω
2 pF	80 Ω
gL/C	6 Ω

Thus the circuit at this frequency and below can be approximated to by 2 pF in series with 6 Ω. The significance of the 6 Ω is best seen by transforming the series R–C circuit into parallel form, as in Section 5.5.1. We have at 1 GHz

$$\omega C_{\text{S}} r_{\text{S}} = 0.075$$

$$C_{\text{P}} \approx C_{\text{S}} = 2 \text{ pF}$$

Hence

$$r_{\text{P}} \approx 6/(0.075)^2 = 1.06 \text{ k}\Omega$$

The effect of the inductance is therefore to contribute a shunt resistance of about 1 kΩ, which may well be of significance in the operation of the circuit.

5.6 ◆ Bode plots

It is often of interest to display the response of a circuit over a wide range of frequencies, covering perhaps several decades. The normal way of doing this is by use of a logarithmic scale for frequency. Associated with such a wide coverage is usually an equally wide coverage in amplitude response, for which a logarithmic scale is also convenient: such a scale was used in Figure 5.10. In addition to convenience, in certain cases use of the log–log plot leads to very simple graphical methods of obtaining an approximate amplitude response curve. Such graphs are known as Bode plots.

The expression for response as a function of frequency can often be expressed as the product of a number of factors of the type $1 + jf/f_0$. An example occurred in Section 5.2.2, when studying the resistance–capacitance circuit, in which the response was given by

$$\mathbf{H} = \left(1 + j\frac{f}{f_0}\right)^{-1} \tag{5.43}$$

with $f_0 = 2\pi RC$. We consider first the modulus of this expression:

$$|\mathbf{H}| = \left[1 + \left(\frac{f}{f_0}\right)^2\right]^{-1/2} \tag{5.44}$$

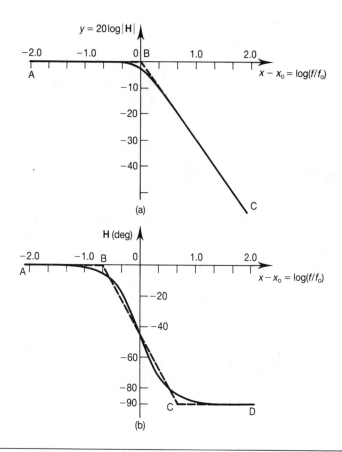

Figure 5.22 Bode plots for $\mathbf{H} = (1 + f/f_0)^{-1}$: (a) gain in decibels; (b) phase.

In Figure 5.22(a) a graph of this expression is shown in the form

$$y = 20\log_{10}|\mathbf{H}|$$

$$= -10\log_{10}\left[1 + \left(\frac{f}{f_0}\right)^2\right] \tag{5.45}$$

$$x = \log_{10} \mathrm{f}$$

In these equations x, y are taken as the conventional symbols for Cartesian axes. With the factor of $\times 20$ in the expression for the ordinate, the resulting unit is customarily said to be the decible. Since $|\mathbf{H}|$ is a voltage ratio and not a power ratio this is an illicit use of the decibel, as discussed in Section 3.9.2. It is, however, hallowed by usage and since in the present context power is not involved confusion should not arise. Further, f in the expression for x has to be presumed divided by unit frequency so that it becomes a numeric. This merely fixes the origin of x.

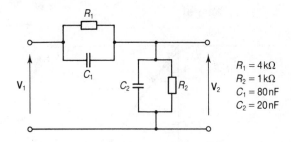

Figure 5.23 Circuit for Example 5.20.

Plotted also in the figure are the asymptotes to which the curve tends in the limits $x \to -\infty$ and $x \to \infty$, corresponding to $f \to 0$ and $f \to \infty$ respectively. As $f \to 0$, $x \to -\infty$ and $y \to 0$, giving the line AB in the figure; as $f \to \infty$, $x \to +\infty$ and $y \to -20 \, (x - \log_{10} f_0)$, giving the line BC. These two lines, meeting at the point $x = x_0 = \log_{10} f_0$, $y = 0$, are seen to form a good, easily drawn approximation to the true curve. For $x - x_0 < -0.5$ and $x - x_0 > 0.5$, corresponding to $f < f_0/\sqrt{10}$ and $f > f_0\sqrt{10}$ respectively, the errors are negligible. The maximum error occurs when $x = x_0$, $f = f_0$: at this point the approximation gives $y = 0$ while the true value is $y = 3$. If this exact value is included a more accurate curve can be sketched in from the approximation provided by the asymptotes.

The phase is plotted in Figure 5.22(b), in the form

$$y = \phi = -\tan^{-1}(f/f_0) \tag{5.46}$$

$$x = \log_{10} f$$

A useful approximation to this curve can also be provided by straight lines. Two of these, AB and CD, represent asymptotes:

$$x - x_0 < -1, \, f < 0.1 f_0, \, y = 0°$$

$$x - x_0 > 1, \, f > 10 f_0, \, y = -90°$$

The third, BC, joins the end points of the other two. It passes through the correct value of 45° at $x = x_0$, $f = f_0$. Maximum errors of about 10° are incurred. The approximate curve can be corrected by use of a table of errors.

The way in which these approximate constructions can be used will be illustrated by an example.

EXAMPLE 5.20 For the circuit of Figure 5.23, derive an expression for the ratio

$$\mathbf{H} = \frac{\mathbf{V}_2}{\mathbf{V}_1}$$

in the form

$$\mathbf{H} = K \frac{1 + jfT_1}{1 + jfT_2}$$

Using the values

$$R_1 = 4 \text{ k}\Omega, \ R_2 = 1 \text{ k}\Omega, \ C_1 = 80 \text{ nF}, \ C_2 = 20 \text{ nF}$$

draw Bode plots to illustrate the variation of the amplitude and phase of **H** over the frequency range from 10 Hz to 100 kHz.

Denoting

$$\frac{1}{\mathbf{Z}_1} = \mathbf{Y}_1 = \frac{1}{R_1} + j\omega C_1$$

$$\frac{1}{\mathbf{Z}_2} = \mathbf{Y}_2 = \frac{1}{R_2} + j\omega C_2$$

we have

$$\mathbf{H} = \frac{\mathbf{V}_2}{\mathbf{V}_1} = \frac{\mathbf{Z}_2}{\mathbf{Z}_1 + \mathbf{Z}_2}$$

$$= \frac{\mathbf{Y}_1}{\mathbf{Y}_1 + \mathbf{Y}_2}$$

Substituting,

$$\mathbf{H} = \frac{1/R_1 + j\omega C_1}{1/R_1 + 1/R_2 + j\omega(C_1 + C_2)}$$

$$= K \frac{1 + jfT_1}{1 + jfT_2}$$

where

$$K = \frac{R_2}{R_1 + R_2}$$

$$T_1 = 2\pi R_1 C_1$$

$$T_2 = 2\pi \frac{R_1 R_2}{R_1 + R_2} (C_1 + C_2)$$

Writing the expression for **H** in the form

$$\mathbf{H} = K \frac{\mathbf{H}_1}{\mathbf{H}_2}$$

with

$$\mathbf{H}_1 = 1 + jfT_1$$

$$\mathbf{H}_2 = 1 + jfT_2$$

we form

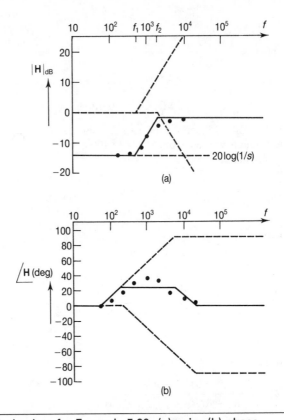

Figure 5.24 Bode plots for Example 5.20: (a) gain; (b) phase.

$$20 \log |\mathbf{H}| = 20 \log K + 20 \log |\mathbf{H}_1| - 20 \log |\mathbf{H}_2|$$

and

$$\angle \mathbf{H} = \angle \mathbf{H}_1 - \angle \mathbf{H}_2$$

Allowing for the fact that the logarithm of the reciprocal of a quantity is just the negative of the logarithm of the quantity, the behaviour of each term is of the same as that of Equation 5.43. The resultant graph can therefore be constructed by addition or subtraction of the component graphs.

Inserting numerical values, we find

$$K = 0.2$$

$$T_1 = 2.01 \times 10^{-3}, \ f_1 = 1/T_1 \approx 500 \text{ Hz}$$

$$T_2 = 5.03 \times 10^{-4}, \ f_2 = 1/T_2 \approx 2 \text{ kHz}$$

The formation of the Bode approximation for the amplitude is shown in Figure 5.24(a), and for the phase in Figure 5.24(b). For comparison, in both figures are shown the values obtained from the exact expressions.

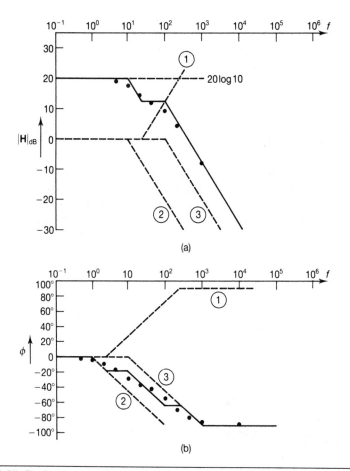

Figure 5.25 Bode plots for Example 5.21: (a) gain; (b) phase.

EXAMPLE 5.21   A system has the response function

$$\mathbf{H} = 10 \,\frac{1 + j0.04f}{(1 + j0.1f)(1 + j0.01f)}$$

Make Bode plots for this response.

Each factor in this response function is of the form $1 + jf/f_0$:

$$\mathbf{H}_1 = 1 + j0.04f = 1 + jf/25$$

$$\mathbf{H}_2 = 1 + j0.1f = 1 + jf/10$$

$$\mathbf{H}_3 = 1 + j0.01f = 1 + jf/100$$

We have

$$20 \log_{10} |\mathbf{H}| = 20 \log_{10} 10 + 20 \log_{10} |\mathbf{H}_1| - 20 \log_{10} |\mathbf{H}_2| - 20 \log_{10} |\mathbf{H}_3|$$

$$\angle \mathbf{H} = \angle \mathbf{H}_1 - \angle \mathbf{H}_2 - \angle \mathbf{H}_3$$

The behaviour of each term can be approximated by the asymptotic forms used above and the final result obtained by addition. The process is illustrated for the amplitude response in Figure 5.25(a) and for phase response in Figure 5.25(b).

5.7 ◆ Conclusions

The sections of this chapter have been primarily studies of particular circuits. Although each circuit may be of importance in its own right, the object of the chapter as a whole is to illustrate applications of AC circuit analysis. In the course of a professional career it is likely that a great variety of circuits will be encountered. Analysis will be necessary as an aid to understanding and design. Many of them will be simple circuits of the types illustrated in this chapter; others will be much more complicated and require the sophisticated methods of analysis introduced in the next chapter.

The principle results found for the circuits studied in the chapter are summarized in the next section.

5.8 ◆ Summary of formulae

Series resistance–capacitance circuit (Figure 5.1)

$$\mathbf{V}_C = \mathbf{E}\,\frac{1}{1 + j\omega CR}$$

$$\mathbf{V}_R = \mathbf{E}\,\frac{1}{1 + 1/j\omega CR}$$

$$\frac{\mathbf{V}_C}{\mathbf{E}} = \frac{1}{[1 + (\omega CR)^2]^{1/2}}\exp(-j\theta)$$

$$\frac{\mathbf{V}_R}{\mathbf{E}} = \frac{\omega CR}{[1 + (\omega CR)^2]^{1/2}}\exp[j(\pi/2 - \theta)]$$

where $\theta = \tan^{-1}(\omega CR)$.

Series inductance–resistance circuit (Figure 5.11)

$$\frac{\mathbf{V}_R}{\mathbf{E}} = \frac{1}{1 + (j\omega L/R)}$$

Hence

$$|V_R/E| = \frac{1}{[1 + (\omega L/R)^2]^{1/2}}$$

$$\angle(V_R/E) = -\tan^{-1}(\omega L/R)$$

Series *R–L–C* circuit (Figure 5.12)

At resonance

$$\omega_0 = \frac{1}{(LC)^{1/2}}$$

$$f_0 = \frac{1}{2\pi(LC)^{1/2}}$$

$$Q = \frac{1}{R}\left(\frac{L}{C}\right)^{1/2} = \frac{\omega_0 L}{R}$$

$$\mathbf{E} = \mathbf{I}\left[R + j\left(\frac{L}{C}\right)^{1/2}\left(\frac{\omega}{\omega_0} - \frac{\omega_0}{\omega}\right)\right]$$

$$= \mathbf{I}R\left[1 + jQ\left(\frac{f}{f_0} - \frac{f_0}{f}\right)\right]$$

$$|\mathbf{I}| = \frac{|\mathbf{E}|/R}{[1 + Q^2(f/f_0 - f_0/f)^2]^{1/2}}$$

$$\angle(\mathbf{I}/\mathbf{E}) = -\tan^{-1}\left[Q\left(\frac{f}{f_0} - \frac{f_0}{f}\right)\right]$$

Bandwidth to $1/\sqrt{2}$

$$\Delta f = \frac{f_0}{Q}$$

Parallel *R–L–C* circuit (Figure 5.15)

$$f_0 = \frac{\omega_0}{2\pi} = \frac{1}{2\pi(LC)^{1/2}}$$

$$Q = R\left(\frac{C}{L}\right)^{1/2} = \frac{R}{\omega_0 L}$$

$$|\mathbf{V}| = \frac{|\mathbf{I}_S|R}{[1 + Q^2(f/f_0 - f_0/f)^2]^{1/2}}$$

$$\angle(\mathbf{V}/\mathbf{I}_S) = -\tan^{-1}\left[Q\left(\frac{f}{f_0} - \frac{f_0}{f}\right)\right]$$

For high *Q* circuits, over the frequency range of interest the approximation

$$\frac{f}{f_0} - \frac{f_0}{f} \approx \frac{2\,\delta f}{f_0}$$

may be used.

Series–parallel equivalence

At a single frequency, the series combination of a reactance X_S and resistance R_S may be replaced by a parallel combination of X_P and R_P, where

$$X_P = X_S \left(1 + \frac{1}{Q^2}\right)$$

$$R_P = R_S(1 + Q^2)$$

and

$$Q = \frac{X_S}{R_S} = \frac{R_P}{X_P}$$

Bode plots

See Figure 5.21.

PROBLEMS

5.1 A microphone gives an output over the range of frequencies 100 Hz–25 kHz. Its internal impedance is resistive, of value 1 kΩ. It is connected to an amplifier of resistive input impedance 10 kΩ through a series capacitor. Estimate a value for the capacitor which will make the response uniform to within 10% over the operating frequency range. With this value of capacitance calculate the phase shift at the upper and lower frequencies.

5.2 In Figure 5.26 is shown the circuit of a voltage divider. Obtain an expression for the ratio **V/E**.

In a particular case $R_1 = R_2 = 10$ kΩ, $C_1 = C_2 = 5$ nF. Determine the frequency for which the output is in phase with the input and calculate the division ratio at that frequency. Sketch graphs to illustrate the variation with frequency of the magnitude and phase of the division ratio.

The source E in the divider delivers 12 V between terminals and its frequency is adjusted to give maximum output when no load is connected. Determine the value of the output when the divider is connected to a 5 kΩ resistive load.

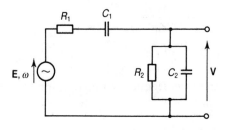

Figure 5.26

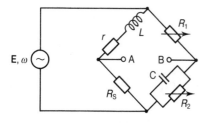

Figure 5.27

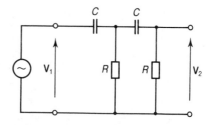

Figure 5.28

5.3 A variable frequency source has a resistive internal impedance of 1 kΩ. The voltage between terminals on open circuit is constant at 10 V as the frequency is altered. The source is connected to a load consisting of 2 kΩ resistance in parallel with 10 mH inductance through a cable of negligible resistance but having 5 mH series inductance. Obtain an expression for the voltage appearing across the load as a function of frequency. Determine the frequency at which this voltage is a maximum and the value of that maximum. Determine also the frequencies for which the phase difference between output and input is 45°.

5.4 In Figure 5.27 is shown the circuit of an AC bridge designed to measure the inductance and resistance of an unknown inductor. The capacitance C and the resistance R_S are fixed standards; the resistances R_1 and R_2 are variable resistance boxes. The bridge is excited by an AC source of angular frequency ω. In use, R_1 and R_2 are adjusted to reduce the voltage between nodes A and B to zero. Derive the balance conditions for the bridge and hence show that knowledge of the frequency is not required.

In a particular case

$$C = 0.1 \ \mu\text{F}, \ R_S = 1 \ \text{k}\Omega, \ R_1 = 5370 \ \Omega, \ R_2 = 15\,230 \ \Omega$$

Determine L and r.

5.5 In Figure 5.28 is shown a two-stage R–C circuit. Obtain an expression for the voltage ratio $\mathbf{V}_2/\mathbf{V}_1$ as a function of frequency. Determine the value of the ratio in magnitude and phase at the frequency for which $\omega RC = 1$. Find also the value of ωRC for which $|\mathbf{V}_2/\mathbf{V}_1| = 0.5$. What is then the difference in phase between output and input?

5.6 For the circuit of Figure 5.29, containing a voltage-controlled dependent current source, show that the voltage ratio $\mathbf{V}_2/\mathbf{V}_1$ can be expressed in the form

$$\frac{\mathbf{V}_2}{\mathbf{V}_1} = -K \frac{1 + j\omega T_1}{1 + j\omega T_2}$$

In a particular case $g = 0.01 \ \text{S}$, $R_1 = 150 \ \Omega$, $R_2 = 5 \ \text{k}\Omega$. Estimate a value for C so that at a frequency of 100 Hz $|\mathbf{V}_2/\mathbf{V}_1|$ is not less than 80% of its value for very high frequencies. With this value of C, calculate $\angle(\mathbf{V}_2/\mathbf{V}_1)$ at 100 Hz.

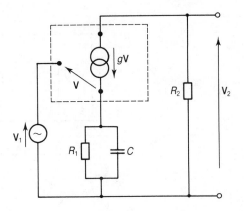

Figure 5.29

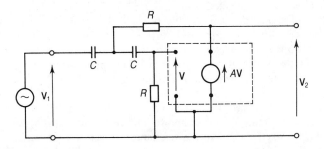

Figure 5.30

5.7 In the circuit of Figure 5.30, the gain A of the voltage-controlled dependent voltage source is unity. Show that

$$\frac{V_2}{V_1} = \left(1 + \frac{1}{j\omega CR}\right)^{-2}$$

5.8 In the circuit of Figure 5.31 the source is variable in frequency but of constant RMS value E. Derive a Thévenin equivalent for that part of the circuit to the left of the terminals A, B, and hence show that the phasor representation of the voltage across the resistor can be expressed in the form

$$V = E\angle(-2\theta)$$

where

$$\tan \theta = \omega(LC)^{1/2}$$

5.9 An inductor is connected in series with a 1 nF capacitor to a variable frequency source giving 0.5 V between terminals on open circuit, independent of frequency. The internal impedance of the source is 50 Ω resistive. The inductor can be regarded as the series combination of an inductance with a resistance. When the frequency is varied, it is found that the voltage across the capacitor rises to a maximum of 4.5 V at 12 kHz. It is also found that the bandwidth to $1/\sqrt{2}$ is

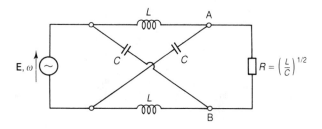

Figure 5.31

1.3 kHz. Check the data for internal consistency and estimate values for the inductance and resistance of the inductor.

5.10 In a series $R-L-C$ circuit the components have the values $L = 50\,\mu\text{H}$, $C = 100\,\text{pF}$ and $R = 5\,\Omega$. Determine the resonant frequency, the Q value and the bandwidth. If the source voltage is $1\,\text{mV}$, independently of frequency, determine the maximum voltage appearing across the capacitor and the frequency at which it occurs.

5.11 The parallel combination of a variable inductor and a $12\,\text{pF}$ capacitor is connected to variable frequency source through a $2.5\,\text{k}\Omega$ resistor. The source delivers $5\,\text{V}$ (RMS) on open circuit, and has an internal impedance of $500\,\Omega$. The resonant frequency of the circuit is to be $45\,\text{MHz}$. Calculate

(a) the value of the inductance,

(b) the Q factor,

(c) the bandwidth to $1/\sqrt{2}$, and

(d) the current in the inductor at $45\,\text{MHz}$.

5.12 It is required to make a parallel $L-C-R$ circuit which resonates at $30\,\text{kHz}$ with a Q factor of 10. An available inductor has inductance $1.8\,\text{mH}$ and a Q factor of 20 at $30\,\text{kHz}$. The available capacitors use a dielectric for which $\tan\delta = 0.02$. Calculate the value of capacitance and resistance required.

5.13 For the circuit of Figure 5.32, obtain an expression for the response function $\mathbf{V}/\mathbf{E}$. By means of careful sketches illustrate the variation of $|\mathbf{V}/\mathbf{E}|$ and $\angle(\mathbf{V}/\mathbf{E})$ with frequency. Show that $|\mathbf{V}/\mathbf{E}|$ exhibits a maximum at a non-zero frequency if $R > (L/2C)^{1/2}$.

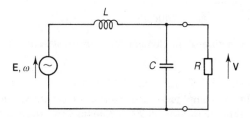

Figure 5.32

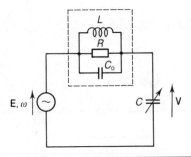

Figure 5.33

In a particular case

$$L = 20 \text{ mH}, \; C = 200 \text{ pF}, \; R = 20 \text{ k}\Omega$$

Calculate the position and magnitude of the peak response. Calculate also the magnitude and phase of the response at 50 kHz.

5.14 In the circuit of Figure 5.33 the elements within the dotted line form an equivalent circuit for an inductor. The properties of an inductor suggest that whereas the inductance L and self-capacitance C_0 will be independent of frequency, the resistance R may not be constant. The capacitor C and the frequency of the source may be varied. Derive an expression for the RMS voltage across the capacitor C. Show that as C is varied, at fixed frequency, the voltage has a maximum value given by

$$V_M = E[1 + Q^2(1 - \omega^2 LC_0)^2]^{1/2}$$

in which $Q = R/\omega L$, and that at this maximum

$$\omega^2 L(C + C_0) = 1$$

For a particular inductor, experiment gave the following values:

$f = \omega/2\pi$ (MHz)	C (pF)	V_M/E
1.0	308	117
1.5	132	92
2.0	69	75

Analyse these results to obtain estimates for

   (i) the value of L and of C_0, and

   (ii) the value of Q at each frequency.

5.15 The transfer function of an element of a control system is given by

$$\mathbf{H} = \frac{10}{(1 + j0.2f)(1 + j0.5f)}$$

Prepare Bode plots of amplitude and phase. From these plots, estimate the frequency for which the phase lags by 90°. Estimate also the gain at this frequency. Compare these estimates with those values obtained by calculation.

6 ◆ The Formal AC Analysis of a General Linear Network

This chapter details the systematic approach to setting up the equations necessary for analysis of a network excited by AC sources. The choice of suitable unknown variables is considered, leading to nodal and loop (mesh) methods of analysis. The use of the computer in network analysis is introduced.

6.1 ◆ Introduction

The circuits hitherto studied have been straightforward to analyse primarily because few components have been involved and suitable unknowns easily chosen. For networks with more unknowns and more involved interconnections a systematic approach is essential. A formulation of the general network problem not only aids solution but is also necessary for the proof of theorems such as those of Thévenin and Norton, application of which can greatly simplify the approach both to analysing and to understanding a particular network. A general formulation is also a necessary preliminary to the writing of computer programs which can aid the process of analysis.

A network is characterized both by the nature and values of the individual components and by the web of interconnections. This web of interconnections, usually described graphically by means of a circuit diagram, is known as the **topology** of the network.

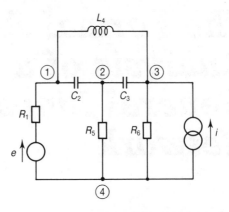

Figure 6.1 Circuit used as an example. Ringed numbers mark nodes.

6.2 ◆ Network description: branches, nodes and loops

The general method of network description is most easily introduced with the aid of an example. The example to be used is shown in Figure 6.1, being a **bridged-T** circuit containing inductance, capacitance and resistance, one constant voltage source and one constant current source.

6.2.1 ◆ Nodes and branches

The terms node and branch have been used in previous examples in connection with circuit diagrams, and are of general application. A branch signifies the existence of a component and a node signifies the junction of different components. In principle each ideal component requires two nodes, but in certain cases it may not be necessary to introduce a node which has only two components joined to it: for example, two resistances in series can be replaced by one equivalent resistance. It is often adequate in AC analysis to describe a series combination of reactance and resistance by a single 'component' of specified inpedance, or a parallel combination by a single admittance. The network of Figure 6.1 may be said to have four nodes and six branches, provided that we treat each source and its associated resistance as a single branch. The form of interconnections between the various branches, the topology, is shown by the **graph** of the circuit, appearing for the present case as Figure 6.2. In order to be able to assign unambiguously directions to current and voltage, it is convenient to assign directions to the branches, as in the figure.

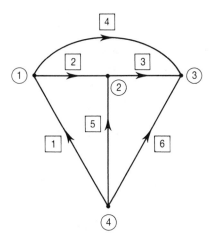

Figure 6.2 The graph of the circuit of Figure 6.1: branches are given direction and number (in squares).

6.2.2 ♦ Loops

Another feature of the graph of Figure 6.2, and of the graph of any network, is the existence of sequences of branches through each of which a closed path can be traced. Such paths are known as **loops**. In a case such as that of Figure 6.2 in which no connections cross it is easy to identify the loops. Each loop can be identified by a number and described by a sequence of (numbered) branches (Figure 6.3). This is done for the present case in Table 6.1.

Some of these loops can be made up from others. Loops are said to be independent if each contains one branch not contained in the others. Thus in Table 6.1 the first three loops contain between them all the branches and each has one branch unique to itself. They therefore form an independent set of loops: the rest can be made up by combinations of these three. It is to be noted that other sets of independent loops could be chosen: the choice is not unique.

Table 6.1

Loop	Branches
1	1–2–5
2	5–3–6
3	4–3–2
4	1–2–3–6
5	4–6–5–2
6	1–4–3–5
7	1–4–6

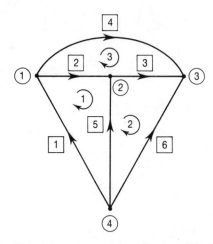

Figure 6.3 Graph for Figure 6.1 with numbered loops (meshes).

6.2.3 ◆ Loops in non-planar networks

Independent loops can easily be chosen for networks which can be laid out in the form of a fish-net, without branches crossing. It is less easy when this is not the case, and a formal way of choosing a set of independent loops is desirable. To do this we firstly consider the concept of a **tree** of a network: a tree is obtained by extracting branches from the graph of the network until removal of one more would leave one node isolated. Some trees for the graph of Figure 6.2 are shown in Figure 6.4, in which the

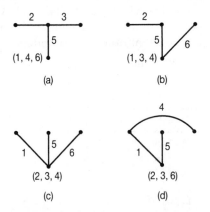

Figure 6.4 Trees for the graph of Figure 6.2. Numbers below indicate branches removed to construct tree.

numbers of the branches removed are displayed in parentheses underneath each tree.

In general, a tree of a network with n nodes will contain $n - 1$ branches. Replacement of any one of the extracted branches will create a loop, and the set of loops thus created from a single tree will necessarily be independent since each will contain a branch unique to itself. The set of loops given in Table 6.1 correspond to the tree shown in Figure 6.4(a). For the general network containing b branches and n nodes, the number, l, of independent loops is thus given by

$$l = b - n + 1 \qquad\qquad\qquad (6.1)$$

6.3 ◆ Kirchhoff's Laws

Kirchhoff's laws were introduced in Section 3.3. Having set up a convention for the direction of each branch in the network, the voltage and current for each branch can be given a sign, enabling the laws to be stated algebraically:

- *Current law*: The algebraic sum of the currents leaving any node of the network is zero.
- *Voltage law*: The algebraic sum of all the voltages round a loop of the network is zero.

The first of these laws rests on the assumption that a node has no physical reality which allows accumulation of electric charge. The second is saying that, as in an electrostatic field, no work is gained or lost in taking a charge round a closed path in the network; as in the field case this implies that, with respect to some origin in the network, a unique voltage can be ascribed to each node. It is to be noted that in the above formulation of the laws the currents and voltages related to the actual, instantaneous values. As has been shown in Section 4.7, in the case of AC analysis the same algebraic formulation applies if these actual currents and voltages are replaced by their phasor representations. In this chapter we are concerned solely with AC analysis, so that all 'voltages' and 'currents' will be assumed to be phasor representations including, contrary to previous practice, those denoted by small letters.

The laws are usually applied in one of two ways, differing in choice of unknowns. Although a complete analysis will determine voltage and current for each branch, labour can be reduced by choosing unknowns to satisfy automatically the set of equations resulting from application of one of the laws. If one node is chosen as origin and the voltages of all the other nodes with respect to this origin are taken as unknowns, the voltage laws are automatically enforced. This choice leads to the method of nodal

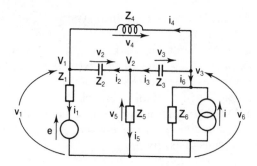

Figure 6.5 Nodal analysis of the circuit of Figure 6.1. Lower case **v**, **i** denote branch variables; upper case **V** denote nodal voltages.

analysis. Alternatively, currents may be chosen which automatically enforce the current laws. This can be done piecemeal, or by choosing as unknowns what are called **loop currents**. These are fictitious currents assumed to circulate round each of a set of independent loops, with the current in a particular branch taken as the algebraic sum of the loop currents flowing through it. This method leads to loop, or mesh, analysis. These methods will be illustrated by the analysis of the network of Figure 6.1.

6.3.1 ♦ Nodal analysis

An obvious choice for the reference node in the circuit of Figure 6.1 is node 4, resulting in the three nodal voltages shown as $\mathbf{V}_1$, $\mathbf{V}_2$, $\mathbf{V}_3$ in Figure 6.5. We express the branch voltages in terms of these unknowns:

$$\begin{aligned}
&\mathbf{v}_1 = \mathbf{V}_1 & &\mathbf{v}_4 = \mathbf{V}_3 - \mathbf{V}_1 \\
&\mathbf{v}_2 = \mathbf{V}_2 - \mathbf{V}_1 & &\mathbf{v}_5 = \mathbf{V}_2 \\
&\mathbf{v}_3 = \mathbf{V}_3 - \mathbf{V}_2 & &\mathbf{v}_6 = \mathbf{V}_3
\end{aligned} \tag{6.2}$$

The nature of the branches provides the following equations:

$$\begin{aligned}
&\mathbf{v}_1 = \mathbf{e} + \mathbf{Z}_1\mathbf{i}_1 & &\mathbf{v}_4 = \mathbf{Z}_4\mathbf{i}_4 \\
&\mathbf{v}_2 = \mathbf{Z}_2\mathbf{i}_2 & &\mathbf{v}_5 = \mathbf{Z}_5\mathbf{i}_5 \\
&\mathbf{v}_3 = \mathbf{Z}_3\mathbf{i}_3 & &\mathbf{v}_6 = \mathbf{Z}_6(\mathbf{i}_6 + \mathbf{i})
\end{aligned} \tag{6.3}$$

in which

$$\begin{aligned}
&\mathbf{Z}_1 = R_1 & &\mathbf{Z}_4 = j\omega L \\
&\mathbf{Z}_2 = 1/j\omega C & &\mathbf{Z}_5 = R_5 \\
&\mathbf{Z}_3 = 1/j\omega C & &\mathbf{Z}_6 = R_6
\end{aligned}$$

Application of the current laws requires that

$$i_1 - i_2 - i_4 \quad = 0 \qquad \text{node 1}$$

$$i_2 - i_3 + i_5 \quad = 0 \qquad \text{node 2} \tag{6.4}$$

$$i_3 + i_4 + i_6 \quad = 0 \qquad \text{node 3}$$

Using Equations 6.2 and 6.3 the currents can be expressed in terms of the nodal voltages. These expressions when substituted in Equations 6.4, give three equations in three unknowns. Denoting by $\mathbf{Y}$ the admittance $1/\mathbf{Z}$ we find

$$\mathbf{Y}_1(\mathbf{V}_1 - \mathbf{e}) - \mathbf{Y}_2(\mathbf{V}_2 - \mathbf{V}_1) - \mathbf{Y}_4(\mathbf{V}_3 - \mathbf{V}_1) = 0$$

$$\mathbf{Y}_2(\mathbf{V}_2 - \mathbf{V}_1) - \mathbf{Y}_3(\mathbf{V}_3 - \mathbf{V}_2) + \mathbf{Y}_5\mathbf{V}_2 = 0$$

$$\mathbf{Y}_3(\mathbf{V}_3 - \mathbf{V}_2) + \mathbf{Y}_4(\mathbf{V}_3 - \mathbf{V}_1) + \mathbf{Y}_6(\mathbf{V}_3 - \mathbf{Z}_6\mathbf{i}) = 0$$

Rearranging, these become

$$(\mathbf{Y}_1 + \mathbf{Y}_2 + \mathbf{Y}_4)\mathbf{V}_1 - \mathbf{Y}_2\mathbf{V}_2 - \mathbf{Y}_4\mathbf{V}_3 = \mathbf{Y}_1\mathbf{e}$$

$$-\mathbf{Y}_2\mathbf{V}_1 + (\mathbf{Y}_2 + \mathbf{Y}_3 + \mathbf{Y}_5)\mathbf{V}_2 - \mathbf{Y}_3\mathbf{V}_3 = 0 \tag{6.5}$$

$$-\mathbf{Y}_4\mathbf{V}_1 - \mathbf{Y}_3\mathbf{V}_2 + (\mathbf{Y}_3 + \mathbf{Y}_4 + \mathbf{Y}_6)\mathbf{V}_3 = \mathbf{i}$$

Inspection of these equations reveals a pattern for the coefficients, which may be seen to be of general application. Considering the first equation, derived from the current balance at node 1, on the left-hand side the coefficient of $\mathbf{V}_1$ is the total admittance of all the components connected to that node, the coefficient of $\mathbf{V}_2$ is the negative of the admittance of the component joining node 1 to node 2 and the coefficient of $\mathbf{V}_3$ is similarly the negative of the admittance of the component joining node 1 to node 3. Exactly similar interpretations can be given to the coefficients on the left-hand sides of the second and third equations. The right-hand sides can also be given a physical interpretation: if all the nodes are joined to the reference node 4, in each case the term on the right-hand side is equal to the current leaving the particular node and flowing to node 4. With this interpretation, Equation 6.5 could have been formulated by inspection of the network, avoiding all the intervening algebra.

The set of equations 6.5 can be written in matrix form

$$\mathbf{YV} = \mathbf{I}$$

in which $\mathbf{Y}$ is the matrix

$$\mathbf{Y} = \begin{bmatrix} \mathbf{Y}_1 + \mathbf{Y}_2 + \mathbf{Y}_4 & -\mathbf{Y}_2 & -\mathbf{Y}_4 \\ -\mathbf{Y}_2 & \mathbf{Y}_2 + \mathbf{Y}_3 + \mathbf{Y}_5 & -\mathbf{Y}_3 \\ -\mathbf{Y}_4 & -\mathbf{Y}_3 & \mathbf{Y}_3 + \mathbf{Y}_4 + \mathbf{Y}_6 \end{bmatrix} \tag{6.6}$$

and $\mathbf{V}$, $\mathbf{I}$ are the column vectors

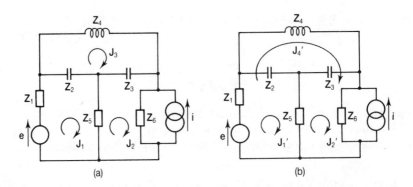

Figure 6.6 Loop analysis of the circuit of Figure 6.1. Upper case **J** denote loop currents.

$$V = \begin{bmatrix} V_1 \\ V_2 \\ V_3 \end{bmatrix} \qquad I = \begin{bmatrix} Y_1 e \\ 0 \\ i \end{bmatrix}$$

The matrix **Y** is termed the definite nodal admittance matrix, definite because it is formed with respect to a particular node chosen as reference, in this case node 4; the components of the column vectors have the significance given to them above. It will be noted that the form of Equation 6.5 and the form of the matrix **Y** depend only on the graph of the network. The nature of the individual branches determines the actual value of the coefficients. Whatever the nature and value of the components in the network of Figure 6.1 the nodal admittance matrix will be of the same form.

6.3.2 ◆ Loop analysis

To carry out a loop analysis for the network of Figure 6.1 we choose loop currents J_1, J_2, J_3 as shown in Figure 6.6(a). We have

$$
\begin{aligned}
i_1 &= -J_1 & i_4 &= -J_3 \\
i_2 &= -J_1 + J_3 & i_5 &= J_1 - J_2 \\
i_3 &= -J_2 + J_3 & i_6 &= J_2
\end{aligned}
\tag{6.7}
$$

The voltage laws require

$$
\begin{aligned}
v_1 + v_2 - v_5 &= 0 \\
v_3 + v_5 - v_6 &= 0 \\
v_2 - v_3 + v_4 &= 0
\end{aligned}
\tag{6.8}
$$

Using Equations 6.3, which still apply, together with Equations 6.7 in Equation 6.8 we find

$$(\mathbf{Z}_1 + \mathbf{Z}_2 + \mathbf{Z}_5)\mathbf{J}_1 - \mathbf{Z}_5\mathbf{J}_2 - \mathbf{Z}_2\mathbf{J}_3 = \mathbf{e}$$

$$-\mathbf{Z}_5\mathbf{J}_1 + (\mathbf{Z}_3 + \mathbf{Z}_5 + \mathbf{Z}_6)\mathbf{J}_2 - \mathbf{Z}_3\mathbf{J}_3 = -\mathbf{i}\mathbf{Z}_6 \qquad (6.9)$$

$$-\mathbf{Z}_2\mathbf{J}_1 - \mathbf{Z}_3\mathbf{J}_2 + (\mathbf{Z}_2 + \mathbf{Z}_3 + \mathbf{Z}_4)\mathbf{J}_3 = 0$$

Superficially, the interpretation of the coefficients in these equations follows a pattern similar to that for the nodal equations: on the left-hand side of the first equation, derived from loop 1, the coefficient of $\mathbf{J}_1$ is the total impedance round loop 1, the coefficient of $\mathbf{J}_2$ is the negative of the impedance common to loops 1 and 2, and the coefficient of $\mathbf{J}_3$ the negative of the impedance common to loops 1 and 3. The right-hand side is the total source voltage developed round the open-circuited loop, measured in the direction of the loop. For the other two equations the same pattern is followed, since for loop 2 the voltage on open circuit is $-\mathbf{Z}_6\mathbf{i}$. However, taken as a recipe for forming the equations, this form is not correct in all cases. To illustrate the problem, the analysis will be repeated using the set of independent loops derived from the tree of Figure 6.4(c) as shown in Figure 6.6(b).

Using the same methods as in the previous analysis, taking the unknown loop currents $\mathbf{J}_1'$, $\mathbf{J}_2'$, $\mathbf{J}_4'$, we form the following equations:

loop 1 $\qquad \mathbf{Z}_1(\mathbf{J}_1' + \mathbf{J}_4') + \mathbf{Z}_2\mathbf{J}_1' + \mathbf{Z}_5(\mathbf{J}_1' - \mathbf{J}_2') - \mathbf{e} = 0$

loop 2 $\qquad \mathbf{Z}_5(\mathbf{J}_2' - \mathbf{J}_1') + \mathbf{Z}_3\mathbf{J}_2' + \mathbf{Z}_6(\mathbf{J}_2' + \mathbf{J}_4' + \mathbf{i}) = 0$

loop 4 $\qquad \mathbf{Z}_1(\mathbf{J}_1' + \mathbf{J}_4') + \mathbf{Z}_4\mathbf{J}_4' + \mathbf{Z}_6(\mathbf{J}_2' + \mathbf{J}_4' + \mathbf{i}) - \mathbf{e} = 0$

On rearrangement these equations give

$$(\mathbf{Z}_1 + \mathbf{Z}_2 + \mathbf{Z}_5)\mathbf{J}_1' - \mathbf{Z}_5\mathbf{J}_2' + \mathbf{Z}_1\mathbf{J}_4' = \mathbf{e}$$

$$-\mathbf{Z}_5\mathbf{J}_1' + (\mathbf{Z}_3 + \mathbf{Z}_5 + \mathbf{Z}_6)\mathbf{J}_2' + \mathbf{Z}_6\mathbf{J}_4' = -\mathbf{i}\mathbf{Z}_6$$

$$\mathbf{Z}_1\mathbf{J}_1' + \mathbf{Z}_6\mathbf{J}_2' + (\mathbf{Z}_1 + \mathbf{Z}_4 + \mathbf{Z}_6)\mathbf{J}_4' = \mathbf{e} - \mathbf{i}\mathbf{Z}_6$$

In the equation for loop 1, the coefficient of $\mathbf{J}_1'$ is, as before, the total impedance in the loop, the coefficient of $\mathbf{J}_2'$ is the negative of the impedance common to loops 1 and 2. The coefficient of $\mathbf{J}_4'$ is, however, the common impedance without change of sign: this is because the loop currents through that common impedance are in the same direction. Taking this into account, the equations can be set up by inspection, but not in quite so straightforward a manner as with the nodal equations.

Equations 6.9 can also be written in matrix form:

$$\mathbf{ZJ} = \mathbf{E}$$

in which $\mathbf{Z}$ is the loop impedance matrix, for the set of loops chosen, $\mathbf{J}$ is

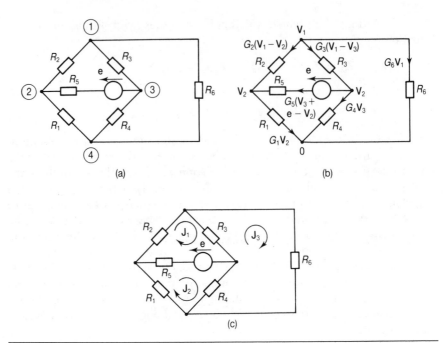

Figure 6.7 Example 6.1: (a) circuit; (b) nodal analysis; (c) loop analysis.

the column vector of the loop currents and **E** the column vector of the loop voltages. The matrix **Z** is given by

$$\mathbf{Z} = \begin{bmatrix} \mathbf{Z}_1 + \mathbf{Z}_2 + \mathbf{Z}_5 & -\mathbf{Z}_5 & -\mathbf{Z}_2 \\ -\mathbf{Z}_5 & \mathbf{Z}_3 + \mathbf{Z}_5 + \mathbf{Z}_6 & -\mathbf{Z}_3 \\ -\mathbf{Z}_2 & -\mathbf{Z}_3 & \mathbf{Z}_2 + \mathbf{Z}_3 + \mathbf{Z}_4 \end{bmatrix} \quad \textbf{(6.10)}$$

As with the nodal admittance matrix, the form is dictated by the network graph and the loops chosen; the actual values of the components are reflected in the values of the elements of the matrix.

EXAMPLE 6.1 For the Wheatstone bridge circuit of Figure 6.7(a) set up the equations for determining the voltages of all nodes with respect to node 4. (As noted above, it is quite irrelevant to the setting up of the equations that the components are all resistors.)

Joining all nodes together, we see that a current of magnitude $\mathbf{e}G_5$ leaves at node 2 and enters at node 3. Hence we can write

$$(G_2 + G_3 + G_6)\mathbf{V}_1 - G_2\mathbf{V}_2 - G_3\mathbf{V}_3 = 0$$

$$-G_2\mathbf{V}_1 + (G_1 + G_2 + G_5)\mathbf{V}_2 - G_5\mathbf{V}_3 = \mathbf{e}G_5$$

$$-G_3\mathbf{V}_1 - G_5\mathbf{V}_2 + (G_3 + G_4 + G_5)\mathbf{V}_3 = -\mathbf{e}G_5$$

It is equally simple, and often advantageous, just to start with the fact that nodal voltages are suitable unknowns and to write in the corresponding currents on the circuit diagram, as shown in Figure 6.7(b). Current balance equations can then be written down directly.

EXAMPLE 6.2   Repeat for the circuit of Figure 6.7(a) the analysis using the loop currents shown in Figure 6.7(c).
   We have

$$(R_2 + R_3 + R_5)\mathbf{J}_1 = R_5\mathbf{J}_2 - R_3\mathbf{J}_3 = \mathbf{e}$$

$$-R_5\mathbf{J}_1 + (R_1 + R_4 + R_5)\mathbf{J}_2 - R_4\mathbf{J}_3 = -\mathbf{e}$$

$$-R_3\mathbf{J}_1 - R_4\mathbf{J}_2 + (R_3 + R_4 + R_6)\mathbf{J}_3 = 0$$

6.3.3 ♦ Nodal analysis for the general network

The interpretations of Equations 6.5 and 6.9 are applicable to any network containing resistance, inductance and capacitance, together with constant voltage or constant current sources. The generalization of Equations 6.5 can be written

$$\sum_{q=1}^{n-1} \mathbf{Y}_{pq}\mathbf{V}_q = \mathbf{I}_p, \qquad p = 1, 2, \ldots, n - 1 \tag{6.11}$$

or in matrix form

$$\mathbf{YV} = \mathbf{I}$$

In these equations node n has been taken as reference node, $\mathbf{Y}_{pp}$ is the total admittance connected to node p and $\mathbf{Y}_{pq}$ is the negative of the admittance joining nodes p and q. These coefficients form the nodal admittance matrix of the network:

$$\mathbf{Y} = \begin{bmatrix} \mathbf{Y}_{11} & \mathbf{Y}_{12} & \mathbf{Y}_{13} & \cdots & \mathbf{Y}_{1,n-1} \\ \mathbf{Y}_{21} & \mathbf{Y}_{22} & \mathbf{Y}_{23} & \cdots & \\ \mathbf{Y}_{31} & & & & \\ \vdots & \vdots & & & \\ \mathbf{Y}_{n-1,1} & \mathbf{Y}_{n-2,2} & & \cdots & \mathbf{Y}_{n-1,n-1} \end{bmatrix} \tag{6.12}$$

It will be noticed that for this type of network the matrix is symmetrical:

$$\mathbf{Y}_{pq} = \mathbf{Y}_{qp} \tag{6.13}$$

The right-hand sides in Equations 6.11, forming the column vector $\mathbf{I}$, are found by considering the currents leaving the network at each node when all nodes are connected to the reference node. The unknown nodal voltages are formed as the vector $\mathbf{V}$.

   The components of the current vector $\mathbf{I}$ can be obtained in a slightly different way, using the Thévenin–Norton equivalences to transform the voltage sources into current sources. In the network chosen as the example

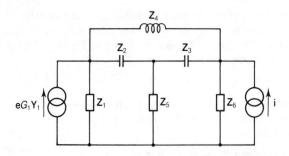

Figure 6.8 Circuit of Figure 6.1 with the voltage source in branch 1 transformed into a parallel current source.

in Section 6.3, Figure 6.1, the combination of **e** in series with R_1 can be transformed into a current source of strength eG_1 in parallel with a conductance G_1, as in Figure 6.8. The nodal voltages obtained by analysing this network will be the same as those obtained by analysing the network of Figure 6.1, although the current through R_1 will be different in the two cases. If all the nodes are joined to the reference node 4, the current leaving node 1 and entering node 4 will be eG_1 which is the correct expression for the right-hand side of the first equation of Equations 6.7. In a similar way, the network of Figure 6.7(a) can be transformed into that of Figure 6.9(a) and thence to that of Figure 6.9(b): the current sources correctly give the right-hand sides for nodes 2 and 3 in the nodal equations determined in Example 6.1.

This approach to finding the components of the current vector **I** can be applied to a general network. A typical node of a network is illustrated in Figure 6.10(a). Applying the equivalence to each branch we arrive at the circuit of Figure 6.10(b) and thence to Figure 6.10(c). If this is done systematically for all nodes we finish with the excitations in the network being represented by current sources injecting current into a source-free

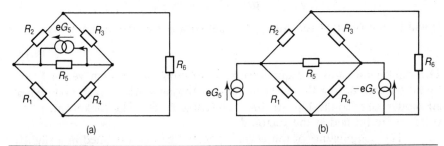

(a)   (b)

Figure 6.9 Transformation of voltage source in the circuit of Figure 6.7(a) to equivalent nodal sources: (a) first step; (b) final form.

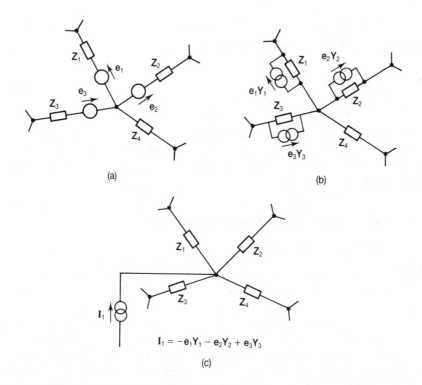

(a)

(b)

$$I_1 = -e_1Y_1 - e_2Y_2 + e_3Y_3$$

(c)

Figure 6.10 Illustrating the general form for transformation of voltage sources to nodal current sources: (a) circuit for one node; (b) applying Norton's theorem; (c) final form.

network. The nodal voltages obtained by analysing this network will be the same as those appertaining to the original network. The currents in the various branches of the original network can then be found using the values obtained for these nodal voltages.

6.3.4 ◆ Loop analysis for the general network

Most of the comments made in the last section apply equally to the case of loop analysis. The general form of equations for a network with l loops can be written

$$\sum_{q=1}^{l} \mathbf{Z}_{pq}\mathbf{J}_q = \mathbf{E}_p \qquad p = 1, 2, \ldots, l \qquad (6.14)$$

In these equations $\mathbf{Z}_{pp}$ is the total impedance in loop p; $\mathbf{Z}_{pq}$ is either the impedance common to loops p and q or its negative, depending on choice of loops; $\mathbf{E}_p$ is the total source voltage round loop p. In matrix form these equations can be written

$$\mathbf{ZJ} = \mathbf{E} \tag{6.15}$$

in which $\mathbf{Z}$ is the loop impedance matrix

$$\mathbf{Z} = \begin{bmatrix} \mathbf{Z}_{11} & \mathbf{Z}_{12} & \mathbf{Z}_{13} & \cdots & \mathbf{Z}_{1l} \\ \mathbf{Z}_{21} & \mathbf{Z}_{22} & \mathbf{Z}_{23} & \cdots & \\ \vdots & \vdots & & & \\ \mathbf{Z}_{l1} & \mathbf{Z}_{l2} & & & \mathbf{Z}_{ll} \end{bmatrix} \tag{6.16}$$

In the present case it is symmetrical since $\mathbf{Z}_{pq}$ and $\mathbf{Z}_{qp}$ arise from the same impedance. Since it is the total source voltage round the loop which is needed it is necessary to transform all current sources into voltage sources when finding the terms $\mathbf{E}_q$.

6.3.5 ◆ Comparison of nodal and loop methods

Both nodal and loop methods of analysis can be used for a particular network. Both methods lead to a set of simultaneous equations, so a choice must be made. One method may commend itself rather than the other because of the nature of the investigation being pursued: why work out currents if it is voltages that are wanted? Another matter influencing choice is the number of equations which have to be solved. In the nodal case there are $n - 1$ equations in the set; in the loop case the number is equal to the number of independent loops, $l = b - n + 1$. In the example of Figure 6.1 used as a basis for description both methods lead to three equations. This is accidental; usually one number will be greater than the other. One extreme case occurs when a number of components are connected in series, in which case there is only one loop and one loop current suffices. At the other extreme lies the cases when many components are in parallel, when there are only two nodes and only one nodal voltage to find. Since the labour of 'manual' solution, as well as the likelihood of error, increases with the number of equations, using the method leading to smaller number of unknowns is likely to be advantageous. Many networks are cross-connected, with branches between most pairs of nodes. If this process is complete, it can be shown that with five or more nodes the nodal method needs fewer unknowns. There is also the fact that neither the choice of a set of independent loops nor the formation of the impedance matrix is necessarily straightforward, and is not unique. Furthermore, the loop currents are somewhat fictitious. With the nodal method there is a choice of reference node to be made, but there is very frequently an obvious choice: electronic circuits usually have a common, earth, line which is a natural reference level. Setting up the nodal admittance matrix is straightforward. Nodal voltages also have a very definite physical significance: in an actual circuit they can, in principle, be measured directly with a voltmeter. To this author, the balance seems clearly in favour of the nodal method, which will be used for general purposes throughout the book.

6.3.6 ◆ Voltage sources in nodal analysis

The nodal admittance matrix is straightforward to set up by inspection of the circuit diagram, but when the excitation is in the form of a voltage source conversion to a current source has to take place before the final equations can be set up. This has been seen in the analysis of the circuit of Figure 6.5 and in the discussion of Section 6.3.3. In 'real' situations this is always possible, since real sources always have some internal impedance and can be characterized in terms of a current source. It is nevertheless desirable for general analysis programs to be able to include ideal voltage sources without prior manipulation. An example occurs in calculating the gain of a voltage amplifier, when it would be desired to calculate the output voltage for a given input voltage. Fortunately, the nodal equations can be simply modified to include such sources in a systematic way by introducing the current through the source as an extra unknown. Consider the network shown in Figure 6.11. Using the notation in the diagram the following nodal equations can be written down by inspection:

$$\mathbf{Y}_1\mathbf{V}_1 - \mathbf{Y}_1\mathbf{V}_2 \qquad\qquad + 0 \times \mathbf{V}_3 \qquad = -\mathbf{I}_1$$

$$-\mathbf{Y}_1\mathbf{V}_1 + (\mathbf{Y}_1+\mathbf{Y}_2+\mathbf{Y}_3)\mathbf{V}_2 - \mathbf{Y}_2\mathbf{V}_3 \qquad = 0$$

$$0 \times \mathbf{V}_1 - \mathbf{Y}_2\mathbf{V}_2 \qquad\qquad + (\mathbf{Y}_2 + \mathbf{Y}_4)\mathbf{V}_3 = \mathbf{I}_3$$

In addition

$$\mathbf{V}_1 = \mathbf{E}_1$$

If we treat as unknowns the three nodal voltages together with the current $\mathbf{I}_1$, these four equations can be written in the matrix form

$$
\begin{bmatrix}
\mathbf{Y}_1 & -\mathbf{Y}_1 & 0 & | & 1 \\
-\mathbf{Y}_1 & \mathbf{Y}_1 + \mathbf{Y}_2 + \mathbf{Y}_3 & -\mathbf{Y}_2 & | & 0 \\
0 & -\mathbf{Y}_2 & \mathbf{Y}_2 + \mathbf{Y}_4 & | & 0 \\
\hline
1 & 0 & 0 & | & 0
\end{bmatrix}
\begin{bmatrix}
\mathbf{V}_1 \\ \mathbf{V}_2 \\ \mathbf{V}_2 \\ \mathbf{I}_1
\end{bmatrix}
=
\begin{bmatrix}
0 \\ 0 \\ \mathbf{I}_3 \\ \mathbf{E}_1
\end{bmatrix}
\qquad \textbf{(6.17)}
$$

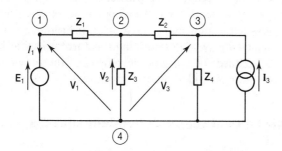

Figure 6.11 Modified nodal analysis (Section 6.3.6): circuit for first example.

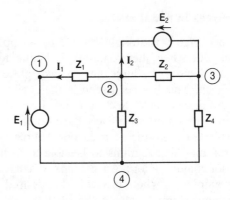

Figure 6.12 Modified nodal analysis: circuit for second example.

That part of the matrix within dotted lines is just the nodal admittance matrix of the network. The new terms follow immediately from the fact that $\mathbf{E}_1$ is connected between nodes 1 and 0, and can also be written in by inspection.

Another example is shown in Figure 6.12, in which neither of the end nodes of one voltage source is connected to the reference node. In this case we add the two unknown source currents $\mathbf{I}_1$ and $\mathbf{I}_2$ to the three nodal voltages. It can be checked that the pattern of equations is expressed by

$$
\begin{bmatrix}
\mathbf{Y}_1 & -\mathbf{Y}_1 & 0 & 1 & 0 \\
-\mathbf{Y}_1 & \mathbf{Y}_1 + \mathbf{Y}_2 + \mathbf{Y}_3 & -\mathbf{Y}_2 & 0 & 1 \\
0 & -\mathbf{Y}_2 & \mathbf{Y}_2 + \mathbf{Y}_4 & 0 & -1 \\
1 & 0 & 0 & 0 & 0 \\
0 & 1 & -1 & 0 & 0
\end{bmatrix}
\begin{bmatrix}
\mathbf{V}_1 \\
\mathbf{V}_2 \\
\mathbf{V}_3 \\
\mathbf{I}_1 \\
\mathbf{I}_2
\end{bmatrix}
=
\begin{bmatrix}
0 \\
0 \\
0 \\
\mathbf{E}_1 \\
\mathbf{E}_2
\end{bmatrix}
\tag{6.18}
$$

In this example too the pattern of 1s and 0s in the added rows and columns follows immediately from the numbers of nodes to which the sources are connected, and could be filled in by inspection. It is worth noticing that if either of $\mathbf{E}_1$ or $\mathbf{E}_2$ were to be made zero, solution of the equations would still yield the current through that source. The inclusion of a '0' source in a branch thus provides a way of determining the current in that branch.

Formal procedures of the type developed in the previous paragraphs are essential for computer-aided analysis programs.

6.4 ◆ Networks containing dependent sources

This chapter has so far been concerned with networks containing resistance, inductance and capacitance together with fixed sources. It has been

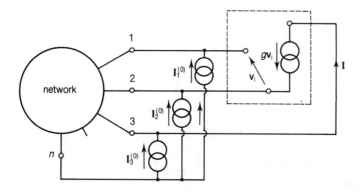

Figure 6.13 Illustrating the treatment of a network containing one voltage-controlled dependent current source.

seen in previous chapters that networks involving dependent sources are important. This section considers how such sources can be included in a general analysis.

We first consider a network containing a single voltage-controlled current source, as well as resistance, inductance and capacitance. Such a network can be thought of as a network of the type considered above connected to the dependent source, as illustrated in Figure 6.13 in which the source is shown connected to nodes 1, 2 and 3 of the part network. This part network is presumed to be source free: external exciting sources are represented by the fixed current sources $I_1^{(0)}$, $I_2^{(0)}$, ..., $I_{n-1}^{(0)}$. Operation of this part network can then be described by a set of equations

$$\sum_{q=1}^{n-1} \mathbf{Y}_{pq}\mathbf{V}_q = \mathbf{I}_p^{(0)}, \qquad p = 1, 2, 3, \ldots, n-1 \tag{6.19}$$

in which $\mathbf{Y}_{pq} = \mathbf{Y}_{qp}$.

For the dependent source we have

$$\mathbf{v}_i = \mathbf{V}_1 - \mathbf{V}_2$$

$$\mathbf{I} = g\mathbf{v}_i = g(\mathbf{V}_1 - \mathbf{V}_2)$$

Assuming that the control terminal of the dependent source takes no current, the effect of that source is to augment the input to node 2 and to reduce the input to node 3 by $\mathbf{I}$. Thus the second of the set of Equations 6.19 becomes

$$\mathbf{Y}_{21}\mathbf{V}_1 + \mathbf{Y}_{22}\mathbf{V}_2 + \ldots + \mathbf{Y}_{2,n-1}\mathbf{V}_{n-1} = \mathbf{I}_2^{(0)} + \mathbf{I} = \mathbf{I}_2^{(0)} + g(\mathbf{V}_1 - \mathbf{V}_2)$$

Collecting terms we have

$$(\mathbf{Y}_{21} - g)\mathbf{V}_1 + (\mathbf{Y}_{22} + g)\mathbf{V}_2 + \mathbf{Y}_{23}\mathbf{V}_3 + \ldots \mathbf{Y}_{2,n-1}\mathbf{V}_{n-1} = \mathbf{I}_2^{(0)} \tag{6.20}$$

The third of the set of equations becomes

$$\mathbf{Y}_{31}\mathbf{V}_1 + \mathbf{Y}_{32}\mathbf{V}_2 + \mathbf{Y}_{33}\mathbf{V}_3 + \ldots + \mathbf{Y}_{3,n-1}\mathbf{V}_{n-1} = \mathbf{I}_3^{(0)} - g(\mathbf{V}_1 - \mathbf{V}_2)$$

or

$$(\mathbf{Y}_{31} + g)\mathbf{V}_1 + (\mathbf{Y}_{32} - g)\mathbf{V}_2 + \mathbf{Y}_{33}\mathbf{V}_3 + \ldots + \mathbf{Y}_{3,n-1}\mathbf{V}_{n-1} = \mathbf{I}_3^{(0)}$$

(6.21)

The remaining equations of the set are unaltered. Equations 6.20 and 6.21 together with those remaining from Equations 6.19 constitute a new set from which to obtain the nodal voltages. This new set can be written in the same form as Equations 6.19:

$$\mathbf{Y'V} = \mathbf{I}$$

(6.22)

with the new nodal admittance matrix

$$\mathbf{Y'} = \begin{bmatrix} \mathbf{Y}_{11} & \mathbf{Y}_{12} & \mathbf{Y}_{13} & \cdots & \mathbf{Y}_{1,n-1} \\ \mathbf{Y}_{21} - g & \mathbf{Y}_{22} + g & \mathbf{Y}_{23} & \cdots & \mathbf{Y}_{2,n-1} \\ \mathbf{Y}_{31} + g & \mathbf{Y}_{32} - g & \mathbf{Y}_{33} & & \\ \mathbf{Y}_{41} & \mathbf{Y}_{42} & \mathbf{Y}_{43} & & \\ \vdots & \vdots & & & \\ \mathbf{Y}_{n-1,1} & \mathbf{Y}_{n-1,2} & \mathbf{Y}_{n-1,3} & & \mathbf{Y}_{n-1,n-1} \end{bmatrix}$$

(6.23)

The basic difference between $\mathbf{Y}$ and $\mathbf{Y'}$ is that whereas $\mathbf{Y}$ is symmetrical, $\mathbf{Y'}$ is not.

EXAMPLE 6.3 For the circuit of Figure 6.14 set up equations from which the nodal voltages $\mathbf{V}_1$, $\mathbf{V}_2$, $\mathbf{V}_3$ can be obtained.

The configuration and node numbering of this example follows exactly that of Figure 6.13 so that the equations can be written down with the aid of Equations 6.22 and 6.23. They may also be obtained by direct analysis. Referring to the circuit we formulate the current balance equation at each node:

node 1 $G_1\mathbf{V}_1 + G_2(\mathbf{V}_1 - \mathbf{V}_2) + j\omega C(\mathbf{V}_1 - \mathbf{V}_3) = \mathbf{I}$

node 2 $G_2(\mathbf{V}_2 - \mathbf{V}_1) + \mathbf{V}_2/j\omega L + G_3(\mathbf{V}_2 - \mathbf{V}_3) = g\mathbf{V}_i = g(\mathbf{V}_1 - \mathbf{V}_2)$

node 3 $j\omega C(\mathbf{V}_3 - \mathbf{V}_1) + G_3(\mathbf{V}_3 - \mathbf{V}_2) + G_4\mathbf{V}_3 = -g\mathbf{V}_i = -g(\mathbf{V}_1 - \mathbf{V}_2)$

Collecting terms we find

$$(G_1 + G_2 + j\omega C)\mathbf{V}_1 - G_2\mathbf{V}_2 - j\omega C\mathbf{V}_3 = \mathbf{I}$$

$$-(G_2 + g)\mathbf{V}_1 + (G_2 + G_3 + g + 1/j\omega L)\mathbf{V}_2 - G_3\mathbf{V}_3 = 0$$

$$(g - j\omega C)\mathbf{V}_1 - (g + G_3)\mathbf{V}_2 + (G_3 + G_4 + j\omega C)\mathbf{V}_3 = 0$$

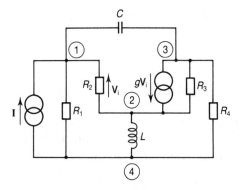

Figure 6.14 Circuit for Example 6.3.

These will be recognized as in agreement with

$$\mathbf{Y'V} = \begin{bmatrix} \mathbf{I} \\ 0 \\ 0 \end{bmatrix}$$

The inclusion of more than one source does not add anything new to the discussion: it merely results in more terms of the nodal admittance matrix being modified. Without going into details, it can be seen that the inclusion of any of the four kinds of dependent sources (Section 1.5.3) will lead to the same types of equations. For example, if in the circuit in Figure 6.12 the voltage source $\mathbf{E}_2$ was controlled by the voltage between node 1 and node 2, the term $\mathbf{E}_2$ in the last row of the right-hand side of Equation 6.18 would be replaced by $A(\mathbf{V}_1 - \mathbf{V}_2)$. On collecting terms, Equation 6.18 would become

$$\begin{bmatrix} \mathbf{Y}_1 & -\mathbf{Y}_1 & 0 & 1 & 0 \\ -\mathbf{Y}_1 & \mathbf{Y}_1 + \mathbf{Y}_2 + \mathbf{Y}_3 & -\mathbf{Y}_2 & 0 & 1 \\ 0 & -\mathbf{Y}_2 & \mathbf{Y}_2 + \mathbf{Y}_4 & 0 & -1 \\ 1 & 0 & 0 & 0 & 0 \\ -A & 1 + A & -1 & 0 & 0 \end{bmatrix} \begin{bmatrix} \mathbf{V}_1 \\ \mathbf{V}_2 \\ \mathbf{V}_3 \\ \mathbf{I}_1 \\ \mathbf{I}_2 \end{bmatrix}$$

$$= \begin{bmatrix} 0 \\ 0 \\ 0 \\ \mathbf{E}_1 \\ 0 \end{bmatrix} \tag{6.24}$$

The importance of formulating equations in this way lies in the fact that once the node connections of the various sources and components are known the equations can be set up by rules suitable for use in a computer program.

6.5 ◆ Solution of the network equations

In the previous sections it has been shown that the analysis of linear networks results finally in a set of simultaneous linear equations. While the mathematical techniques for the solution of such equations are outside the scope of this book, it is relevant to consider the various forms solutions can take. We shall consider that nodal variables have been used and that a set of equations in the form of Equations 6.11 have been found:

$$\sum_{q=1}^{n-1} \mathbf{Y}_{pq}\mathbf{V}_q = \mathbf{I}_p, \qquad p = 1, 2, \ldots, n-1$$

These equations are linear in the Vs and Is, and consequently a solution exists of the form

$$\mathbf{V}_p = \sum_{q=1}^{n-1} \mathbf{M}_{pq}\mathbf{I}_q, \qquad p = 1, 2, \ldots, n-1 \tag{6.25}$$

The coefficients $\mathbf{M}_{pq}$ may be found in a variety of ways: the formal one making use of determinants, while important for theoretical investigations, may or may not be the most suitable. If we denote by the determinant

$$\Delta = \begin{vmatrix} \mathbf{Y}_{11} & \mathbf{Y}_{12} & \cdots & \mathbf{Y}_{1,n-1} \\ \mathbf{Y}_{21} & \mathbf{Y}_{22} & \cdots & \mathbf{Y}_{2,n-1} \\ \vdots & \vdots & & \\ \mathbf{Y}_{n-1,1} & \mathbf{Y}_{n-1,2} & & \mathbf{Y}_{n-1,n-1} \end{vmatrix} \tag{6.26}$$

then

$$\mathbf{M}_{pq} = \Delta_{pq}/\Delta \tag{6.27}$$

in which Δ_{pq} is the cofactor of $\mathbf{Y}_{pq}$ in Δ.

A number of different situations can be envisaged. The simplest is that all the terms in the matrices $\mathbf{Y}$ and $\mathbf{I}$ are given as (complex) numbers. Numerical methods of solution can then be used to determine the values of the nodal voltages: a computer program can be profitably used for this task. Less restrictively, it may be that although the terms of the matrix $\mathbf{Y}$ are given numerically we wish to find the effect of varying one or more sources: we then have to determine numerically all the coefficients $\mathbf{M}_{pq}$ in order to derive algebraic expressions for the nodal voltages in terms of individual current sources. Finally, we may be interested in the effects of varying one or more components, calling for a totally algebraic solution.

It is clear that much depends on the set of equations: not only does the labour of solution increase rapidly with the number of equations but the significance of results containing numbers of unknown coefficients is not easy to assess. Consideration of the totally algebraic solutions leads to the mathematical theory of networks and network functions, of importance, for example, in the design of filters. Very frequently, however, the design process starts with simple, idealized circuits which can be easily analysed. Realization of such circuits inevitably introduces other elements, the effect of which is best determined using computer-aided analysis programs. Such a process is equivalent to continual refinement based on measurement of an actual circuit.

6.6 ◆ Computer-aided analysis

The previous sections have shown how a systematic analysis in terms of nodal variables can be carried out, leading to a set of simultaneous equations. In cases when the nodal admittance matrix takes numerical form the solution can be with profit handled by a computer program. However, the preparation of the nodal admittance matrix in numerical form is in itself a considerable labour, a labour which has to be repeated for every frequency. It has been seen that simple rules exist by which the terms in the nodal admittance matrix can be calculated once the configuration of the network is given. If the configuration can be specified in a form acceptable by the computer, the whole process of forming the equations can be handled in the same program. This is the way in which network analysis programs work.

6.6.1 ◆ Input of network data

The information from the circuit diagram required to assemble the nodal admittance matrix is the nature of each component, its value and the end nodes to which it is connected.

A circuit of the form of that of Figure 6.1 is shown in Figure 6.15, differing only in that the inductor is represented by inductance in series with resistance. (This makes little difference to the previous analysis: it is merely necessary to redefine Z_4.) In order to specify each separate component it is necessary to introduce more nodes than in the previous analysis, as shown in the figure. Using this node numbering Table 6.2 can be drawn up. In this table the end nodes have been distinguished with +, −: for a simple component the distinction is irrelevant, but for a source some convention is needed to specify direction. The table contains enough information to reconstruct the circuit and, once frequency is specified, forms an adequate base from which to set up the appropriate nodal equations, modified as in Section 6.3.5. It will be noted that, with a

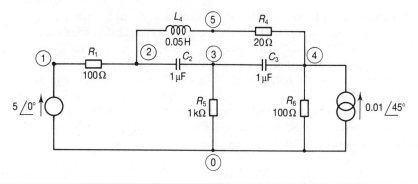

Figure 6.15 Circuit used to illustrate the input of circuit data in tabular form, Section 6.6.1.

Table 6.2

Component	Type	End Nodes (+)	End Nodes (−)	Values
R1	Resistance	1	2	100
R4	Resistance	4	5	20
R5	Resistance	3	0	1000
R6	Resistance	4	0	100
C2	Capacitance	2	3	1×10^{-6}
C3	Capacitance	3	4	1×10^{-6}
L4	Inductance	2	5	0.05
V1	Voltage source	1	0	$5\angle 0°$
I1	Current source	4	0	$0.01\angle 45°$

subsidiary definition of symbols in the first column, the 'type' column can be omitted. The entry of such a table can form the input data for a network analysis program.

6.6.2 ◆ Circuit simulation

A table of the form of Table 6.2 can be drawn up for any network, and contains a complete description of the circuit. The only additional information for an AC analysis is specification of frequency, or frequencies.

The context of this chapter has been AC analysis of linear circuits, but since the circuit information in the table is complete it may be expected that programs can be devised to determine response to other forms of excitation. In this sense the program can simulate the circuit: provided that the circuit has been correctly specified, allowing for component and wiring imperfections, the computed results will be those which would be determined by setting up the circuit and making measurements.

Indeed, it may be possible to do better than physical measurement in the sense that attaching instruments to a circuit will modify its action in some degree. Clearly such a network analysis program can be a very valuable tool.

It is to be noted that components other than those occurring in Table 6.2 could be specified, including non-linear components. The table defines the topology of the network: provided that a definition of the voltage–current relationship for the component is added, circuit simulation is possible.

6.6.3 ♦ SPICE

SPICE is a well-known circuit simulation program which, although designed for circuits containing many semiconductor devices, can be used for the linear circuits forming the main subject matter of this book. More details about SPICE and how to use it will be found in Chapter 15.

6.7 ♦ Summary

Nodal analysis (Section 6.3.3)
Nodal voltages are defined by choosing one node as a reference and then taking the voltage of each node with respect to that reference. A linear network can be analysed in terms of the set of nodal voltages, leading to a set of equations of the form

$$\sum_{q=1}^{n-1} \mathbf{Y}_{pq}\mathbf{V}_q = \mathbf{I}_p, \qquad p = 1, 2, \ldots, n-1$$

The coefficients $\mathbf{Y}_{pq}$ form the nodal admittance matrix appropriate to the chosen reference node. For a network containing only resistance, inductance and capacitance and in which the sources are all independent, $\mathbf{Y}_{pq} = \mathbf{Y}_{qp}$ is the negative of the admittance joining node p to node q, $\mathbf{Y}_{pp}$ is the total admittance connected to node p, and the current $\mathbf{I}_p$ is the current which flows into node p when all the nodes are joined together.

If the network contains linear dependent sources $\mathbf{Y}_{pq}$ is not necessarily equal to $\mathbf{Y}_{qp}$.

Loop (mesh) analysis (Section 6.3.4)
Loops are closed paths through the network. A set of independent loops can be chosen by ensuring that each loop contains a component not contained in any other, and each component is in one loop. The number of independent loops is given by the formula $l = b - n + 1$.

A linear network can be analysed in terms of the set of loop currents considered to circulate round each of a set of independent loops. This leads to a set of equations of the form

$$\sum_{q=1}^{l} \mathbf{Z}_{pq} \mathbf{I}_q = \mathbf{E}_p, \qquad p = 1, 2, \ldots, l$$

The coefficients form the loop impedance matrix for the chosen set of loops. For a network containing only resistance, inductance and capacitance and in which the sources are all independent, $\mathbf{Z}_{pq} = \mathbf{Z}_{qp}$ is either the impedance common to loops p and q or its negative, depending on choice of loops, $\mathbf{Z}_{pp}$ is the total impedance in loop p, and the voltage $\mathbf{E}_p$ is the total voltage round loop p when the loops are all open circuited.

Modified nodal analysis (Section 6.3.6)
The presence of voltage sources can be allowed for explicitly in the formulation of nodal equations.

Solution of nodal equations (Section 6.5)
Solution of the set of nodal equations can be expressed in the form

$$\mathbf{V}_p = \sum_{q=1}^{n-1} \mathbf{M}_{pq} \mathbf{I}_q, \qquad p = 1, 2, \ldots, n-1$$

Circuit table (Section 6.6.1)
A circuit diagram can be presented in tabular form.

PROBLEMS

6.1 Enumerate the numbers of branches, nodes and independent loops for the network for which the graph is described by

(a) a pyramid with square base ABCD and vertex E,

(b) a cube, and

(c) a hexagon ABCDEF with links AD, BE, CF.

6.2 For each of the cases of Problem 6.1 determine a set of independent loops.

6.3 For the network shown in Figure 6.16 set up the equations for the nodal voltages, taking node 3 as the reference node.

6.4 For the network of Figure 6.16 set up equations for the loop currents in the independent loops e_1–$\mathbf{Z}_1$–$\mathbf{Z}_2$, $\mathbf{Z}_2$–$\mathbf{Z}_3$–$\mathbf{Z}_4$–e_2, e_2–$\mathbf{Z}_4$–$\mathbf{Z}_5$.

6.5 Show how solution of the network of Figure 6.16 can be reduced to a single mesh problem by use of Thévenin's theorem.

6.6 The source e_2 in the network of Figure 6.16 is replaced by a current-controlled dependent voltage source $r\mathbf{i}$, the control current $\mathbf{i}$ being that flowing in the

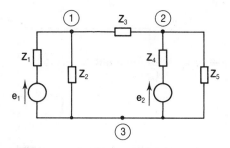

Figure 6.16

impedance Z_2. In the expression ri it is assumed that when i is directed into the reference node then the voltage is directed away from that node. Reformulate the nodal equations found in Problem 6.3.

Comment on the resulting nodal admittance matrix.

6.7 In Figure 6.17 is shown a twin-T network. Derive, by inspection or otherwise, equations for the nodal voltages, taking node 5 as the reference node.

Show that the graph of this network is that of Problem 6.1(a).

6.8 The twin-T network of Figure 6.17 contains four independent loops. Show that the following loops form an independent set:

Loop	Current	Branches
1	J_1	1–4–6
2	J_2	1–2–7
3	J_3	6–5–8
4	J_4	7–3–8

Using this set of loops, set up the equations for the loop currents.

6.9 For the network of Figure 6.18, set up the modified nodal equations for the variables V_1, V_2, V_3, V_4, I_1, I_4, taking node 5 as reference.

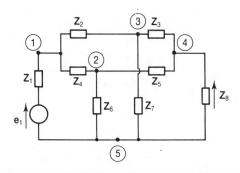

Figure 6.17

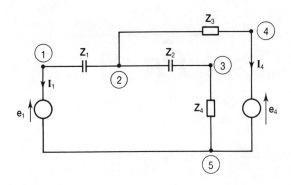

Figure 6.18

6.10 In the network shown in Figure 6.18, the source e_4 is replaced by a dependent voltage source controlled by the nodal voltage V_3, so that $e_4 = AV_3$. Derive the modified nodal equations with the same variables as in Problem 6.9.

6.11 In the circuit of Figure 6.17, the impedances are as follows:

Z_1 1 kΩ resistor

Z_2, Z_3 10 mH inductor

Z_4, Z_5 100 Ω resistor

Z_6 0.1 μF capacitor

Z_7 300 Ω resistor

Z_8 1 kΩ resistor

The voltage source e_1 is AC, giving 5 V at 100 Hz. Draw up a tabular description of the network on the lines of Table 6.2. (In order to do this, relabel the reference node to node 6, and insert a new node 5 between e_1 and Z_1.)

7 ◆ Three-phase Circuits

The distribution of AC power at high levels takes place using the three-phase method. In this chapter network analysis is applied to the study of three-phase circuits. The concepts of balanced and unbalanced systems are introduced.

7.1 ◆ Introduction

It has been repeatedly commented that network analysis is applied in all branches of electrical engineering. Details depend on the precise form of the application. In this and the next chapter applications to two areas of importance in the field of AC power are considered. The present chapter concerns the use of the three-phase method of AC power transmission.

    The generation and distribution of AC power at high levels, for example from a power station, uses a three-wire system rather than the more obvious two wires. By taking pairs among the three wires, three different voltages can be measured. The characteristic of a three-phase system is that these three voltages are all equal in magnitude, but successively differ in phase by 120°. It is convenient to identify the three wires by the letters R, Y, B after a commonly used system of colour coding. Since we are concerned with AC, we can use phasor representations for the voltages. If we denote the phasor voltages between pairs of

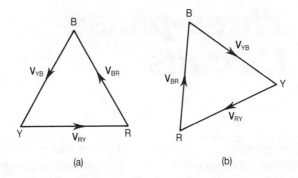

Figure 7.1 Phasor diagrams for three-phase supply: (a) positive sequence;
(b) negative sequence.

wires RY, YB, BR by $\mathbf{V}_{RY}$, $\mathbf{V}_{YB}$, $\mathbf{V}_{BR}$ respectively, and since the
magnitudes are equal, the relation between them must be of the form
illustrated in the phasor diagrams of Figures 7.1(a) and 7.1(b). In Figure
7.1(a) in the order $\mathbf{V}_{RY}$, $\mathbf{V}_{YB}$, $\mathbf{V}_{BR}$ each voltage lags on the previous one
by 120°: this is said to be a positive phase sequence. In Figure 7.1(b), in
the order given, each voltage leads the one before by 120°: this constitutes
a negative phase sequence. Placing the phasors in the form of an
equilateral triangle emphasizes that necessarily (Kirchhoff's voltage law)
the sum of the three voltages must be zero:

$$\mathbf{V}_{RY} + \mathbf{V}_{YB} + \mathbf{V}_{BR} = 0 \tag{7.1}$$

The magnitude

$$|\mathbf{V}_{RY}| = |\mathbf{V}_{YB}| = |\mathbf{V}_{BR}| = V_L \tag{7.2}$$

is termed the voltage between lines, or the **line voltage**.

7.2 ◆ Source configuration

In terms of conventional two-terminal voltage sources, a three-phase
voltage source can be modelled in two convenient forms, a mesh or delta
connection and a star connection.

7.2.1 ◆ Mesh or delta connection

In this form, three voltage sources with phasor representations $\mathbf{V}_{RY}$, $\mathbf{V}_{YB}$,
$\mathbf{V}_{BR}$ are connected as shown in Figure 7.2(a). The phasor diagram is as in
Figure 7.1, depending on the phase sequence. Although it is frequently
convenient to maintain symmetry by using three sources, it is to be noted
that any one may be removed without altering the situation: for example,
by Equation 7.1, if the source representing $\mathbf{V}_{BR}$ is removed the voltage

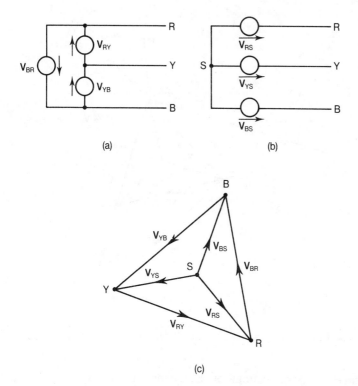

Figure 7.2 Equivalent source configurations: (a) delta connection; (b) star connection; (c) phasor diagram showing relation between star and delta voltages.

between R and B must be fixed correctly. If the sources were real sources with internal impedances this would not necessarily be true, but on the other hand a circuit simulation program would reject any attempt to use the three ideal sources.

7.2.2 ◆ Star connection

A star connection of three sources is shown in Figure 7.2(b). In Figure 7.2(c) is shown the phasor diagram for the case when the voltages $\mathbf{V}_{RS}$, $\mathbf{V}_{YS}$, $\mathbf{V}_{BS}$ are equal in magnitude and in a positive phase sequence. From this diagram it is seen that the voltages $\mathbf{V}_{RY}$, $\mathbf{V}_{YB}$, $\mathbf{V}_{BR}$ are also in a positive sequence, as in Figure 7.1(a). (If the source voltages are in a negative phase sequence, then $\mathbf{V}_{RY}$, $\mathbf{V}_{YB}$, $\mathbf{V}_{BR}$ are also in a negative sequence.)

The magnitude

$$|\mathbf{V}_{RS}| = |\mathbf{V}_{YS}| = |\mathbf{V}_{BS}| = V_P \tag{7.3}$$

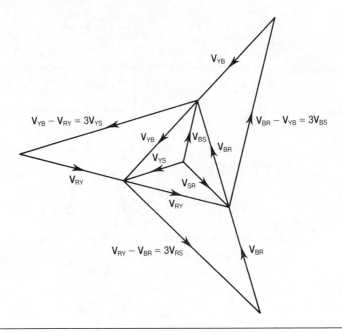

Figure 7.3 Phasor diagram illustrating Equation 7.7, Section 7.2.2.

is termed the **phase voltage** of the system. It can be seen from the phasor diagram that it is related to the line voltage by

$$V_L = V_P\sqrt{3} \tag{7.4}$$

It is usual to define the origin of phase by taking some convenient voltage as zero phase: there is no absolute origin. A frequently useful choice is to take $\mathbf{V}_{RS}$, when we have

$$\mathbf{V}_{RS} = V_P\angle 0° = V_P(1 + j \times 0)$$
$$\mathbf{V}_{YS} = V_P\angle -120° = V_P\tfrac{1}{2}(-1 - j\sqrt{3}) \tag{7.5}$$
$$\mathbf{V}_{BS} = V_P\angle -240° = V_P\angle + 120° = V_P\tfrac{1}{2}(-1 + j\sqrt{3})$$

From these equations, or by inspection of the phasor diagram, we find

$$\mathbf{V}_{RY} = \mathbf{V}_{RS} - \mathbf{V}_{YS} = V_P\sqrt{3}\angle 30° = V_P\tfrac{1}{2}\sqrt{3}(\sqrt{3} + j)$$
$$\mathbf{V}_{YB} = \mathbf{V}_{YS} - \mathbf{V}_{BS} = V_P\sqrt{3}\angle -90° = V_P\sqrt{3}(-j) \tag{7.6}$$
$$\mathbf{V}_{BR} = \mathbf{V}_{BS} - \mathbf{V}_{RS} = V_P\sqrt{3}\angle +150° = V_P\sqrt{3}\tfrac{1}{2}(-\sqrt{3} + j)$$

For future use we note either from Equations 7.5 and 7.6 or from the phasor diagram of Figure 7.3 that

$$\mathbf{V}_{RY} - \mathbf{V}_{BR} = 3\mathbf{V}_{RS}$$
$$\mathbf{V}_{YB} - \mathbf{V}_{RY} = 3\mathbf{V}_{YS} \tag{7.7}$$
$$\mathbf{V}_{BR} - \mathbf{V}_{YB} = 3\mathbf{V}_{BS}$$

7.3 ◆ Currents in three-phase circuits

The connection of a three-phase source to a load results in currents I_R, I_Y, I_B flowing along the wires: these are referred to as the **line currents**. We have

$$I_R + I_Y + I_B = 0 \tag{7.8}$$

These currents, as well as the voltages, are easily monitored in a system. The load can take many forms, but two standard forms are the star and delta configurations.

7.3.1 ◆ Load in delta configuration

In Figure 7.4 is shown a three-phase source supplying a delta-connected load. The load is characterized by the impedances or admittances of its three components: we take the admittances to be Y_{RY}, Y_{YB}, Y_{BR}. In this circuit the voltage across each component is known so that the current in each can be calculated. The line currents follow by summation. We have

$$I_R = I_{RY} - I_{BR} = V_{RY}Y_{RY} - V_{BR}Y_{BR}$$

$$I_Y = I_{YB} - I_{RY} = V_{YB}Y_{YB} - V_{RY}Y_{RY} \tag{7.9}$$

$$I_B = I_{BR} - I_{YB} = V_{BR}Y_{BR} - V_{YB}Y_{YB}$$

7.3.2 ◆ Load in star configuration

The general form for a star-connected load is shown in Figure 7.5. In star-connected circuits it is convenient to use the star equivalent for the three-phase source and to measure all voltages with respect to the star point of the source arrangement. In the circuit of Figure 7.5 the simplest way to proceed is to calculate the voltage between the star point of the load, O, and the star point of the source, S. Once this is done it is a straightforward matter to calculate the line currents. Denoting this voltage by V_{OS}, we have

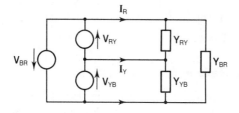

Figure 7.4 Circuit showing load in delta connection.

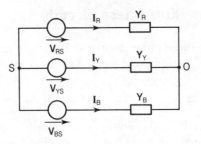

Figure 7.5 Circuit showing load in star connection.

$$I_R = (V_{RS} - V_{OS})Y_R$$
$$I_Y = (V_{YS} - V_{OS})Y_Y \qquad (7.10)$$
$$I_B = (V_{BS} - V_{OS})Y_B$$

Substituting in Equation 7.8 we find

$$V_{RS}Y_R + V_{YS}Y_Y + V_{BS}Y_B = V_{OS}(Y_R + Y_Y + Y_B)$$

and therefore

$$V_{OS} = \frac{V_{RS}Y_R + V_{YS}Y_Y + V_{BS}Y_B}{Y_R + Y_Y + Y_B} \qquad (7.11)$$

The application of the formulae of this and the previous section will now be considered.

7.4 ◆ Balanced and unbalanced loads

It is frequently the case that the load on a three-phase system is symmetrical in the three phases. In such a case the three impedances in a delta-connected load, or those of a star-connected load, will all be the same. Such, a load is described as **balanced**. In this case of a balanced load the line currents will all be equal in magnitude although differing in phase by 120°. When the load is not symmetrical it is described as **unbalanced**. Practical systems are usually designed to have balanced loads. It will be shown in the following sections that in this case of balanced loads a simplified analysis is possible.

7.4.1 ◆ Balanced loads in delta connection

Consider the case when the three impedances in the delta-connected load of Figure 7.4 are all equal. In terms of admittances we write

$$Y_{RY} = Y_{YB} = Y_{BR} = Y_D = |Y_D|\angle\phi \qquad (7.12)$$

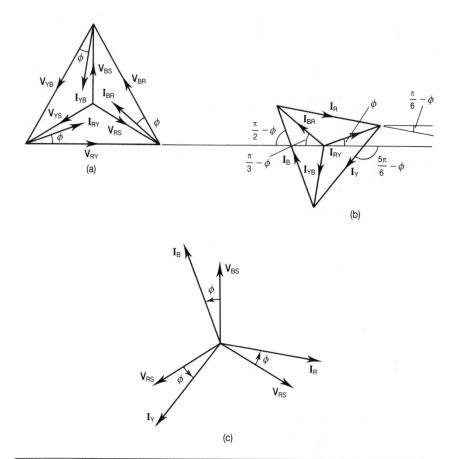

Figure 7.6 Phasor diagrams for currents with load in delta connection:
(a) delta currents in relation to delta voltages; (b) delta currents in relation to
line currents; (c) line currents in relation to star voltages.

Equations 7.9 then become

$$\mathbf{I}_R = \mathbf{I}_{RY} - \mathbf{I}_{BR} = (\mathbf{V}_{RY} - \mathbf{V}_{BR})\mathbf{Y}_D$$

$$\mathbf{I}_Y = \mathbf{I}_{YB} - \mathbf{I}_{RY} = (\mathbf{V}_{YB} - \mathbf{V}_{RY})\mathbf{Y}_D \qquad (7.13)$$

$$\mathbf{I}_B = \mathbf{I}_{BR} - \mathbf{I}_{YB} = (\mathbf{V}_{BR} - \mathbf{V}_{YB})\mathbf{Y}_D$$

These equations are illustrated by the phasor diagrams of Figure 7.6. The
currents $\mathbf{I}_{RY}$, $\mathbf{I}_{YB}$, $\mathbf{I}_{BR}$ are all equal in magnitude and are disposed in the
same phase sequence as the line voltages. The line currents $\mathbf{I}_R$, $\mathbf{I}_Y$, $\mathbf{I}_B$ are
also equal and in the same phase sequence, and can be disposed as the
three sides of an equilateral triangle. This indicates that Equation 7.8 is
satisfied. If we denote

$$|\mathbf{I}_{RY}| = |\mathbf{I}_{YB}| = |\mathbf{I}_{BR}| = I_D$$

and

$$|\mathbf{I}_R| = |\mathbf{I}_Y| = |\mathbf{I}_B| = I_L$$

we see from the diagram that

$$I_L = \sqrt{3}I_D \qquad\qquad (7.14)$$

The results illustrated by the phasor diagram can be obtained algebraically as follows. Taking $\mathbf{V}_{RY}$ as reference, we have

$$\mathbf{V}_{YB} = \mathbf{V}_{RY} \exp(-j2\pi/3)$$

$$\mathbf{V}_{BR} = \mathbf{V}_{RY} \exp(j2\pi/3)$$

Hence

$$\mathbf{V}_{RY} - \mathbf{V}_{BR} = \mathbf{V}_{RY}\,[1 - \exp(j2\pi/3)]$$

$$= \mathbf{V}_{RY} \exp(j\pi/3)(-2j)\sin(\pi/3)$$

$$= \sqrt{3}\mathbf{V}_{RY} \exp(-j\pi/6)$$

In a similar fashion we find

$$\mathbf{V}_{YB} - \mathbf{V}_{RY} = \sqrt{3}\mathbf{V}_{RY} \exp(-j5\pi/6)$$

$$\mathbf{V}_{BR} - \mathbf{V}_{YB} = \sqrt{3}\mathbf{V}_{RY} \exp(j\pi/2)$$

Hence, using Equations 7.12 and 7.13, we find

$$\mathbf{I}_R = \sqrt{3}\mathbf{V}_{RY}|\mathbf{Y}_D| \exp[j(\phi - \pi/6)]$$

$$\mathbf{I}_Y = \sqrt{3}\mathbf{V}_{RY}|\mathbf{Y}_D| \exp[j(\phi - 5\pi/6)] \qquad (7.15)$$

$$\mathbf{I}_B = \sqrt{3}\mathbf{V}_{RY}|\mathbf{Y}_D| \exp[j(\phi + \pi/2)]$$

The various phase angles will be found to agree with those in Figure 7.6. The magnitude of the line current is given by

$$I_L = \sqrt{3}V_L|\mathbf{Y}_D| \qquad\qquad (7.16)$$

The results expressed by Equations 7.13 can alternatively be expressed in terms of the phase voltages. Substituting from Equations 7.7, we find

$$\mathbf{I}_R = 3\mathbf{V}_{RS}\mathbf{Y}_D$$

$$\mathbf{I}_Y = 3\mathbf{V}_{YS}\mathbf{Y}_D \qquad\qquad (7.17)$$

$$\mathbf{I}_B = 3\mathbf{V}_{BS}\mathbf{Y}_D$$

EXAMPLE 7.1 Three loads each of impedance $16 + j12\ \Omega$ are connected in the delta configuration to a three-phase supply of 415 V between lines. Calculate the magnitude of the line current.

In this case Equation 7.16 can be applied directly:

$$\mathbf{Z}_D = 16 + j12 = 20\angle 36.9°$$

$$\mathbf{Y}_D = 1/\mathbf{Z}_D = 0.05\angle -36.9°$$

Hence

$$\mathbf{I}_L = \sqrt{3} \times 415 \times 0.05 = 36.0 \text{ A}$$

7.4.2 ◆ Balanced loads in star connection

In this case, referring to Figure 7.5,

$$\mathbf{Y}_R = \mathbf{Y}_Y = \mathbf{Y}_B = \mathbf{Y}_S = |\mathbf{Y}_S|\angle \phi \qquad (7.18)$$

so that Equation 7.11 reduces to

$$\mathbf{V}_{OS} = \mathbf{V}_{RS} + \mathbf{V}_{YS} + \mathbf{V}_{BS} = 0$$

Thus in the case of a balanced load there is no voltage difference between the star points of source and load. Putting $\mathbf{V}_{OS}$ equal to zero in Equations 7.10, expressions for the line currents are obtained:

$$\mathbf{I}_R = \mathbf{V}_{RS}|\mathbf{Y}_S|\exp(j\phi)$$

$$\mathbf{I}_Y = \mathbf{V}_{RS}|\mathbf{Y}_S|\exp[j(\phi - 2\pi/3)] \qquad (7.19)$$

$$\mathbf{I}_B = \mathbf{V}_{RS}|\mathbf{Y}_S|\exp[j(\phi + 2\pi/3)]$$

These equations show the result mentioned earlier that

$$\mathbf{I}_Y = \mathbf{I}_R \exp(-j2\pi/3)$$

$$\mathbf{I}_B = \mathbf{I}_R \exp(j2\pi/3)$$

The phasor diagram is similar to that of Figure 7.6(c). The magnitude of the line currents is given by

$$I_L = V_P|\mathbf{Y}_S|$$

$$= \frac{1}{\sqrt{3}}V_L|\mathbf{Y}_S| \qquad (7.20)$$

Comparison of the expressions of Equations 7.19 with those of Equations 7.17 shows that the star-connected load and the delta-connected load behave in exactly the same way if

$$\mathbf{Y}_S = 3\mathbf{Y}_D$$

or

$$\mathbf{Z}_D = 3\mathbf{Z}_S \qquad (7.21)$$

This is consistent with the star–delta formulae presented in Section 4.14.

EXAMPLE 7.2 The three loads of Example 7.1 are connected to the same three-phase supply in star. Calculate the line current.

The result follows from application of Equation 7.20, giving

$$I_\text{L} = \frac{1}{\sqrt{3}} \times 415 \times 0.05 = 12.0 \text{ A}$$

EXAMPLE 7.3 The balanced load of a 415 V three-phase supply consists of three star-connected impedances in parallel with three delta-connected impedances. Each of the impedances in the star is equal to $8 + j6 \ \Omega$; those in the delta connection are each equal to $12 + j16 \ \Omega$. Calculate the magnitude of the line current.

Since the load is balanced we may calculate the current in any one line in order to find the magnitude. Consider $\mathbf{I}_\text{R}$: the current provided by the supply is made up of the component going to the star-connected load and that going to the delta connection. To combine these two contributions it will be necessary to allow for the difference in phase. We accordingly obtain a complete expression for each component. Consider first the star connection:

$$\mathbf{Z}_\text{S} = 8 + j6 = 10\angle 36.9°$$

Therefore

$$\mathbf{Y}_\text{S} = 0.1\angle -36.9°$$

Applying Equation 7.19, taking $\mathbf{V}_\text{RS}$ as the real reference phasor, we have

$$\mathbf{I}_\text{R} = (415/\sqrt{3}) \times 0.1\angle -36.9° = 24.0\angle -36.9° = 19.2 - j14.4$$

To find the component for the delta connection we can use Equation 7.17. We have

$$\mathbf{Z}_\text{D} = 12 + j16 = 20.0\angle 53.1°$$

Therefore

$$\mathbf{Y}_\text{D} = 0.05\angle -53.1°$$

and

$$\mathbf{I}_\text{R} = 3(415/\sqrt{3}) \times 0.05\angle -53.1° = 35.9\angle -53.1 = 21.6 - j28.7$$

The line current at the supply is therefore given by

$$\mathbf{I}_\text{R} = 19.2 - j14.4 + 21.6 - j28.7$$
$$= 40.8 - j43.1$$
$$= 59.3\angle -46.6°$$

Hence

$$I_\text{L} = 59.3 \text{ A}$$

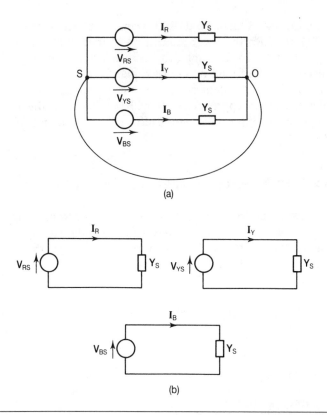

(a)

(b)

Figure 7.7 Illustrating the derivation of a single-phase equivalent circuit for a balanced load: (a) adding a zero-impedance link between star points; (b) separation into three similar circuits.

7.5 ◆ Single-phase equivalent circuits for balanced three-phase

The methods by which the line current has been determined in the previous sections have basically involved one phase only, indicating that such problems can be solved by a single-phase equivalent circuit. Consider the star-connected source and load shown in Figure 7.5. Since the two star points have zero voltage between them, the situation will be unchanged if they are joined by a conductor, as shown in Figure 7.7(a). Assuming this conductor to be of zero resistance this circuit can be separated into three single-phase circuits, as shown in Figure 7.7(b). The values obtained by calculation on any one of these circuits can be applied to the others by allowing for the change in phase.

The delta connection can be treated in the same way: the currents I_{RY}, I_{YB}, I_{BR} flowing in the three branches of the circuit of Figure 7.4 will, even for an unbalanced load, be those flowing in the three separate circuits shown in Figure 7.8. In the balanced case they will all be equal in magnitude, so that $\mathbf{I}_D$, for example, can be calculated from any one circuit. The line current is then found by use of Equation 7.16.

Either a delta equivalent circuit or a star equivalent can be used for such calculations, as is appropriate. It will be shown in the next section that the single-phase equivalent also suffices for calculation of power or volt–amps reactive.

7.6 ◆ Power in balanced three-phase circuits

Since both power and volt–amps reactive are conserved, as discussed in Sections 4.9 and 4.10, we may calculate values for the supply by summing the relevant quantities for each branch of the load. In the case of the balanced star-connected load of Section 7.4.2 we have

$$\mathbf{V}_{RS}\mathbf{I}_R^* = \mathbf{V}_{RS}(\mathbf{V}_{RS}\mathbf{Y}_R)^*$$
$$= |\mathbf{V}_{RS}|^2|\mathbf{Y}_S|\exp(-j\phi)$$
$$= V_P I_L \exp(-j\phi)$$

Similarly

$$\mathbf{V}_{YS}\mathbf{I}_Y^* = \mathbf{V}_{BS}\mathbf{I}_B^* = V_P I_L \exp(-j\phi)$$

Thus for the total power we have

$$P = 3V_P I_L \cos(\phi)$$
$$= \sqrt{3}V_L I_L \cos(\phi) \tag{7.22}$$

and for the volt–amps reactive

$$Q = \sqrt{3}V_L I_L \sin(-\phi) \tag{7.23}$$

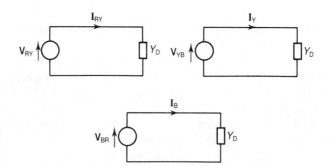

Figure 7.8 Separation into similar circuits with balanced load in delta connection.

From these equations we also have

$$(P^2 + Q^2)^{1/2} = \sqrt{3}V_L I_L \tag{7.24}$$

[It is to be noted that ϕ has been defined as the phase angle of the admittance: for an inductive or lagging load $\phi < 0$, for a capacitative or leading load $\phi > 0$.]

The case of the delta-connected load can be dealt with in a similar way. We have

$$\mathbf{V}_{RY}\mathbf{I}_{RY}^* = |\mathbf{V}_{RY}|^2 |\mathbf{Y}_D| \exp(-j\phi)$$

$$= V_L^2 |\mathbf{Y}_D| \exp(-j\phi)$$

$$= V_L I_D \exp(-j\phi)$$

Similarly

$$\mathbf{V}_{YB}\mathbf{I}_{YB}^* = \mathbf{V}_{BR}\mathbf{I}_{BR}^* = V_L I_D \exp(-j\phi)$$

In terms of $\mathbf{I}_L$, using Equation 7.16, we find the same expressions as before

$$P = \sqrt{3}V_L I_L \cos(\phi)$$

$$Q = \sqrt{3}V_L I_L \sin(-\phi)$$

EXAMPLE 7.4 Determine the power taken from the supply in the cases of Example 7.1 and Example 7.2.

From the values given in Example 7.1 we have

$$V_L = 415, \ I_L = 36.0, \ \phi = -36.9°$$

Hence

$$P = \sqrt{3} \times 415 \times 36.0 \cos(36.9°) = 20.7 \text{ kW}$$

$$Q = \sqrt{3} \times 415 \times 36.0 \sin(36.9°) = 15.5 \text{ kVAR}$$

In this example we can proceed in an alternative fashion.

$$|\mathbf{I}_{RY}| = |\mathbf{V}_{RY}\mathbf{Y}_D| = 415 \times 0.05 \text{ A}$$

This current flows through an impedance $16 + j12 \ \Omega$. Hence the power dissipated is given by

$$P_{RY} = 16|\mathbf{I}_{RY}|^2 = 6889 \text{ W}$$

and

$$Q_{RY} = 12|\mathbf{I}_{RY}|^2 = 5167 \text{ VAR}$$

For the supply therefore

$$P = 3 \times 6889 = 20.7 \text{ kW}$$

$$Q = 3 \times 5167 = 15.5 \text{ kVAR}$$

In Example 7.2 the phase angle is again $-36.9°$ so that

$$P = \sqrt{3} \times 415 \times 12.0 \cos(36.9°) = 6.9 \text{ kW}$$

$$Q = \sqrt{3} \times 415 \times 12.0 \sin(36.9°) = 5.2 \text{ kVAR}$$

Treating individual branches these results should also be given by

$$P = 3 \times 16 \times (12.0)^2 = 6.9 \text{ kW}$$

$$Q = 3 \times 12 \times (12.0)^2 = 5.2 \text{ kVAR}$$

EXAMPLE 7.5 Calculate the power supplied to the combined load in Example 7.3.

It was shown in the calculations for Example 7.3 that the line current I_R was 59.3 A, lagging on the phase voltage V_{RS} by 46.6°. Applying Equation 7.22, the power supplied is given by

$$P = \sqrt{3} \times 415 \times 59.3 \times \cos(46.6°)$$

$$= 29.3 \text{ kW}$$

Also

$$Q = \sqrt{3} \times 415 \times 59.3 \times \sin(46.6°)$$

$$= 31.0 \text{ kVAR}$$

We can check these results by summing the individual contributions. For the star-connected load the line current was found to be 24.0 A, passing through the impedance of $8 + j6 \ \Omega$. Hence for the star load

$$P_{STAR} = 3 \times 8 \times (24.0)^2 = 13.82 \text{ kW}$$

$$Q_{STAR} = 3 \times 6 \times (24.0)^2 = 10.37 \text{ kVAR}$$

For the delta-connected load

$$I_D = V_L |Y_D| = 415 \times 0.05$$

Hence, for the impedance $12 + j16 \ \Omega$,

$$P_{DELTA} = 3 \times 12 \times I_D^2 = 15.50 \text{ kW}$$

$$Q_{DELTA} = 3 \times 12 \times I_D^2 = 20.67 \text{ kVAR}$$

Summing these contributions we find, correctly,

$$P = 29.3 \text{ kW}, \ Q = 31.0 \text{ kVAR}$$

7.7 ◆ The 'two-wattmeter method' of power measurement

A standard method of power measurement in three-phase circuits uses two wattmeters in the configuration shown in Figure 7.9. The three-phase supply may be taken to consist only of the two voltage sources V_{RY} and

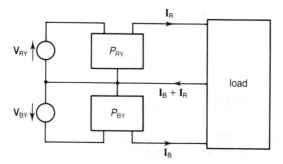

Figure 7.9 Circuit connections and notation for the two-wattmeter method of power measurement.

$\mathbf{V}_{BY}$, as discussed in Section 7.2.1. Clearly, since the correct voltages are maintained in this arrangement, the power delivered to the load is found by summing the powers delivered by these two sources, as indicated by the wattmeters. In what follows, the load will be assumed balanced, although the two-wattmeter method is equally valid for unbalanced loads (provided that connection is by three wires only). Treating the Y line as common, for each source the power delivered can be calculated in the form

$$P_{RY} = \text{Re}\,(\mathbf{V}_{RY}\mathbf{I}_R^*)$$

$$P_{BY} = \text{Re}\,(\mathbf{V}_{BY}\mathbf{I}_B^*)$$

From Equations 7.15 the line current is seen to lead the voltage $\mathbf{V}_{RY}$ by the angle $\phi - \pi/6$. Hence

$$P_{RY} = V_L I_L \cos\,(\phi - \pi/6) \tag{7.25}$$

To find P_{BY} we note that

$$\mathbf{V}_{BY} = -\mathbf{V}_{YB} = -\mathbf{V}_{RY}\exp\,(-j2\pi/3)$$

and that, from Equations 7.15, $\mathbf{I}_B$ leads the voltage $\mathbf{V}_{RY}$ by an angle $\phi + \pi/2$. Therefore

$$\mathbf{V}_{BY}\mathbf{I}_B^* = -V_L\exp\,(-j2\pi/3) \times I_L\exp\,[-j(\phi + \pi/2)]$$

$$= V_L I_L\exp\,[-j(\phi + \pi/6)]$$

Hence

$$P_{BY} = V_L I_L \cos\,(\phi + \pi/6) \tag{7.26}$$

To check that the total power is given correctly we form

$$P_{RY} + P_{BY} = V_L I_L \cos\,(\phi - \pi/6) + \cos\,(\phi + \pi/6)]$$

$$= V_L I_L \times 2\cos\phi\cos\,(\pi/6)$$

$$= \sqrt{3}V_L I_L \cos\phi$$

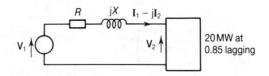

Figure 7.10 Form of circuit for Example 7.7.

EXAMPLE 7.6 Calculate the readings the two wattmeters would give if the load were that of Example 7.3.

In Example 7.3 the combined loads give a line current of 59.3 A, with a lagging phase angle of 46.6°. Hence

$$P_{RY} = 415 \times 59.3 \times \cos(16.6°) = 23.58 \text{ kW}$$

$$P_{BY} = 415 \times 59.3 \times \cos(76.6°) = 5.70 \text{ kW}$$

The total power is therefore 29.3 kW, as found earlier.

It is to be noticed that the reading of one of the wattmeters can be negative: if, for example, the load is purely reactive so that $\phi = \pi/2$, then

$$P_{RY} = V_L I_L \cos(\pi/2 - \pi/6) = \tfrac{1}{2} V_L I_L$$

$$P_{BY} = V_L I_L \cos(\pi/2 + \pi/6) = -\tfrac{1}{2} V_L I_L$$

7.8 ◆ Balanced loads specified by power and power factor

In problems involving power distribution it is common to specify loads by the power and power factor of the load, rather than by impedance or admittance. Such specification often makes it convenient to perform calculations in terms of power and volt–amps reactive rather than to convert the load specification to impedance form. The technique will be illustrated by an example.

EXAMPLE 7.7 A three-phase system supplies at a line voltage of 132 kV a load of 60 MW at power factor 0.85 lagging. Each conductor carrying the supply has an impedance of $10 + j55$ Ω. Calculate the line current and the line voltage at the supply end of the lines.

The problem will first be solved with the aid of a single-phase equivalent circuit, shown in Figure 7.10. Taking the voltage at the load, V_2, as the phasor reference we have

$$\mathbf{V}_2 = V_2 \angle 0 = (132/\sqrt{3}) \times 10^3 = 76.21 \text{ kV}$$

The current $\mathbf{I}$ will have one component in phase with $\mathbf{V}_2$ and one lagging by $\pi/2$:

$$\mathbf{I} = I_1 - jI_2 = I_L(\cos\phi - j\sin\phi)$$

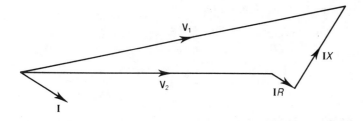

Figure 7.11 Phasor diagram of Example 7.7.

where

$$\cos \phi = 0.85$$

and

$$V_2 I_{\mathrm{L}} \cos \phi = (60/3) \times 10^6$$

We find

$$I_{\mathrm{L}} = 308.7 \text{ A}$$

$$\sin \phi = 0.527$$

Using these values in the circuit of Figure 7.10 we have

$$\mathbf{V}_1 = \mathbf{V}_2 + (10 + \mathrm{j}55) \times 308.7(0.85 - \mathrm{j}0.527)$$

$$= \mathbf{V}_2 + (37.5 + \mathrm{j}41.5) \times 308.7$$

$$= 87\,790 + \mathrm{j}12\,811$$

The line voltage at the supply end is therefore equal to

$$\sqrt{3} \times |\mathbf{V}_1| = \sqrt{3} \times 88\,720 = 153.7 \text{ kV}$$

The phasor diagram is shown in Figure 7.11.

The alternative method proceeds as follows. To find the line current we use Equation 7.22:

$$\sqrt{3} \times 132 \times 10^3 \times I_{\mathrm{L}} \times 0.85 = 60 \times 10^6$$

$$I_{\mathrm{L}} = 308.7 \text{ A}$$

We then consider power and volt–amps reactive: for the load

$$P = 60 \times 10^6$$

$$Q = 60 \times 10^6 \sin \phi / \cos \phi = 37.20 \times 10^6$$

For the line impedances

$$P = 3 \times 10 \times (308.7)^2 = 2.86 \times 10^6$$

$$Q = 3 \times 55 \times (308.7)^2 = 15.73 \times 10^6$$

Hence we find for the totals

$$P = 62.86 \text{ MW}$$

$$Q = 52.93 \text{ MVAR}$$

Applying Equation 7.24 we have

$$\sqrt{3}V_L I_L = (P^2 + Q^2)^{1/2} = 82.18 \times 10^6$$

Hence

$$V_L = 82.18 \times 10^6/\sqrt{3} \times 308.7 = 153.7 \text{ kV}$$

7.9 ◆ Power factor correction

In a situation such as that occurring in Example 7.7 it is common practice to increase the power factor towards unity by the addition of a capacitance in parallel with the main load. This has the effect of reducing the line current and thus reducing transmission losses. The process is illustrated in the following example.

EXAMPLE 7.8 Determine the value of the capacitances which when connected in delta to the load of Example 7.7 will give a power factor of unity.

To correct the power factor to unity it is necessary to compensate for the reactive, lagging, component denoted by I_2 in the discussion of Example 7.7. In the single-phase equivalent circuit of Figure 7.10 this can be achieved by placing a capacitance in parallel with the load, as in Figure 7.12. The capacitance takes a leading current of magnitude $V_2 \omega C_2$, in which

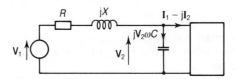

Figure 7.12 Single-phase equivalent circuit showing the addition of a capacitor to correct power factor, Example 7.8.

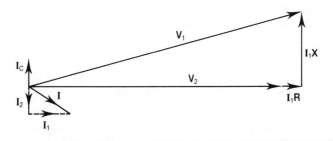

Figure 7.13 Phasor diagram for Example 7.8.

ω is the angular frequency of the supply. The phasor diagram is shown in Figure 7.13. Exact correction will be obtained when

$$V_2 \omega C_S = 308.7 \times 0.527$$
$$\omega C_S = 308.7 \times 0.527 \times \sqrt{3}/132 \times 10^3$$
$$= 2.13 \times 10^{-3} \text{ S}$$

Assuming a supply frequency of 50 Hz

$$C_S = 6.8 \ \mu\text{F}$$

The value required in delta connection will be $C_S/3$ so that

$$C_D = 2.3 \ \mu\text{F}$$

The calculation can be conveniently carried out in terms of volt–amps reactive. The load was shown to take 37.2 MVAR. A single member of a delta connection of capacitors each of value C_D would take

$$- V_L \times V_L \omega C_D \ \text{(VAR)}$$

Hence for correction

$$3 \times \omega C_D \times (132 \times 10^3)^2 = 37.2 \times 10^6$$
$$\omega C_D = 7.12 \times 10^{-4}$$
$$C_D = 2.3 \ \mu\text{F}$$

7.10 ◆ Unbalanced circuits

It is not the purpose of this book to detail the techniques used in the general treatment of unbalanced circuits. However, considering the matter as a circuit problem, we may apply the formulae of Sections 7.3.1 and 7.3.2. In the two following examples these formulae will be applied to an unbalanced delta-connected load and to an unbalanced star.

EXAMPLE 7.9 The load on a 440 V three-phase system consists of a resistance of 100 Ω between R and Y, an impedance of $200 + \text{j}200 \ \Omega$ between Y and B, and an impedance of $80 + \text{j}60 \ \Omega$ between B and R. Assuming that the phase sequence R–Y–B is positive, calculate the magnitudes of the line currents.

In this problem it is convenient to take the voltage V_{RY} as zero phase. We have

$$\mathbf{V}_{RY} = 440\angle 0°$$
$$\mathbf{V}_{YB} = 440\angle -120°$$
$$\mathbf{V}_{BR} = 440\angle +120°$$

In terms of impedances

$$\mathbf{Z}_{RY} = 100\angle 0°$$
$$\mathbf{Z}_{YB} = 200 + \text{j}200 = 200\sqrt{2}\angle 45°$$
$$\mathbf{Z}_{BR} = 80 + \text{j}60 = 100\angle 36.9°$$

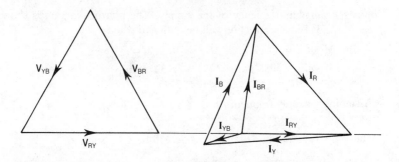

Figure 7.14 Phasor diagram for unbalanced load in delta connection, Example 7.9: (a) voltage diagram; (b) current diagram.

Hence

$$\mathbf{I}_{RY} = 4.4\angle0° = 4.4 + j \times 0$$

$$\mathbf{I}_{YB} = 1.556\angle-165° = 1.503 - j0.403$$

$$\mathbf{I}_{BR} = 4.4\angle83.1° = 0.529 + j4.368$$

Applying Equations 7.9 we find

$$\mathbf{I}_{R} = 3.871 - j4.368 = 5.836\angle-48.5°$$

$$\mathbf{I}_{Y} = -5.903 - j0.403 = 5.917\angle-176.1°$$

$$\mathbf{I}_{B} = 2.031 + j4.771 = 5.185\angle66.9°$$

and hence the required magnitudes. The phasor diagram is shown in Figure 7.14.

EXAMPLE 7.10 A star-connected load consists of impedances of 100 Ω, 200 + j200 Ω and 80 + j60 Ω connected respectively between R, Y and B and the star point of the load. This load is supplied by a 440 V three-phase supply. Assuming a positive phase sequence, determine the magnitudes of the line currents.

Supply voltage is always quoted as line voltage, so that the phase voltage is $V_P = 440/\sqrt{3} = 254$ V. Following Equations 7.5, we write

$$\mathbf{V}_{RS} = 254\angle0°$$

$$\mathbf{V}_{YS} = 254\angle-120°$$

$$\mathbf{V}_{BS} = 254\angle+120°$$

We have

$$\mathbf{Y}_{R} = 0.01\angle0° = 1 \times 10^{-2}$$

$$\mathbf{Y}_{Y} = 0.25\sqrt{2} \times 10^{-2}\angle-45° = 0.25 \times 10^{-2}(1 - j)$$

$$\mathbf{Y}_{B} = 0.01\angle-36.9° = 10^{-2}(0.8 - j0.6)$$

Hence

$$\mathbf{V}_{RS}\mathbf{Y}_R + \mathbf{V}_{YS}\mathbf{Y}_Y + \mathbf{V}_{BS}\mathbf{Y}_B = 2.54(1\angle 0° + 0.25\sqrt{2}\angle -165° + 1.0\angle 83.1°)$$

$$= 3.03\angle 49.2°$$

$$\mathbf{Y}_R + \mathbf{Y}_Y + \mathbf{Y}_B = (2.05 - j0.85) \times 10^{-2}$$

$$= 2.22 \times 10^{-2}\angle -22.5°$$

We substitute in Equations 7.11 to obtain

$$\mathbf{V}_{OS} = 136.3\angle 71.7° = 42.8 + j129.4$$

Hence

$$\mathbf{V}_{RS} - \mathbf{V}_{OS} = 211.2 - j129.4 = 247.7\angle -31.5°$$

$$\mathbf{V}_{YS} - \mathbf{V}_{OS} = -169.8 - j349.4 = 388.5\angle -115.9°$$

$$\mathbf{V}_{BS} - \mathbf{V}_{OS} = -169.8 + j90.6 = 192.4\angle 151.9°$$

Finally

$$\mathbf{I}_R = 2.48\angle -31.5°$$

$$\mathbf{I}_Y = 1.37\angle -160.9°$$

$$\mathbf{I}_B = 1.92\angle 115.0°$$

The phasor diagram appears in Figure 7.15.

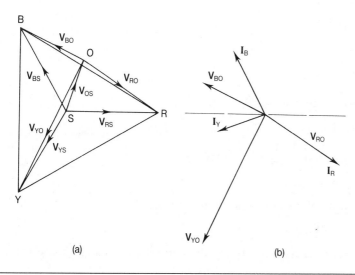

(a)   (b)

Figure 7.15 Phasor diagram for unbalanced load in star connection, Example 7.10: (a) voltage diagram; (b) current diagram.

7.11 ◆ Summary of formulae

phase voltage $= \dfrac{1}{\sqrt{3}} \times$ line voltage

For a three-phase supply with R–Y–B in a positive sequence, phasor diagram Figure 7.1,

$\mathbf{V}_{YB} = \mathbf{V}_{RY} \exp(-j2\pi/3)$

$\mathbf{V}_{BR} = \mathbf{V}_{RY} \exp(j2\pi/3)$

$|\mathbf{V}_{RY}| = |\mathbf{V}_{YB}| = |\mathbf{V}_{BR}| = V_L$

For the equivalent star connection of sources

$\mathbf{V}_{RS} = \dfrac{1}{\sqrt{3}} \mathbf{V}_{RY} \exp(-j\pi/6)$

$\mathbf{V}_{YS} = \mathbf{V}_{RS} \exp(-j2\pi/3)$

$\mathbf{V}_{BS} = \mathbf{V}_{RS} \exp(j2\pi/3)$

$|\mathbf{V}_{RS}| = |\mathbf{V}_{YS}| = |\mathbf{V}_{BS}| = V_P = \dfrac{1}{\sqrt{3}} V_L$

For balanced loads, positive phase sequence

$\mathbf{I}_Y = \mathbf{I}_R \exp(-j2\pi/3)$

$\mathbf{I}_B = \mathbf{I}_R \exp(j2\pi/3)$

$|\mathbf{I}_R| = |\mathbf{I}_Y| = |\mathbf{I}_B| = I_L$

For a delta-connected load

$\mathbf{I}_{RY} = \dfrac{1}{\sqrt{3}} \mathbf{I}_R \exp(j\pi/6)$

$\mathbf{I}_{YB} = \mathbf{I}_{RY} \exp(-j2\pi/3)$

$\mathbf{I}_{BR} = \mathbf{I}_{RY} \exp(j2\pi/3)$

$|\mathbf{I}_{RY}| = |\mathbf{I}_{YB}| = |\mathbf{I}_{BR}| = I_D = \dfrac{1}{\sqrt{3}} I_L$

For a star of equal admittances $\mathbf{Y}_S$

$\mathbf{I}_R = \mathbf{V}_{RS} \mathbf{Y}_S$

For a delta connection of admittances $\mathbf{Y}_D$

$\mathbf{I}_R = 3\mathbf{V}_{RS} \mathbf{Y}_D$

then ·If ϕ is the angle by which the current $\mathbf{I}_R$ leads the phase voltage $\mathbf{V}_{RS}$

power $= P = \sqrt{3} V_L I_L \cos\phi$ (W)

volt–amps reactive $= Q = \sqrt{3} V_L I_L \sin(-\phi)$ (VAR)

Two-wattmeter method

Regarding line Y as common, for a balanced load,

$$P_{RY} = V_L I_L \cos(\phi - \pi/6)$$
$$P_{BY} = V_L I_L \cos(\phi + \pi/6)$$

PROBLEMS

In the following problems the voltage quoted for a three-phase supply is the voltage between lines.

7.1 Three $50\,\Omega$ resistors are used to form a three-phase load. The load is connected to a 415 V three-phase supply. Find the line current and power when the resistors are (a) connected in star and (b) connected in delta.

7.2 A load consists of three equal impedances, each of value $25 + j10\,\Omega$, connected in star. The load is connected to a 415 V, three-phase supply. Draw a single-phase equivalent circuit and hence determine the line current. Determine also the power and power factor.

7.3 Each impedance of a balanced, delta-connected load consists of a $100\,\Omega$ resistance in parallel with an inductive reactance of $180\,\Omega$. The load is connected to a three-phase, 415 V supply. Determine the line current, power and power factor.

7.4 The star-connected load of Problem 7.2 and the delta-connected load of Problem 7.3 are both connected to the same 415 V, three-phase supply. Determine the line current, power and power factor.

7.5 A high-voltage, three-phase load can be represented by three star-connected impedances, each impedance being equal to $20\,\Omega$ resistance in series with 0.5 H inductance. The load is connected to a 2.2 kV, 50 Hz, three-phase supply. Determine the line current, power and power factor.

7.6 To the system of Problem 7.5 is connected an additional load consisting of three $4\,\mu F$ capacitors in delta configuration. Determine the line current, power and power factor.

7.7 With the supply and load of Problem 7.5 determine the value of the equal capacitors which when connected in the delta configuration will increase the power factor to 0.95 lagging.

7.8 The power consumed from a 415 V, three-phase supply by a load of three $20\,\Omega$ resistors in the delta connection is measured by the two-wattmeter method. Determine the readings of the wattmeters.

7.9 The resistors of Problem 7.8 are replaced by capacitors of the same magnitude of impedance. Determine the readings of the wattmeters.

7.10 Determine the readings of the wattmeters if the two-wattmeter method were used to measure the power delivered in the system of Problem 7.4.

7.11 Power is delivered to a factory at 132 kV by a 50 Hz, three-phase supply. The factory presents a balanced load consuming 100 MW at 0.8 lagging power factor. Determine the line current.

Each of the three lines connecting the generating station to the factory has an impedance of $2 + j10\,\Omega$. Determine the voltage at the sending end and the power lost in the lines.

7.12 In the system of Problem 7.11 the power factor at the factory is to be corrected to 0.95 lagging by adding capacitors in delta configuration. Determine the value of the required capacitors. Determine also the new values of the voltage at the sending end and the power lost in the lines.

7.13 A three-phase load consists of a resistor, an inductor and a capacitor connected in delta configuration. The impedance of each component has the same magnitude of $50\,\Omega$. The load is connected to a 415 V, three-phase supply. Determine the line currents.

7.14 A load consisting of three unequal resistors in the star configuration is connected to a 415 V, three-phase supply. The resistors have values of $10\,\Omega$, $10\,\Omega$ and $7.5\,\Omega$. Calculate the voltage between the neutral (star point) of the supply and the star point of the load.

7.15 Each arm of a balanced, three-phase, star-connected load consists of 63.7 mH inductance in series with $10\,\Omega$ resistance. The load is connected to a 415 V, 50 Hz, three-phase supply through wires each of resistance $1\,\Omega$. Calculate the line current and the power dissipated in the load.

The resistance in one arm of the load drops to zero. Calculate the voltage between the star point of the load and the neutral of the supply. Hence determine the line currents.

8 ♦ *Magnetically Coupled Inductors:*
Transformers

In this chapter analysis is applied to circuits involving inductors coupled through a common magnetic flux. The AC power transformer is introduced as a particular case of close coupling.

8.1 ♦ Introduction

The element consisting of two magnetically coupled inductors finds use in a very wide range of electrical and electronic circuits. Such an element is usually referred to as a **transformer**. This name arises from the frequent use of the device to change from one level to another the voltage at which power is delivered. It can also be used to effect the power matching between source and load discussed in Section 4.15.

8.2 ♦ Circuit equations

The equations relating to coupled inductors were given in Section 1.3.4. In phasor form, using the notation in Figure 8.1, Equations 1.6 become

$$\mathbf{V}_1 = j\omega L_1 \mathbf{I}_1 + j\omega M \mathbf{I}_2$$

$$\mathbf{V}_2 = j\omega M \mathbf{I}_1 + j\omega L_2 \mathbf{I}_2 \tag{8.1}$$

As pointed out in Section 1.3.4, the sign of M can only be made definite when some convention has been established by which the directions of voltages and currents on the two sides can be related. The dots shown in Figure 8.1 are one such convention: with currents entering at the dots and

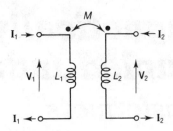

Figure 8.1 Coupled inductors, showing phasor variables.

voltages being measured in the sense shown in the figure, M will be positive. The dots only have physical significance when the two sides of the coupled pair are in some way interconnected.

Equations 8.1 also represent the circuit of Figure 8.2, which is therefore equivalent to that of Figure 8.1, except that the isolation of the two sides has been omitted. In the circuit of Figure 8.2, one or more of the inductances may be negative, either because M is negative or because, if positive, it lies between L_1 and L_2. If this is the case, the equivalence is still valid for the purpose of calculations but the circuit could not be realized in practice.

The coefficients L_1, L_2, M characterizing the coupled inductors can in principle be calculated from the geometry of the coils. It can be shown from such calculations that necessarily

$$M^2 \leq L_1 L_2$$

This result can also be proved from considerations of energy: this proof appears in Appendix B, Section B.2. The ratio

$$k = \frac{M}{(L_1 L_2)^{1/2}} \leq 1 \tag{8.2}$$

is termed the coupling coefficient.

EXAMPLE 8.1 The two coils in the circuit of Figure 8.1 are connected in series. Show that the inductance of the combination can have either of the values $L_1 + L_2 \pm 2M$, depending on the sense of the connection.

The two possible connections are shown in Figure 8.3. In the circuit of Figure 8.3(a) we have

$$\mathbf{V} = \mathbf{V}_1 + \mathbf{V}_2$$

$$\mathbf{I} = \mathbf{I}_1 = \mathbf{I}_2$$

Adding Equations 8.1 we find

$$\mathbf{V} = \mathbf{V}_1 + \mathbf{V}_2 = j\omega(L_1\mathbf{I}_1 + M\mathbf{I}_2 + M\mathbf{I}_1 + L_2\mathbf{I}_2)$$

Hence

$$\mathbf{V} = j\omega(L_1 + L_2 + 2M)\mathbf{I}$$

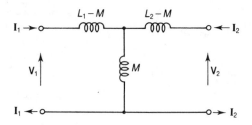

Figure 8.2 Equivalent T-circuit.

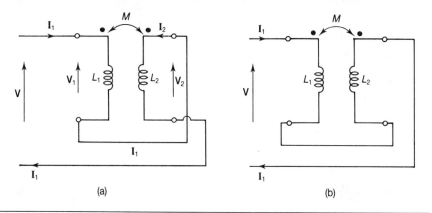

Figure 8.3 Example 8.1, showing the two possible connections: (a) aiding; (b) opposing.

so that the combined coils have an inductance $L_1 + L_2 + 2M$.

With the connection of Figure 8.3(b) we now have

$$\mathbf{V} = \mathbf{V}_1 - \mathbf{V}_2$$

$$\mathbf{I} = \mathbf{I}_1 = -\mathbf{I}_2$$

In this case we find

$$\mathbf{V} = j\omega(L_1 + L_2 - 2M)\mathbf{I}$$

corresponding to an inductance $L_1 + L_2 - 2M$.

8.3 ♦ Referred impedance

We consider in this section how the connection of an impedance across one coil affects the impedance seen across the other. The situation is shown in the circuit of Figure 8.4. We have

$$\mathbf{V}_1 = j\omega L_1 \mathbf{I}_1 + j\omega M \mathbf{I}_2$$

$$\mathbf{V}_2 = j\omega M \mathbf{I}_1 + j\omega L_2 \mathbf{I}_2$$

$$\mathbf{V}_2 = -\mathbf{Z}_2 \mathbf{I}_2$$

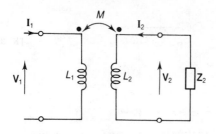

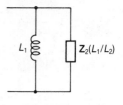

Figure 8.4 Coupled inductors with load Z_2 on secondary.

Figure 8.5 Equivalent circuit for input to primary in the case of perfect coupling.

From the last two equations we find

$$(Z_2 + j\omega L_2)I_2 = -j\omega M I_1$$

Hence

$$V_1 = \left[j\omega L_1 + \frac{(\omega M)^2}{Z_2 + j\omega L_2} \right] I_1$$

The impedance seen between terminals 1 and 2 is therefore

$$Z_1 = j\omega L_1 + \frac{(\omega M)^2}{Z_2 + j\omega L_2} \tag{8.3}$$

This is equivalent to the primary inductance L_1 in series with the 'referred' impedance

$$\frac{(\omega M)^2}{Z_2 + j\omega L_2}$$

In the case when the coupling is perfect, $L_1 L_2 = M^2$, Equation 8.3 reduces to

$$Z_1 = \frac{j\omega L_1 Z_2}{Z_2 + j\omega L_2}$$

which may be rearranged in the form

$$\frac{1}{Z_1} = \frac{1}{j\omega L_1} + \frac{L_2}{L_1}\frac{1}{Z_2} \tag{8.4}$$

The impedance Z_1 is therefore equivalent to the parallel combination of an inductance L_1 and an impedance equal to Z_2 multiplied by L_1/L_2, as shown in Figure 8.5.

EXAMPLE 8.2 The primary of a pair of coupled coils has an inductance of 1 mH; the inductance of the secondary is 100 μH and the coefficient of coupling $k = 0.79$. Determine, at a frequency of 100 kHz, the impedance presented by the primary when a 50 Ω resistance is connected across the secondary coil.

For this calculation we use Equation 8.3. We first calculate the value of M:

$$M = 0.79 \times \sqrt{(10^{-3} \times 10^{-4})^{1/2}} = 250 \; \mu\text{H}$$

We then have

$$\omega M = 2\pi \times 10^5 \times 250 \times 10^{-6} = 50\pi = 157.10 \; \Omega$$

$$\omega L_2 = 2\pi \times 10^5 \times 10^{-4} = 20\pi = 62.83 \; \Omega$$

$$\omega L_1 = 2\pi \times 10^5 \times 10^{-3} = 200\pi = 628.3 \; \Omega$$

$$\mathbf{Z}_2 + j\omega L_2 = 50 + j62.83$$

$$= 80.30\angle 51.49°$$

Hence

$$\frac{(\omega M)^2}{\mathbf{Z}_2 + j\omega L_2} = 307.4\angle -51.49° = 191.4 - j240.5$$

and

$$\mathbf{Z}_1 = j628.3 + 191.4 - j240.5 = 191.4 + j387.8$$

Thus $\mathbf{Z}_1$ may be represented as a resistance of 191 Ω in series with a reactance of 387.8 Ω. At 100 kHz the latter corresponds to an inductance of 617 μH.

EXAMPLE 8.3 Show that in the circuit of Example 8.2 the addition of a suitable capacitor in parallel with the primary can make the impedance seen at the primary purely resistive. Determine the value of the capacitor and of the resulting resistance.

The impedance of the primary was calculated in Example 8.2 to be

$$\mathbf{Z}_1 = 191.4 + j387.7 = 432.4\angle 63.73°$$

For cancellation of the reactive component by means of a capacitor in parallel we need to put this into the form of an admittance:

$$\mathbf{Y}_1 = 2.313 \times 10^{-3}\angle -63.73° = (1.024 - j2.074) \times 10^{-3}$$

This will become purely resistive if a capacitance of susceptance 2.074×10^{-3} Ω is placed in parallel: hence

$$\omega C = 2.074 \times 10^{-3}$$

$$C = 3.30 \text{ nF}$$

The impedance then presented by the primary will be

$$1/1.024 \times 10^{-3} = 977 \; \Omega$$

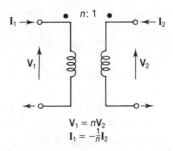

$$V_1 = nV_2$$
$$I_1 = -\frac{1}{n}I_2$$

Figure 8.6 The ideal transformer.

8.4 ◆ The ideal transformer

If it were possible to obtain perfect coupling, so that

$$M = (L_1 L_2)^{1/2}$$

then Equations 8.1 could be rearranged into the form

$$\mathbf{V}_1 = j\omega(L_1 L_2)^{1/2}[(L_1/L_2)^{1/2}\mathbf{I}_1 + \mathbf{I}_2]$$
$$\mathbf{V}_2 = j\omega L_2[(L_1/L_2)^{1/2}\mathbf{I}_1 + \mathbf{I}_2] \tag{8.5}$$

and we would then have

$$\frac{\mathbf{V}_2}{\mathbf{V}_1} = \left(\frac{L_1}{L_2}\right)^{1/2} \tag{8.6}$$

Thus the ratio of the two voltages would be fixed by coil geometry only: such a transformer would give a voltage ratio fixed independently of currents flowing. Although perfect coupling cannot be achieved the concept gives rise to an ideal component known as the **ideal transformer**. This may be imagined as a pair of coils of very high self-inductance with perfect coupling between them. If we denote $(L_1/L_2)^{1/2}$ by n, Equation 8.6 becomes

$$\mathbf{V}_1 = n\mathbf{V}_2 \tag{8.7}$$

If L_1 and L_2 become very large with the voltages $\mathbf{V}_1$, $\mathbf{V}_2$, finite then from Equations 8.5 we must have

$$\mathbf{I}_1 = -\frac{1}{n}\mathbf{I}_2 \tag{8.8}$$

Equations 8.7 and 8.8 define this ideal transformer, shown symbolically in Figure 8.6.

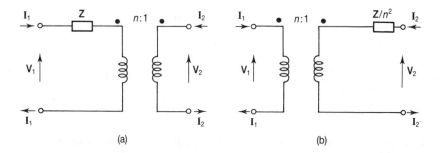

Figure 8.7 Equivalence for series connection with ideal transformer: (a) and (b) are equivalent as far as the external terminals are concerned.

8.4.1 ◆ Some circuit transformations

In dealing with circuits containing ideal transformers two circuit transformations are often found useful. The first is shown in Figure 8.7. The equations for the circuit of Figure 8.7(a) are

$$V_1 = ZI_1 + nV_2$$

$$I_2 = -nI_1$$

The first of these can be rearranged to give

$$V_2 = \frac{1}{n}V_1 + \frac{Z}{n^2}I_2$$

which also describes the situation in the circuit of Figure 8.7(b).

The second transformation is shown in Figure 8.8. For the circuit of Figure 8.8(a) we have

$$V_1 = nV_2$$

$$I_1 = \frac{1}{Z_1}V_1 - \frac{1}{n}I_2$$

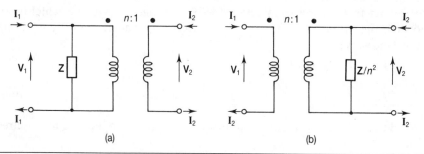

Figure 8.8 Equivalence for parallel connection with ideal transformer: (a) and (b) are equivalent as far as the external terminals are concerned.

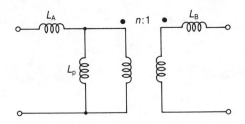

Figure 8.9 A general equivalent circuit for coupled inductors using inductors and ideal transformer.

Rearrangement gives

$$\mathbf{I}_2 = - n\mathbf{I}_1 + (n^2/\mathbf{Z})\mathbf{V}_2$$

which is also correct for the circuit of Figure 8.8(b).

8.5 ◆ Equivalent circuits for transformers

One equivalent circuit for a pair of coupled coils is that already shown in Figure 8.2. This circuit, however, is not satisfactory for many purposes. Transformers are often used primarily as voltage changers, a function which is basically that of an ideal transformer. It then becomes desirable to have an equivalent circuit which clearly expresses the ways in which the real transformer differs from an ideal. The circuit of Figure 8.2 does not fulfil this requirement and, as mentioned earlier, can involve unrealizable components. A more appropriate circuit can be derived. On physical grounds we can justify an equivalent circuit of the form shown in Figure 8.9. In this circuit the inductance L_A represents magnetic flux arising from the current in the primary winding which does not link any turn on the secondary winding; that flux linking the secondary is represented by the inductance L_P; the ideal transformer $n{:}1$ represents the ratio of turns on the primary to turns on the secondary. In a similar way the inductance L_B represents flux arising from current in the secondary winding which does not link the primary. If the circuits are to be equivalent then, by equating impedances, we must have

$$L_1 = L_A + L_P$$

$$L_2 = L_B + \frac{1}{n^2}L_P \tag{8.9}$$

In addition, the correct output:input voltage ratio on open circuit requires that

$$M = \frac{1}{n}L_P \tag{8.10}$$

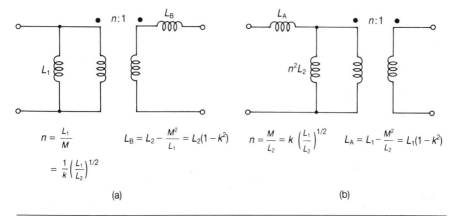

$$n = \frac{L_1}{M} \qquad L_B = L_2 - \frac{M^2}{L_1} = L_2(1-k^2) \qquad n = \frac{M}{L_2} = k\left(\frac{L_1}{L_2}\right)^{1/2} \qquad L_A = L_1 - \frac{M^2}{L_2} = L_1(1-k^2)$$

$$= \frac{1}{k}\left(\frac{L_1}{L_2}\right)^{1/2}$$

(a) (b)

Figure 8.10 Reduced forms of the circuit of Figure 8.9.

We thus have only three equations while there are four unknowns. This problem arises because measurement of voltage determines flux–turns and cannot separate flux from turns. Any sensible choice can be made: in Figure 8.10(a) is shown the circuit which results from taking L_A to be zero; in Figure 8.10(b), L_B has been taken to be zero. Both circuits are correct equivalents and no external measurements could distinguish between them. From the formulae given in the figures it can be seen that as the coupling between the coils becomes closer, k tending to unity, both circuits reduce to the ideal transformer in parallel with an inductance. In Figure 8.11(a) is shown the result of applying the transformation of Figure 8.7 to the circuit of Figure 8.10(a). For comparison, Figure 8.11(b) shows the deceptively similar circuit of Figure 8.10(b): in these circuits the order of the series and shunt components is all important.

> **EXAMPLE 8.4** Derive an equivalent circuit in the form of Figure 8.10(a) for the coupled coils of Example 8.2, assuming that L_1 represents the coil of higher inductance.
>   Since $M = k(L_1L_2)^{1/2}$ the transformer ratio is given by
>
> $$n = \frac{L_1}{M} = \frac{1}{k}\left(\frac{L_1}{L_2}\right)^{1/2} = 4.00$$
>
> $$L_B = (1 - k^2)L_2 = 0.376 \times 100 = 37.6 \,\mu H$$
>
> $$L_P = 1 \,mH$$

8.5.1 ◆ Close coupling

Many uses require a transformer with close coupling. To achieve this, the coils are wound on a core of magnetic material forming a closed magnetic circuit. When the coupling coefficient k is close to unity the values of

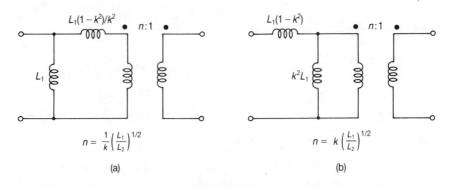

$$n = \frac{1}{k}\left(\frac{L_1}{L_2}\right)^{1/2}$$

(a)

$$n = k\left(\frac{L_1}{L_2}\right)^{1/2}$$

(b)

Figure 8.11 Alternative forms of equivalent circuit.

comparable components in the two circuits of Figure 8.11 become closely equal: they will be seen to differ only by a factor of k^2, and for all practical purposes may be assumed to be the same. The fact that the position of the shunt arm does not seem to matter is because the reactance of the series arm is in this case much smaller than that of the shunt arm. Transformers can be designed for instrumentation in which the coupling coefficient is unity to within 1 part in 10^5.

8.5.2 ◆ Representation of losses

Loss in a transformer arises from both resistance of the windings and magnetic loss in the core. The former can be directly represented as in Figures 8.12(a) and 8.12(b). If coupling is close, as considered in Section 8.5.1, and the two resistances are of the same small order of magnitude as the series reactance, then with negligible error they may be combined, as shown in Figure 8.12(c). This would not be correct if the resistances were large. Losses in the magnetic core may be represented by a resistance in parallel with the shunt inductance, as shown in Figure 8.12(d).

8.6 ◆ AC power transformers

Power transformers are usually close-coupled, and the circuit given in Figure 8.13 is a suitable equivalent. The parameters in this circuit are often determined by carrying out two standard tests, known as the open- and short-circuit tests. In the open-circuit test the working voltage is applied to one winding, leaving the other unconnected. Measurement is made of the supply voltage, current, power consumed and voltage appearing across the second winding. This test is conducted at rated voltage in order to ensure that the magnetic state of the core and the consequent magnetic loss will be the same as in normal operation. For the short-circuit test the terminals

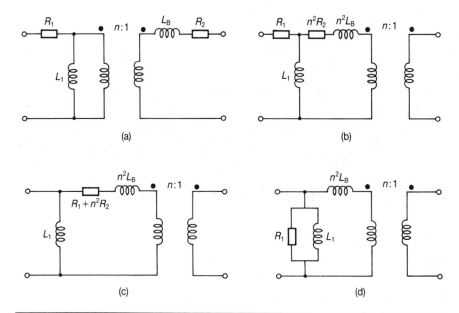

Figure 8.12 Incorporation of losses into the equivalent circuit: (a) winding resistance; (b) reduced form for (a); (c) approximation valid if $R_1 \ll \omega L_1$; (d) allowance for iron losses.

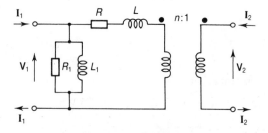

Figure 8.13 Equivalent circuit for close-coupled power transformer.

of one winding are joined together and a suitable voltage applied to the other winding. Voltage, current and power are measured. This test is usually done with approximately full-load current.

8.6.1 ◆ Analysis of open- and short-circuit tests

- *Open-circuit test*: Referring to the circuit of Figure 8.13 and assuming that the supply is connected to end 1, for this test $I_2 = 0$. Hence

$$|\mathbf{I}_1/\mathbf{V}_1|^2 = 1/R_1^2 + 1/X_1^2$$

$$n = |\mathbf{V}_1/\mathbf{V}_2|$$

$$P_1 = |\mathbf{V}_1|^2/R_1$$

$$X_1 = \omega L_1$$

From these equations n, R_1, X_1 and hence L_1 can be found.

- *Short-circuit test*: In this test, assuming that the supply is still connected to end 1, $\mathbf{V}_2 = 0$. The impedance seen by the supply therefore consists of the series combination of R and X in parallel with R_1, X_1. It has been argued in Section 8.5.1 that the impedance of R_1 and X_1 is much greater than that of the series arm, so that in this test the supply may be assumed to be loaded only by R in series with X. Hence

$$|\mathbf{V}_1/\mathbf{I}_1|^2 = R^2 + X^2$$

$$P_1 = R|\mathbf{I}_1|^2$$

$$X = \omega L$$

From these equations R and L can be found.

It is to be noted that the tests need not, and often are not, carried out from the same side. For very high-voltage transformers it is usual to supply the low-voltage side in the open-circuit test, and the high-voltage side in the short-circuit test. The interpretation of results is similar.

EXAMPLE 8.5 A low-voltage, 50 Hz, transformer on test gave the following results:

open circuit: input 240 V, 2.6 A, 384 W, output 110 V

short circuit: input 23.3 V, 20 A, 225 W

Both tests were carried out on the same winding. Derive the components of an equivalent circuit of the form of Figure 8.13.

Since the open-circuit test is conducted at rated voltage, the tests have both been carried out on the 240 V side. From the open-circuit test we have

$$n = 240/110$$

$$240^2 R_1 = 384; \quad R_1 = 150 \ \Omega$$

$$1/R_1^2 = 1/X_1^2 = (2.6/240)^2; \quad X_1 = 117 \ \Omega$$

From the short-circuit test we have

$$20^2 R = 225, \quad R = 0.56 \ \Omega$$

$$(23.3/20)^2 = R^2 + X^2, \quad X = 1.02 \ \Omega$$

Thus

$$n = 2.18$$
$$L_1 = 0.372 \text{ H}$$
$$L = 3.2 \text{ mH}$$
$$R_1 = 150 \text{ }\Omega$$
$$R = 0.56 \text{ }\Omega$$

8.6.2 ◆ Regulation and efficiency

The main use of transformers in AC power systems is that of changing the voltage at which power is delivered. In such systems it is desirable the voltage supply changes only by a small amount when a load is connected. The regulation is defined in terms of the voltage at no load and that at full load by

$$\text{regulation} = \frac{\text{no-load voltage} - \text{full-load voltage}}{\text{no-load voltage}}$$

It is usually expressed as a percentage.

A second criterion of performance is the efficiency, defined as

$$\text{efficiency} = \frac{\text{output power}}{\text{input power}}$$

The power loss in a transformer consists of the core loss, measured on the open-circuit test, and the ohmic loss in the windings, represented by the series resistance in the equivalent circuit. The core loss does not alter appreciably with the current taken by the load whereas the ohmic loss increases as the square of the current. The maximum loss is therefore known once the full-load current is specified: for this reason transformers are rated in volt–amps, being the product of rated voltage and full-load current on the same side. Thus if in Example 8.5 the full-load current on the low-voltage side were given at 25 A, the transformer would be rated at $110 \times 25 = 2.75 \text{ kV A}$.

EXAMPLE 8.6 The transformer of Example 8.5 is used to supply a resistive load with a current of 50 A on the low-voltage side. The primary is fed at 240 V. Determine the efficiency and the regulation.

It is convenient in this problem to transfer the series impedance to the low-voltage side, giving the equivalent circuit of Figure 8.14(a), in which

$$R_2 = 0.56/2.18^2 = 0.118 \text{ }\Omega$$

$$X_2 = 1.02/2.18^2 = 0.215 \text{ }\Omega$$

Since the load is resistive, the current will be in phase with the voltage $\mathbf{V}_3$ across the load: we therefore take this to be the reference phasor, leaving the relative phase of the supply voltage to be determined. The phasor

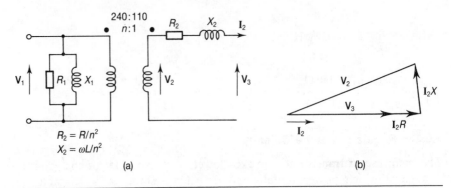

Figure 8.14 Example 8.6: (a) equivalent circuit; (b) phasor diagram.

diagram is shown in Figure 8.14(b). Using the notation in the diagram, we have

$$\mathbf{V}_2 = \mathbf{V}_3 + (0.118 + j0.215)\mathbf{I}_2$$

in which $|\mathbf{V}_2| = 110$, $\mathbf{V}_3 = V_3 \angle 0$, $\mathbf{I}_2 = 50\angle 0$. Hence

$$(110)^2 = (V_3 + 0.118 \times 50)^2 + (0.215 \times 50)^2$$

$$V_3 + 5.9 = \sqrt{(12\,100 - 115.6)} = 109.5$$

$$V_3 = 103.6 \text{ V}$$

The regulation at this current is therefore 5.8%. The losses consist of 384 W in the core together with $0.118 \times 50^2 = 295$ W in the series resistor. The power output is $103.6 \times 50 = 5180$ W. Hence

$$\text{efficiency} = 5180/(5180 + 384 + 295) = 88\%$$

EXAMPLE 8.7 The transformer in Example 8.6 is supplied under the same conditions as in that example but delivers a current of 50 A to a load having a lagging angle of 20°. Determine the new values of regulation and efficiency.

The load current is no longer in phase with the load voltage, so that, keeping the load voltage as phasor reference, we have

$$\mathbf{I}_2 = 50[\cos(20°) - j\sin(20°)] = 50(0.940 - j0.342)$$

Hence

$$\mathbf{V}_2 = \mathbf{V}_3 + 50(0.118 + j0.215)(0.940 - j0.342)$$

$$= \mathbf{V}_3 + 9.2 + j8.1$$

Since $|\mathbf{V}_2| = 110$, we finally have

$$V_3 + 9.2 = 109.7$$

$$V_3 = 100.5$$

The regulation is thus found to be $9.5/110 = 8.5\%$. The losses in the new condition are the same as before, but the output power is now $100.5 \times 50 \times 0.940 = 4724$ W. The efficiency then becomes

$$4724/(4724 + 384 + 295) = 87\%$$

The change in load power factor results in worse performance, but the regulation is affected more than the efficiency.

8.7 ◆ Summary

Connecting the coils of a coupled pair in series gives a combined inductance

$$L_1 + L_2 \pm 2M$$

The impedance seen at the primary when the secondary is loaded with an impedance $\mathbf{Z}_2$ is

$$\mathbf{Z}_1 = j\omega L_1 + \frac{(\omega M)^2}{\mathbf{Z}_2 + j\omega L_2}$$

The ideal transformer, Figure 8.6, is defined by the relations

$$\mathbf{V}_1 = n\mathbf{V}_2$$

$$\mathbf{I}_1 = -\frac{1}{n}\mathbf{I}_2$$

Impedances associated with an ideal transformer in both parallel and series configurations can be transferred from the primary side to the secondary side by allowing a multiplying factor of n^2: see Figures 8.7 and 8.8.

Equivalent circuits for coupled coils have been derived; see Figures 8.2, 8.9, 8.10 and 8.11.

A circuit representation of losses is shown in Figure 8.12

An equivalent circuit for an AC power transformer is shown in Figure 8.13.

PROBLEMS

8.1 The primary of a pair of coupled inductors has inductance 1 mH, the secondary 0.4 mH. What is the maximum value of the mutual inductance? The actual mutual inductance is measured at 0.5 mH. What is the coupling coefficient?

8.2 Assuming that the inductors of Problem 8.1 do not have any resistance, calculate the inductance seen at the primary when the secondary is short circuited.

8.3 In the equivalent circuit of Figure 8.2 the inductance M can be either positive or negative. Show that the impedance presented at the primary terminals when an impedance $\mathbf{Z}$ is connected across the secondary terminals does not depend on the sign of M.

8.4 Two identical inductors form primary and secondary of a coupled pair. Measured on an AC bridge, the inductance of the primary is $80\,\mu\text{H}$ when the secondary is on open circuit and $5\,\mu\text{H}$ when the secondary is short circuited. Calculate the mutual inductance and coupling coefficient.

8.5 The primary of a pair of coupled inductors has inductance $40\,\text{mH}$, the secondary $10\,\text{mH}$. The coupling coefficient is 0.7. Calculate the impedance presented at the primary, at a frequency of $2\,\text{kHz}$, when the component connected across the secondary is (a) a $100\,\Omega$ resistor and (b) a $0.22\,\mu\text{F}$ capacitor.

8.6 If in Problem 8.5(b) the components are lossless (that is, they have no associated resistance) show that at some frequency the impedance presented at the primary becomes (a) indefinitely large and (b) zero. Determine the values of these frequencies.

8.7 Design calculations for a high-frequency transformer make use of the equivalent circuit of Figure 8.10(a). It is concluded that values $n = 5$, $L_1 = 100\,\mu\text{H}$, $L_B = 6\,\mu\text{H}$ are required. Determine the values of the primary and secondary inductors and the coupling coefficient for a suitable transformer.

8.8 Transformers are often used to ensure maximum power transfer from a source to a load. Show that the power transferred from a source of resistive internal impedance R_1 to a resistive load R_2 can be maximized by use an ideal transformer of ratio $n = (R_1/R_2)^{1/2}$.

8.9 In the situation described in Problem 8.8, the source impedance is $600\,\Omega$ and the load impedance $120\,\Omega$. A transformer is designed for which the primary inductance is $150\,\text{mH}$, the secondary inductance $30\,\text{mH}$ and the coupling coefficient 0.98. The source is set to $10\,\text{V}$ on open circuit at a frequency of $2\,\text{kHz}$. Calculate the voltage appearing across the load and the ratio of power actually transferred to the maximum possible.

The following problems relate to closely coupled transformers for which the equivalent circuit of Figure 8.13 applies. In these problems the assumption of Section 8.6.1, that the current in the shunt arm can be ignored when calculating volt drop, should be made.

8.10 A small transformer is used to supply a $12\,\text{V}$, $10\,\text{A}$ projector lamp from $240\,\text{V}$ mains. When the lamp is not connected the ouput gives $12\,\text{V}$. The resistance of the primary winding is $10\,\Omega$, the resistance of the secondary winding is $0.05\,\Omega$ and the leakage reactance referred to the primary side is $15\,\Omega$. In order to ensure that the lamp is run at its rated voltage, the primary is supplied from a source variable up to $270\,\text{V}$. Estimate the input voltage which will be required.

8.11   A 50 Hz power transformer has an open-circuit voltage ratio of 240:110. The values of the elements in the equivalent circuit of Figure 8.13, referred to the high-voltage side, are such that

$$R_1 = 150 \ \Omega, \ \omega L_1 = 120 \ \Omega, \ R = 0.6 \ \Omega, \ \omega L = 1.0 \ \Omega$$

The primary is connected to the 240 V mains and a load connected to the low-voltage side takes a current of 25 A at a power factor of 0.8 lagging. Estimate the output voltage, the power lost in the transformer and the supply current.

8.12   An isolating transformer provides 120 V from the 240 V mains. Tests gave the following results:

	Input voltage (V)	*Input current* (A)	*Input power* (W)	*Output volts* (V)
Open circuit on secondary	240	0.40	45	120
Short circuit on secondary	14.7	6.0	46	–

Deduce an equivalent circuit in the form of Figure 8.13. Estimate the secondary voltage when the primary is supplied at 240 V and the current drawn from the secondary is 10 A at a power factor of 0.75 lagging. Estimate also the efficiency of the transformer under these conditions.

9 ◆ *Transient Analysis 1*

In this chapter the problem of determining the response of a network to an applied excitation of any form is investigated. Equations are set up and solved in various simple cases. It is shown that solutions can be obtained by straightforward numerical methods. Ways in which the general problem can be tackled are outlined.

9.1 ◆ Introduction

Previous chapters have dealt with analysis under two special forms of excitation: DC, time-independent, and AC, sinusoidal waveforms. In general excitation may involve many other forms of variation in time: this chapter considers methods of determining the response to any prescribed waveform. Since the type of excitation applied to a circuit can be completely arbitrary, it is clear that a discussion of transient response will be about methods rather than response to particular excitation waveforms. However, certain situations of a general nature can be envisaged. For example, when a particular network has been excited and the excitation then switched off, practical experience shows the network returns eventually to a quiescent state: in what way does this return proceed? Such behaviour is said to refer to the **natural response** of the network, and characterizes the network rather than the excitation. Another common situation is switching on: an excitation such as DC or sinusoidal AC is applied to a quiescent network. How is the steady state reached? Such a situation is typified by the 'unit step' or 'step function' waveform introduced in Section 2.2.4. Less obvious perhaps, but of frequent occurrence, is the 'impulse' type of excitation in which an intense signal of very short duration is applied to the network, after which a particular form of the natural response takes place. This **impulse response** is in fact characteristic of the network. It can be shown that knowledge of the impulse response enables the response to any form of excitation to be calculated. In the following sections some particular networks will be studied, the equations set up and the response in situations such as those just discussed investigated.

9.2 ◆ Some first-order circuits

The simplest circuits which have appeared in foregoing chapters are the series $R-C$ circuit, Section 5.2, and the series $L-R$ circuit, Section 5.3. These will now be considered from the point of view of excitation under transient conditions.

9.2.1 ◆ The series *R–C* circuit

This circuit, shown in Figure 9.1, has been encountered in Section 5.2, where the excitation $e(t)$ was a waveform sinusoidal in time. In this analysis we will take the instantaneous voltage across the capacitor, $v(t)$, as the unknown. Since the charge on the capacitor is Cv and the current $i(t)$ is building up that charge (Section 1.3.3), we have

$$i(t) = \frac{\mathrm{d}}{\mathrm{d}t}(Cv) = C\frac{\mathrm{d}v}{\mathrm{d}t} \tag{9.1}$$

We also have

$$v + iR = e$$

Hence

$$RC\frac{\mathrm{d}v}{\mathrm{d}t} + v = e \tag{9.2}$$

In mathematical terms this is a linear, constant coefficient, first-order differential equation, reflecting the fact that v and e occur only to the first power, that RC is constant and that only the first derivative of v is involved.

9.2.2 ◆ The series *L–R* circuit

The circuit is shown in Figure 9.2. Using the notation in the figure we have

$$e = v_\mathrm{L} + v$$

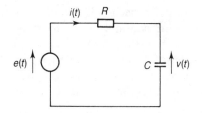

Figure 9.1 The series R–C circuit: variables for transient analysis.

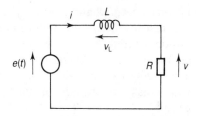

Figure 9.2 The series R–L circuit: variables for transient analysis.

$$v = iR$$

$$v_{L} = L\frac{di}{dt}$$

Hence

$$e = L\frac{di}{dt} + Ri \qquad\qquad (9.3)$$

As in Section 9.2.1, this is a linear, constant coefficient differential equation of the first order. The physical reason for both these equations being of the first order is that in each circuit only one store of energy is present, in one case a store of electrostatic energy, in the other a store of magnetic energy. Some cases in which solutions to these differential equations can be simply obtained will now be considered.

9.2.3 ◆ The natural response

As discussed in Section 9.1, natural response refers to the network behaviour after excitation has been removed. In this case we therefore place the excitation function $e(t)$ in Equation 9.2 equal to zero. The equation then becomes

$$RC\frac{dv}{dt} + v = 0$$

Rearranging, we have

$$\frac{dv}{dt} = -\frac{1}{RC}v$$

Hence

$$v = A\exp(-t/RC) \qquad\qquad (9.4)$$

in which A is a constant. This solution reflects the fact that after $e(t)$

becomes zero whatever energy is left stored in the capacitor is dissipated in the resistor at a rate determined by the circuit configuration: the magnitude is determined by the past excitation, but the form of decay is characteristic of the circuit. The natural response of the circuit is therefore a decaying exponential. The rate of decay depends on the parameter RC, which has dimensions of time and is termed the **time constant** of the circuit.

A similar result is obtained for Equation 9.3. Putting $e(t)$ equal to zero in that equation, we have

$$L\frac{di}{dt} + Ri = 0$$

whence

$$\frac{di}{dt} = -\frac{R}{L}i$$

and

$$i(t) = A\exp(-Rt/L) \tag{9.5}$$

In this result, A is again a constant which can have any suitable value. The natural response of the $R-L$ circuit is also a decaying exponential. The rate of decay depends on the quantity L/R, which is termed the time constant of the L/R circuit.

EXAMPLE 9.1 In a DC power supply a voltage of 250 V appears across a capacitance of 8 μF, as shown in Figure 9.3. When the supply is switched off, by opening the switch S, this capacitor is left isolated except for a parallel resistance of 1 MΩ. Determine the voltage on the capacitor after 30 s has elapsed.

The time constant in this case is $1 \times 10^6 \times 8 \times 10^{-6} = 8$ s. In Equation 9.4 the constant A is found by requiring that the equation gives the correct voltage at the time the supply is switched off. This voltage must be 250 V, since no time has elapsed for the charge on the capacitor to drain away. Hence A must have the value 250, and

$$v(t) = 250\exp(-t/8)$$

After 30 s the voltage is found to be 5.9 V.

EXAMPLE 9.2 The parallel combination of a 5 H inductance and 6 Ω resistance forms part of a circuit carrying a steady current of 2 A, as shown in Figure 9.4. The current flow from the battery is interrupted by opening the switch S. Calculate the time taken for the current in the inductor to fall to 50 mA. Calculate also the maximum voltage appearing across the resistor.

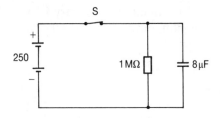

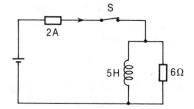

Figure 9.3 Circuit for Example 9.1.	**Figure 9.4** Circuit for Example 9.2.

After the battery is disconnected the circuit consists of just L and R in parallel, which is the same as that of Figure 9.2 with $e(t)$ equal to zero. Hence Equation 9.5 applies. Assuming that the instant of disconnection is taken as the time origin, the constant A in Equation 9.5 must have the value 2 for the initial value of current to be correct. The time constant L/R is equal to $5/6$ s, giving

$$i(t) = 2\exp(-6t/5)$$

For $i = 0.05$ we find $t = 3.1$ s. Since the current through the inductance can only return through the resistance, the voltage across the resistance is a maximum at the moment of disconnection. The value is therefore $2 \times 6 = 12$ V.

9.3 ◆ Continuity

In Example 9.1 it was assumed that the voltage, and hence charge, on the capacitor was unchanged by the act of opening the switch in the circuit of Figure 9.3. The validity of this assumption was based on the conservation of charge and the fact that the change in circuit configuration did not cause charge to drain away in the time of opening the switch. Alternatively, it might be said that there was no mechanism whereby the energy stored in the capacitor could be dissipated in the short switching interval. In the absence of such a mechanism, which is the normal state of affairs, the situations before and after switching are connected by continuity of the voltage across the capacitor.

In Example 9.2 a similar assumption was made about the current in the inductor: it was assumed that the current immediately after the opening of the switch had the same value as it had immediately before. Once again, the switching action does not provide a mechanism by which energy stored in the inductor in magnetic form can be dissipated. Alternatively, the high rate of change of current caused by a sudden jump would have to be accompanied by a high voltage across the inductor, for which the circuit makes no provision. In this circuit, the connection between the states before and after switching is provided by continuity of current in the inductance.

Application of these two principles of continuity is necessary in any case where the excitation function is discontinuous, as it must be, for example, when a 'starting situation' is considered.

9.4 ◆ Response to a step function

We consider in this section the response to a step function in the cases of the $R-C$ and $L-R$ circuits for which the equations have been obtained in Sections 9.21 and 9.2.2.

9.4.1 ◆ Step response of an $R-C$ circuit

We consider the case when the excitation is a step of magnitude E, as shown in Figure 9.5(a), and we firstly assume that the capacitor is initially uncharged. From inspection of the circuit the capacitor voltage must eventually reach the step voltage E, and $v(t) = E$ is seen to be a possible solution of Equation 9.2. It does not, however, fit near $t = 0$ when, applying the principle of continuity, the capacitor voltage must be zero. The natural response corresponds to zero excitation so that, since the equation is linear, the sum

$$v(t) = E + A \exp(-t/RC)$$

is also a solution of Equation 9.2 with $e = E$. We can now choose the constant A to make $v(t)$ zero at time zero:

$$0 = E + A$$

Hence the required solution is

$$v(t) = E[1 - \exp(-t/RC)] \tag{9.6}$$

This is shown in Figure 9.5(b). It may be noted that for times much smaller then the time constant, RC, the exponential function can be replaced by

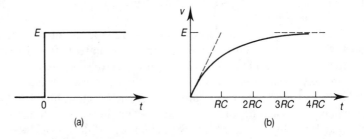

(a)   (b)

Figure 9.5 Step response of $R-C$ circuit (Equation 9.6, Section 9.4.1): (a) input; (b) voltage across capacitor.

$$\exp\left(-t/RC\right) \approx 1 - t/RC$$

in which case Equation 9.6 becomes

$$v(t) \approx Et/RC \tag{9.7}$$

This approximation is indicated by the pecked line in Figure 9.5b. In physical terms, this approximate expression is valid while the voltage across the capacitor is much smaller than the drive voltage E: when this is the case the current is approximately constant at the value E/R.

It might be that, because of previous excitation, at the moment the step was applied the capacitor voltage was other than zero, say v_0. Since the equation is linear we may superimpose the solution just obtained with the natural response corresponding to the decay of the initial voltage. This gives

$$v(t) = E[1 - \exp\left(-t/RC\right)] + v_0 \exp\left(-t/RC\right) \tag{9.8}$$

EXAMPLE 9.3 A capacitance of 25 μF is charged from a battery of 50 V in series with a resistance of 10 kΩ. Assuming that the capacitor is initially uncharged, calculate the time required for the capacitor to reach 90% of its final voltage.

In this case the time constant RC is $25 \times 10^{-6} \times 10^4 = 0.25$ s. To reach $0.9E$ we must have, from Equation 9.6,

$$E \exp\left(-t/RC\right) = 0.1E$$

Hence

$$t = RC \ln\left(10\right) = 2.3RC = 0.58 \text{ s}$$

9.4.2 ♦ Step response of an *L–R* circuit

If in Equation 9.3 the voltage $e(t)$ is constant and equal to E, the problem is similar to that just solved: the final current will be E/R so that we look for a solution in the form

$$i(t) = \frac{E}{R} + A \exp\left(-\frac{Rt}{L}\right)$$

Assuming that no current flows before the step is applied, continuity of current in the inductance requires

$$i(0) = 0 = \frac{E}{R} + A$$

Hence $A = -L/R$ and

$$i(t) = \frac{E}{R}\left[1 - \exp\left(-\frac{Rt}{L}\right)\right] \tag{9.9}$$

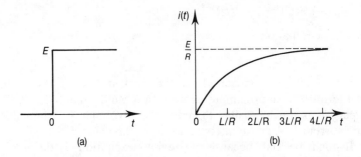

Figure 9.6 Step response of *R–L* circuit (Equation 9.9, Section 9.4.2): (a) input; (b) current.

The graphical form of this solution is shown in Figure 9.6: it has the same general form as the response for the *R–C* circuit. If at the time of application of the step the current were i_0, the solution would then be

$$i(t) = \frac{E}{R}\left[1 - \exp\left(-\frac{Rt}{L}\right)\right] + i_0\exp\left(-\frac{Rt}{L}\right) \tag{9.10}$$

EXAMPLE 9.4 An induction coil can be represented by an inductance of 2 H in parallel with a resistance of 10 kΩ. It is connected to a 12 V battery through a resistance of 10 Ω and a rotary circuit breaker. Contact is made in the circuit breaker for a period of 0.1 s. Determine the maximum voltage appearing across the resistance.

The circuit is shown in Figure 9.7(a). When contact is made, that part of the circuit comprising the battery and the two resistances can be replaced by its Thévenin equivalent, as shown in Figure 9.7(b). In this form Equation 9.9 can be applied to work out the current in the inductance:

$$i(t) = (E'/R')[1 - \exp(-R't/L)]$$

With the values given, to an accuracy within 0.1% we may take $E' = 12$, $R' = 10$. After 0.1 s, just prior to the moment of interruption, the current

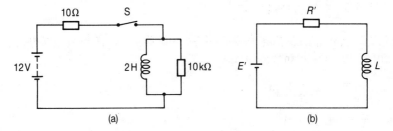

Figure 9.7 Example 9.4: (a) circuit; (b) equivalent circuit after closure of switch.

through the inductance is found to be 0.47 A. Immediately after the interruption, this current will flow through the 10 Ω resistance, giving a voltage of 4.7 kV. Subsequently, the voltage will decay from this peak value according to Equation 9.5, with a time constant of $2/10\,000 = 0.2$ ms.

9.5 ◆ Impulse response

The concept of an impulse was introduced in Section 9.1. The basic characteristic of an impulse is its short duration. As an example of a possible waveform, consider the excitation waveform shown in Figure 9.8(a). Applied to the linear R–C circuit of Figure 9.1 we can regard this waveform as the sum of the two step voltages shown in Figure 9.8(b). We can then superimpose the solutions corresponding to the two step waveforms. The complete solution then takes the form

$$v(t) = v_1(t) + v_2(t)$$

in which

$$\begin{aligned} v_1(t) &= 0 & t < 0 \\ &= E[1 - \exp(-t/RC)] & t > 0 \end{aligned}$$

and

$$\begin{aligned} v_2(t) &= 0 & t < \tau \\ &= -E\{1 - \exp[-(t - \tau)/RC]\} & t > \tau \end{aligned}$$

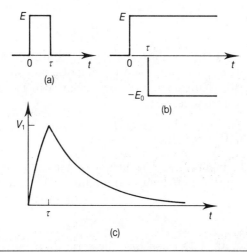

(a)

(b)

(c)

Figure 9.8 Response of R–C circuit to pulse of length τ: (a) input pulse; (b) decomposition into two steps; (c) voltage across capacitor.

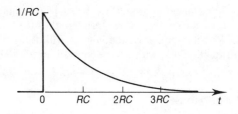

Figure 9.9 Impulse response: limiting form for Figure 9.8(c) when $\tau \to 0$, $E\tau = 1$.

Hence the response to the excitation of Figure 9.8(a) is given by

$$v(t) = E[1 - \exp(-t/RC)] \qquad\qquad 0 < t < \tau$$

$$= E\exp(-t/RC)[\exp(\tau/RC) - 1] \qquad t > \tau$$

This is illustrated in Figure 9.8(c).

Consider now the case when $\tau \ll RC$. The exponential in the square brackets can be replaced by its approximate expression, giving

$$\exp(\tau/RC) - 1 \approx \tau/RC$$

For $t > \tau$ we then have

$$v(t) = E\frac{\tau}{RC}\exp(-t/RC) \tag{9.11}$$

Since in this period the excitation is zero, the response has the form of the natural response. Clearly, if τ is small the response will only be significant if E is large. If we choose to increase E continually as τ becomes smaller in a way such that

$$E\tau = 1$$

then in the limit $\tau \to 0$ the waveform is said to constitute a **unit impulse**. The response to this impulse is termed the **impulse response**. In this book the impulse response, characteristic of the circuit, will be denoted by the symbol $h(t)$. In the present case

$$h(t) = \frac{1}{RC}\exp\left(-\frac{t}{RC}\right) \tag{9.12}$$

This is illustrated in Figure 9.9. The height of the step can be physically interpreted by writing the coefficient in Equation 9.11 in the form

$$\frac{E}{R}\tau\frac{1}{C}$$

In the limit considered, $E \gg v$, so that the term E/R corresponds to the current charging the capacitor during the impulse. The amount of charge transferred is $E\tau/R$, producing a voltage $E\tau/RC$ across the capacitor. The special shape of the excitation of Figure 9.8(a) has been chosen so that an exact solution can be used, but further investigation shows that the exact shape of the waveform is unimportant: it is only the total charge transferred that matters. It will be noticed that with this excitation the voltage on the capacitor is no longer continuous before and after the impulse: this is not a violation of the continuity principle of Section 9.3 since with the impulse a finite amount of energy can be transferred in an infinitesimal time interval.

A voltage impulse applied to the series L–R circuit of Figure 9.2 can be treated in the same way. Since Equation 9.3 is mathematically of the same form as Equation 9.2 an equation of the form of Equation 9.12 will result:

$$h(t) = \frac{1}{L} \exp\left(-\frac{Rt}{L}\right) \tag{9.13}$$

This time $h(t)$ represents the current in the circuit: the dimensions of the expression will be found to be correct if it is remembered that the '1' represents a unit voltage impulse with dimensions of volts $\times$ time. In this case the momentarily infinite voltage causes the current in the inductor to jump to a finite value.

9.6 ◆ Sine wave excitation

Chapter 4 was concerned with the special case of sinusoidal excitation. The validity of the treatment depended on the assumption that in a circuit such as that of Figure 9.1 excitation by a sinusoidal voltage was consistent with a sinusoidal current. We consider here the situation when the exciting waveform starts at a definite time, typical of switching on, and show that the current waveform eventually does becomes sinusoidal. In this case we have

$$e(t) = 0 \qquad\qquad t < 0$$
$$= \hat{E} \cos \omega t \qquad t > 0$$

Equation 9.2 then becomes

$$RC\frac{dv}{dt} + v = \hat{E} \cos \omega t \tag{9.14}$$

Phasor analysis, as carried out in Section 5.2.2, gives

$$\mathbf{V}_c = \frac{\mathbf{E}}{[1 + (\omega CR)^2]^{1/2}} \exp(-j\theta)$$

in which $\mathbf{E} = \hat{E}/\sqrt{2}$ and $\theta = \tan^{-1} \omega CR$. This corresponds to the physical voltage

$$v_c(t) = \frac{\hat{E}}{[1 + (\omega CR)^2]^{1/2}} \cos(\omega t - \theta) \tag{9.15}$$

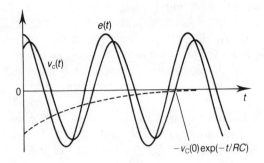

Figure 9.10 The R–C circuit with excitation $\hat{E}\cos(\omega t)$ applied at $t = 0$, showing steady-state and transient components of capacitor voltage.

Substitution will confirm that this function satisfies Equation 9.14. In order to satisfy the condition $v = 0$ at $t = 0$ we add to this function a term representing the natural response:

$$v(t) = v_c(t) + A\exp(-t/RC) \tag{9.16}$$

The initial condition is satisfied by making

$$A = -v_c(0) = -\frac{\hat{E}}{[1 + (\omega CR)^2]^{1/2}}\cos\theta$$

The resulting solution is illustrated in Figure 9.10. Any other initial value of v can be accommodated by choice of A, but in all cases the natural response term decays to zero as time progresses, leaving the sinusoidal waveform given by the phasor analysis.

9.7 ◆ The series L–C–R circuit

The circuit, which has been studied before in Section 5.4, is shown in Figure 9.11. We note that the principle of continuity will require both i and v to be continuous, providing that $e(t)$ is finite. We have

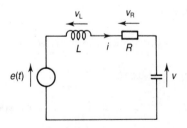

Figure 9.11 The series L–C–R circuit: variables for transient analysis.

$$e = v_\text{L} + v_\text{R} + v$$

$$v_\text{L} = L\frac{di}{dt}$$

$$i = C\frac{dv}{dt}$$ (9.17)

$$v_\text{R} = iR$$

Hence

$$LC\frac{d^2v}{dt^2} + RC\frac{dv}{dt} + v = e$$ (9.18)

Like Equation 9.2, this is a linear, constant coefficient differential equation but is of the second order.

9.7.1 ◆ The natural response

To investigate the natural response we put $e = 0$ in Equation 9.18:

$$LC\frac{d^2v}{dt^2} + RC\frac{dv}{dt} + v = 0$$ (9.19)

For such a linear, constant coefficient equation, the natural response can always be found by trial of the function $\exp(mt)$. Substituting the expression $v = A\exp(mt)$ in Equation 9.19, we find that the equality is satisfied at all times if

$$m^2 LC + mRC + 1 = 0$$ (9.20)

This quadratic equation yields two values for m given by

$$m = \frac{1}{2LC}\{- RC \pm [(RC)^2 - 4LC]^{1/2}\}$$

Three cases are to be distinguished according as R^2 is greater than, less than or equal to $4L/C$.

(i) $R^2 > 4L/C$: In this case the roots given by Equation 9.20 are both real, and are both negative: $m = -\alpha_1, -\alpha_2$ where

$$\alpha_1 = \frac{R}{2L} - \left[\left(\frac{R}{2L}\right)^2 - \frac{1}{LC}\right]^{1/2}$$

$$\alpha_2 = \frac{R}{2L} + \left[\left(\frac{R}{2L}\right)^2 - \frac{1}{LC}\right]^{1/2}$$ (9.21)

The natural response will be a linear combination of the expressions $\exp(-\alpha_1 t)$ and $\exp(-\alpha_2 t)$:

$$v(t) = A\exp(-\alpha_1 t) + B\exp(-\alpha_2 t)$$ (9.22)

(ii) $R^2 < 4L/C$: In this case the roots are complex, forming a conjugate complex pair, $m = -\alpha \pm j\beta$, with the real part negative:

$$\alpha = \frac{R}{2L}; \; \beta = \left[\frac{1}{LC} - \left(\frac{R}{2L}\right)^2\right]^{1/2} \tag{9.23}$$

The solution of Equation 9.19 is a linear combination of the functions $\exp[(-\alpha \pm j\beta)t]$, but since a real expression is required for the natural response we use the relation

$$\exp(j\beta t) = \cos(\beta t) + j\sin(\beta t)$$

To arrive at

$$v(t) = \exp(-\alpha t)[A\cos(\beta t) + B\sin(\beta t)] \tag{9.24}$$

(iii) $R^2 = 4L/C$: As may be checked by substitution, in this special case the natural response takes the form

$$v(t) = \exp(-\alpha t)(A + Bt) \tag{9.25}$$

It will be observed that unless the circuit is completely lossless, $R \equiv 0$, the response in all cases dies away exponentially with time.

To find the actual expression for the voltage under any given initial condition, we have to find the constants A and B in terms of the values of capacitor voltage and inductor current. The procedure is similar in any of the three cases considered above. Detailed algebra will be presented for case (ii), Equation 9.24. Assuming a capacitor voltage v_0 and inductor current i_0, from Equation 9.17 we have to satisfy at $t = 0$

$$v = v_0, \; i = C(\mathrm{d}v/\mathrm{d}t) = i_0$$

The first of these conditions, from Equation 9.24, requires

$$A = v_0 \tag{9.26}$$

Differentiating Equation 9.24 we find

$$\frac{\mathrm{d}v}{\mathrm{d}t} = \exp(-\alpha t)\left[(\beta B - \alpha A)\cos(\beta t) - (\alpha B + \beta A)\sin(\beta t)\right] \tag{9.27}$$

Hence the second condition requires

$$i_0 = C(\beta B - \alpha A)$$

giving

$$B = (i_0 + C\alpha v_0)/C\beta \tag{9.28}$$

EXAMPLE 9.5 A coil has the equivalent circuit of inductance 2 H in series with resistance 100 Ω. It is connected across a charged 10 μF capacitor at a moment when the capacitor voltage is 12 V. Determine the nature of the response and obtain an expression for the current in the inductor.

In this case $4L/C$ has the value 8×10^5, which is greater than $(100)^2$. Hence case (ii) applies. Inserting values into Equation 9.23 we find

$$\alpha = 25, \ \beta = 222.2$$

Equations 9.26 and 9.28 with $v_0 = 12$, $i_0 = 0$, then give

$$A = 12, \ B = \alpha A/\beta = 1.35$$

In the present case, Equation 9.27 reduces to

$$\frac{dv}{dt} = -\exp(-\alpha t)(\alpha B + \beta A)\sin(\beta t)$$

The required expression is therefore

$$i(t) = -C\frac{A}{\beta}(\alpha^2 + \beta^2)\exp(-\alpha t)\sin(\beta t)$$

$$= -0.27\exp(-\alpha t)\sin(\beta t)$$

9.7.2 ◆ The step response

In this section the response of the initially quiescent circuit to a step input will be determined when, again, case (ii) applies: this is the most interesting case, and solution in the other cases follows similar lines. One solution of the equation

$$LC\frac{d^2v}{dt^2} + RC\frac{dv}{dt} + v = E$$

is $v = E$, clearly the correct response to the step after a long time has elapsed. We therefore try the solution

$$v(t) = E + \exp(-\alpha t)[A\cos(\beta t) + B\sin(\beta t)]$$

valid for $t > 0$. As argued above, with this excitation both v and i must be continuous at $t = 0$. Hence v must be zero at the commencement of the step, and in addition, since $i = C\ dv/dt$, dv/dt must also be zero. The condition $v = 0$ at $t = 0$ gives $A = -E$. Evaluating dv/dt, we find that $B = \alpha A/\beta$. Hence the required solution is

$$v(t) = E\left\{1 - \frac{1}{\beta}\exp(-\alpha t)[\beta\cos(\beta t) + \alpha\sin(\beta t)]\right\} \qquad \textbf{(9.29)}$$

This is illustrated in Figure 9.12, from which it is seen that the step excites a damped sine wave which dies away to leave the final steady value, E.

9.7.3 ◆ Connection with the steady-state sinusoidal response

The response of the series $L-C-R$ circuit to steady-state sinusoidal excitation was discussed in Section 5.4, where it was seen that the response

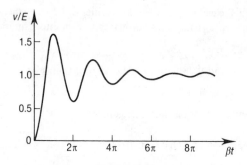

Figure 9.12 Step response of L–C–R circuit, showing voltage across capacitor in the case $Q = 3$ (Equation 9.29, Section 9.7.2).

was characterized by two parameters, the resonant frequency and the value of the Q factor:

$$2\pi f_0 = \omega_0 = \frac{1}{(LC)^{1/2}}$$

$$Q = \frac{1}{R}\left(\frac{L}{C}\right)^{1/2} = \frac{\omega_0 L}{R}$$

In terms of these parameters the expressions for α, β given in Equation 9.23 become

$$\alpha = \frac{R}{2L} = \frac{\omega_0}{2Q}$$

$$\beta = \omega_0\left(1 - \frac{1}{4Q^2}\right)^{1/2}$$

The significance of the Q factor was discussed in Section 5.4.2, where it was seen that if Q was larger than about 5, certain simplifications could be made. If in the present case $Q > 5$, we see that the natural response is a damped sine wave whose frequency is very close to the resonant frequency and which decays according to $\exp(-\omega_0 t/2Q)$. The decay therefore occurs more slowly as Q becomes larger. More precisely, the response will decay by a factor $1/e$ ($= 0.368$) in time

$$t = \frac{2Q}{\omega_0} = \frac{Q}{\pi}\frac{1}{f_0}$$

or Q/π cycles.

EXAMPLE 9.6 The equivalent circuit of a piezo-electric resonator is a series L–C–R circuit. Measurements on a particular resonator show the natural response takes the form of a decaying sinusoid of frequency 10 MHz and that the peak value of the response dies away to 10% of its initial value in 1 ms. Estimate the Q factor of the resonator. Theory suggests a value of 2 pF for the capacitance: determine the other components.

In the period of 1 ms there are 10^4 cycles of the sine wave. With such a slow decay the peak value in the neighbourhood of time t may be taken as $A \exp(-\omega_0 t/2Q)$. Hence over a 1 ms period

$$\exp(-\omega_0 t/2Q) = 0.1$$

This gives

$$\omega_0 t/2Q = \ln(10) = 2.30$$

whence

$$Q = 2\pi \times 10^7 \times 10^{-3}/2 \times 2.30 = 1.37 \times 10^4$$

To find the other component values we have

$$L = 1/C(2\pi \times 10^7)^2 = 127 \ \mu\text{H}$$

and

$$R = \omega_0 L/Q = 0.58 \ \Omega$$

9.8 ◆ The parallel *L–C–R* circuit

The circuit is shown in Figure 9.13, excited by a current source in a parallel configuration. Using the notation in the figure, we have the equations

$$i_L + i_C + \frac{v}{R} = I$$

$$i_C = C\frac{\mathrm{d}v}{\mathrm{d}t} \tag{9.30}$$

$$v_L = L\frac{\mathrm{d}i_L}{\mathrm{d}t}$$

Hence

$$v = L\frac{\mathrm{d}}{\mathrm{d}t}\left(I - \frac{v}{R} - C\frac{\mathrm{d}v}{\mathrm{d}t}\right)$$

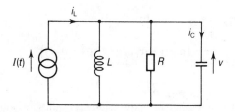

Figure 9.13 The parallel *L–C–R* circuit: variables for transient analysis.

The differential equation for $v(t)$ is therefore

$$LC\frac{d^2v}{dt^2} + \frac{L}{R}\frac{dv}{dt} + v = L\frac{dI}{dt} \tag{9.31}$$

This equation is of the same form as Equation 9.18, the main difference being that the coefficient of dv/dt is L/R rather than RC. To consider the natural response we put the right-hand side equal to zero. A solution can be found by an analysis similar to that of Section 9.7.1. The function $\exp(mt)$ is a solution of Equation 9.31 if

$$LCm^2 + \frac{L}{R}m + 1 = 0$$

Hence

$$m = -\frac{1}{2RC} \pm \left[\left(\frac{1}{2RC}\right)^2 - \frac{1}{LC}\right]^{1/2} \tag{9.32}$$

The formulae of Section 9.7.1 apply with appropriate expressions for α and β.

For the case when I is a step function, the right-hand side of Equation 9.31 is zero once the step has been applied. We then have to determine the initial conditions. The capacitor ensures that the voltage v rises from zero at a finite rate, so that initially no current flows down the resistor; the inductor current must also build up slowly from its initial value of zero so that initially all the source current I must go into the capacitor. We thus have

$$v = 0, \; dv/dt = I/C \text{ at } t = 0$$

Assuming this time that the roots of Equation 9.31 are real, we denote the values of m by $-\alpha_1$, $-\alpha_2$. The solution must then take the form

$$v(t) = A\exp(-\alpha_1 t) + B\exp(-\alpha_2 t)$$

Applying the initial conditions we find

$$0 = A + B$$

$$\frac{I}{C} = -\alpha_1 A - \alpha_2 B$$

Hence the step response is given by

$$v(t) = \frac{I}{C}\frac{1}{\alpha_2 - \alpha_1}[\exp(-\alpha_1 t) - \exp(-\alpha_2 t)] \tag{9.33}$$

EXAMPLE 9.7 Obtain an expression for the response of a parallel L–C–R circuit to a current impulse of strength K, for the case when the roots of Equation 9.32 are complex.

The simplest way to proceed is as follows: the response after the impulse must be a form of the natural response; the precise form is then determined by applying the correct initial conditions. By considering the circuit it is seen that the result of the impulse is to deposit an amount of charge K on the capacitor, thereby raising its voltage to K/C; since the voltage is finite, no charge will be lost through the resistor, and the current through the inductor will remain at zero. The natural response takes the form

$$v(t) = \exp(-\alpha t)[A \cos(\beta t) + B \sin(\beta t)]$$

in which, from Equation 9.32,

$$\alpha = \frac{1}{2RC}, \beta = \left[\frac{1}{LC} - \left(\frac{1}{2RC}\right)^2\right]^{1/2}$$

The constants A, B are to be determined by applying the conditions

$$t = 0: v = K/C; i_L = 0$$

Since i_L and I are both zero just after $t = 0$, from Equations 9.31 we have

$$i_C + \frac{v}{R} = 0$$

$$i_C = C\frac{dv}{dt}$$

The condition $i_L = 0$ therefore corresponds to

$$\frac{dv}{dt} = -\frac{1}{RC}v = -2\alpha\frac{K}{C}$$

Applying these conditions

$$\frac{K}{C} = A$$

$$-2\alpha\frac{K}{C} = -\alpha A + \beta B$$

$$B = -\frac{\alpha}{\beta}\frac{K}{C}$$

The required expression is therefore

$$v = \frac{K}{C}\exp(-\alpha t)[\cos(\beta t) - \frac{\alpha}{\beta}\sin(\beta t)]$$

9.9 ♦ Calculation of transient response by numerical methods

As an alternative to the analytical solutions obtained in the previous sections, responses under prescribed initial conditions can also be obtained by numerical methods. We will consider first the $L-R$ circuit studied in Section 9.2.2 and then the series $L-C-R$ circuit of Section 9.6.

9.9.1 ◆ Step response for the *L–R* circuit

In the former case, the step response requires solution of Equation 9.3:

$$L\frac{di}{dt} + Ri = E$$

We consider the response at a series of discrete times of the form

$$t = n\tau$$

in which n is an integer and τ a short interval over which the current changes only slightly. Writing $i(n\tau) = i_n$, we can replace the differential by the expression

$$\left(\frac{di}{dt}\right)_{t=n\tau} \approx (i_{n+1} - i_n)/\tau = (E - Ri_n)/L$$

Rearranging, we have

$$i_{n+1} = i_n\left(1 - \frac{R}{L}\tau\right) + \frac{\tau}{L}E \tag{9.34}$$

Starting from an initial value for i_0 this equation allows calculation of i_1 and then of successive values i_n. The accuracy of these values depends on the value chosen for the interval τ: in this case it needs to be small compared with the time constant L/R, but the smaller it is, the more points that need to be calculated. Equation 9.34 is in the form of an algorithm suitable for programming into a computer. The values shown in Table 9.1 have been obtained using such a program, taking unity for the numerical values of L, R and E. The table shows values at intervals of one-quarter of the time constant, or 0.25: column 2 was obtained with $\tau = 0.025$, column 3 with $\tau = 0.0125$. Column 4 gives the values calculated from Equation 9.9. The numerical solutions in both columns 2 and 3 are

Table 9.1

1	2	3	4
0	0	0	0
0.25	0.224	0.222	0.221
0.50	0.397	0.395	0.393
0.75	0.532	0.530	0.528
1.00	0.637	0.634	0.632
1.25	0.718	0.716	0.713
1.50	0.781	0.779	0.777
1.75	0.830	0.828	0.826
2.00	0.868	0.866	0.865

Table 9.2

1	2	3
0	0	0
1	0.113	0.123
2	0.428	0.437
3	0.839	0.836
4	1.234	1.211
5	1.517	1.476
6	1.636	1.586
7	1.583	1.537
8	1.395	1.386
9	1.136	1.137
10	0.879	0.910
11	0.687	0.741
12	0.618	0.673

seen to be correct to better than 1% of the final value: the improvement obtained by doubling the number of points is apparent but marginal in this example.

9.9.2 ♦ Step response of the series *L–C–R* circuit

The equations in this case are given by Equations 9.17. Rearrangement of these equations gives

$$\frac{di}{dt} = \frac{1}{L}(e - v - iR)$$

$$\frac{dv}{dt} = \frac{1}{C}i$$

Using similar approximations to that used in Section 9.9.1, these two equations lead to

$$i_{n+1} = \frac{\tau}{L}(E - v_n - Ri_n) + i_n$$

$$v_{n+1} = \frac{\tau}{C}i_n + v_n \qquad \qquad \textbf{(9.35)}$$

Initial values for i_0, v_0 will allow the calculation of i_1, v_1 and then succeeding values. In this case, the time interval should be small compared with $(LC)^{-1/2}$. Table 9.2 shows the results of computing the response to a unit step from Equations 9.35, compared with the exact solution given by Equation 9.29. Values of the capacitor voltage are given at intervals of $1/6\beta$ for the case $Q = 3$, $\alpha = 1/6$. The time step used was $1/60\beta$. Column 1 gives the time in intervals; column 2 the result of Equations 9.35; column 3 shows the exact solution, Equation 9.29. There are differences of the order of 5% of the final value of unity.

As discussed below, more complicated circuits can be treated in the same way. The simple method of numerical integration used requires a small step length for accurate results: larger step lengths can be used with the more sophisticated methods of integration to be found in texts on numerical analysis.

9.10 ♦ General networks

The examples in the earlier sections of this chapter illustrate one way in which the response of a network to a transient can be determined. In carrying out such an analysis the correct variables to choose as unknowns are the currents through inductors and the voltages across capacitors. They relate to the energy stored in these components, and are known as 'state variables'. Use of these variables will always result in differential equations of the type encountered earlier in this chapter, and it is these variables which must be continuous in the presence of finite excitation.

9.10.1 ◆ Form of equations

The equations for determining transient response are set up by applying Kirchhoff's laws and the component relations in much the same way as was done in Sections 6.2 and 6.3. It is, however, convenient to regard each separate inductance, capacitance and resistance as a separate branch of the network, and likewise to regard as a separate branch each voltage source and each current source. The currents through the inductances and the voltages across the capacitances will, for simplicity, be referred to as the 'inductive currents' and 'capacitative voltages' respectively. The number of inductive branches will be denoted by m_L, the number of capacitative branches by m_C and the total number of branches by b. We can classify the equations three groups by type and number:

  (i) Kirchhoff nodal equations, $n - 1$;

 (ii) Kirchhoff loop equations, $b - n + 1$;

(iii) resistive and source branches, $b - m_L - m_C$;

(iv) inductive branches, m_L;

 (v) capacitative branches, m_C.

Consider the situation when m_L inductive currents and the m_C capacitative voltages are assigned values: this is equivalent to replacing each inductance by a known current source, and each capacitance by a known voltage source. The network is then a resistance-only network, and each branch variable can be determined in terms of the actual and assumed sources. This process effectively makes use of the first three groups of equations, $2b - m_L - m_C$ in number, and implies that algebraic expressions can be obtained for, among others, the inductive voltages and the capacitative currents in terms of the inductive currents and the capacitative voltages. The inductive branches lead to m_L equations, each of the form

$$L\frac{di}{dt} = v$$

and the capacitative branches to m_C equations, each of the form

$$C\frac{dv}{dt} = i$$

It has just been shown that the right-hand sides of all these equations can be expressed algebraically in terms of the same inductive currents and capacitative voltages that appear on the left-hand sides. Presuming this to be done, the result is a set of $m_L + m_C$ linear, simultaneous, differential equations in $m_L + m_C$ unknowns. Application of the continuity conditions, Section 9.3, will provide initial values for the variables, leading to a unique solution. It is found that in all cases that the natural response of a network made up of resistance, inductance and capacitance is a linear sum of terms, each of which is either a decaying exponential or a damped sine wave.

Although this approach will lead to a solution of a transient problem, a method involving the Laplace transform is found to be most suitable to network problems. Not only does it yield solutions to specific problems in a straightforward manner, but it provides a close connection to the formulation already used for AC problems. The method is discussed in Chapter 10.

9.10.2 ◆ Numerical solution

A numerical procedure of the same type as considered earlier can be seen to be possible in the general case. As pointed out in the last section, if numerical values are given for the inductive currents and the capacitative voltages, all the other branch variables can be calculated. A stepwise method of solution can then be envisaged. First, initial values of the inductive currents and capacitative voltages are used to determine values for the inductive voltages and the capacitative currents. Knowledge of the voltage across an inductor means that di/dt is known, and by use of the approximation

$$\frac{di}{dt} \approx \frac{1}{\tau}(i_1 - i_0) = \frac{1}{L}v_0$$

for each inductive current an updated value, i_1, is found. A similar expression

$$\frac{dv}{dt} \approx \frac{1}{\tau}(v_1 - v_0) = \frac{1}{C}i_0$$

enables an updated value to be found for each capacitative voltage. The whole process is then repeated.

The first step in the procedure, finding values for the inductive voltages and capacitative currents, is done by use of a standard network analysis programme. The details of a complete implementation are not considered in this book, but in Chapter 15 the use of the transient facility in the SPICE network analysis program is demonstrated.

EXAMPLE 9.8 Set up the equations appropriate to transient response for the circuit of Figure 9.14. Solution is not required.

Using the notation in the diagram, we can set up the following equations:

$$I - i_1 - i_2 = 0$$

$$i_2 - i_3 - i_4 = 0$$

$$v_1 - v_2 - v_3 = 0$$

$$v_3 - v_4 = 0$$

$$i_4 = G_4 v_4$$

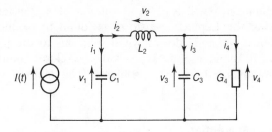

Figure 9.14 Circuit for Example 9.8.

$$C_1(dv_1/dt) = i_1$$

$$C_3(dv_3/dt) = i_3$$

$$L_2(di_2/dt) = v_2$$

From the first five equations we find

$$i_1 = I - i_2$$

$$i_3 = i_2 - G_4v_3$$

$$v_2 - v_1 - v_3$$

Hence

$$\frac{dv_1}{dt} = \frac{1}{C_1}(I - i_2)$$

$$\frac{dv_3}{dt} = \frac{1}{C_3}(i_2 - G_4v_3)$$

$$\frac{di_2}{dt} = \frac{1}{L_2}(v_1 - v_3)$$

In this form, these three simultaneous equations in the variables v_1, v_3, i_2 form the basis for analytical solution. If the differentials on the left-hand sides are replaced by the finite-step approximations as in Sections 9.9.1 and 9.9.2, a set of equations suitable for numerical solution are obtained.

9.11 ◆ Summary

The method of setting up equations from which to determine transient response has been discussed, and examples have been given for first- and second-order systems. The method has been shown to lead to linear, constant coefficient, differential equations. The particular cases of natural response, step response and impulse response have been introduced.

A method of obtaining solution by numerical means has been demonstrated in simple cases.

Extension to the case of a general network has been considered in outline, for both analytical and numerical solutions.

Formulae

Series R–C circuit: response to voltage step

$$v_C(t) = E[1 - \exp(-t/RC)]$$

Series R–L circuit: response to voltage step

$$i(t) = (E/R)[1 - \exp(-Rt/L)]$$

Series L–C–R circuit: response to voltage step
In the following the roots of the quadratic equation

$$LCm^2 + RCm + 1 = 0$$

are denoted by $-\alpha_1$, $-\alpha_2$ if the roots are real, and by $-\alpha \pm j\beta$ if the roots are complex.

$$v_C(t) = E\left\{1 - \frac{1}{\alpha_2 - \alpha_1}[\alpha_2 \exp(-\alpha_1 t) - \alpha_1 \exp(-\alpha_2 t)]\right\}$$

$$v_C(t) = E\left\{1 - \frac{1}{\beta}\exp(-\alpha t)[\beta \cos(\beta t) + \alpha \sin(\beta t)]\right\}$$

Parallel L–C–R circuit: response to current step
In the following the roots of the quadratic equation

$$LCm^2 + (L/R)m + 1 = 0$$

are denoted by $-\alpha_1$, $-\alpha_2$ if the roots are real, and by $-\alpha \pm j\beta$ if the roots are complex.

$$v(t) = I\frac{1}{C}\frac{1}{\alpha_2 - \alpha_1}[\exp(-\alpha_1 t) - \exp(-\alpha_2 t)]$$

$$v(t) = I\frac{1}{C\beta}\exp(-\alpha t)\sin(\beta t)$$

Connection with steady-state parameters

$$\alpha = \frac{R}{2L} = \frac{\omega_0}{2Q}$$

$$\beta = \omega_0\left(1 - \frac{1}{4Q^2}\right)^{1/2}$$

PROBLEMS

9.1

   (a) A transistor radio requires a power supply delivering 20 mA at 9 V. The power supply is smoothed by a 10 000 μF capacitor. When the power is switched off the capacitor is left to discharge through the radio. How long does it take for the voltage to decay by 50%, assuming that the radio behaves as a simple resistance?

(b) A high-voltage generator produces 100 kV across a 0.05 μF capacitor. The capacitor is provided with a shunt leakage resistor of 500 MΩ. The power is turned off. How long is it before the voltage has dropped to 50 V?

9.2 An electromagnet has an inductance of 12 H and a resistance of 20 Ω. When carrying a current of 5 A the terminals are joined together. How long does it take for the current to drop to (a) 1 A and (b) 1 mA? Determine the rate of decay of the current at the moment of joining the terminals.

9.3 A charged capacitor, capacitance C_1, is connected through a switch and resistance R to a second, uncharged capacitor of capacitance C_2. Show that after the switch is closed the voltage across C_2 satisfies the equation

$$\frac{dv_2}{dt} + \frac{1}{R}\frac{C_1 + C_2}{C_1 C_2}v_2 = \frac{1}{RC_2}E$$

in which E is the initial voltage across C_1.

In a particular case $C_1 = C_2 = 10\ \mu$F, $R = 2$ kΩ, $E = 250$ V. Determine the final voltage. How long after the switch is closed is it before the voltage is within 5% of its final value?

9.4 In the general problem of Problem 9.3, $C_1 = C_2 = 2C$. Derive an expression for the current through the resistance, and hence obtain a value for the total energy dissipated before equilibrium is reached. Show that this lost energy corresponds exactly to the loss of stored energy in the capacitors.

9.5 The equivalent circuit of an inductor is an inductance of 50 H in series with a 10 Ω resistance. The inductor is connected to a 200 V supply. Calculate the final current and the time it takes for the current to reach 90% of that final value.

9.6 In the situation described in Problem 9.5, an extra resistance of 10 Ω is placed in series with the inductor before it is connected to the supply. The extra resistance is shorted automatically by a switch when the current reaches 5 A. Calculate

(a) the time taken for the switch to operate, and

(b) the time taken for the current to reach 18 A.

9.7 In the circuit shown in Figure 9.15 with the switch in position 1, the current finally reaches a value I. In this state the switch is operated. Show that the current $i_1(t)$ satisfies the equation

$$L\frac{di_1}{dt} + (r + 2R)i_1 = I\frac{R^2}{r + R}\exp(-rt/L)$$

Show that the function

$$A\exp(-rt/L) + B\exp[-(r + 2R)t/L]$$

satisfies this equation, and hence obtain an expression for $i_1(t)$. Obtain also an expression for the voltage across the resistance R.

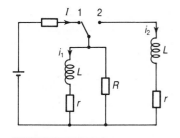

Figure 9.15

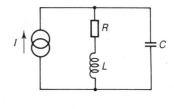

Figure 9.16

9.8 The behaviour of a series R–C circuit when an excitation $\hat{E}\cos(\omega t)$ was applied at $t = 0$ was investigated in Section 9.6. Modify the analysis so that, with respect to the same time origin, the excitation is $\hat{E}\cos(\omega t + \phi)$, to obtain an expression for the voltage across the capacitor, assumed to be initially uncharged. For what value of ϕ is no transient excited?

9.9 In an L–C–R circuit $L = 10$ H, $C = 5\ \mu$F, $R = 3$ kΩ. These components and a switch, initially open, are connected in series, and the capacitor is charged to 50 V. The switch is then closed. Obtain expressions for the voltage across the capacitor and for the current in the circuit. Show that the current reaches a maximum value, and determine that value and when it occurs.

9.10 The inductance, capacitance and resistance in Problem 9.9 are connected in a loop. The terminals of the capacitor are connected through a switch to a 50 V DC power supply. After a long time the voltage across the capacitor reaches a steady value of 50 V. The switch is then opened. Obtain an expression for the voltage across the capacitor.

9.11 In the circuit of Figure 9.16

$$L = 6.4\ \text{mH},\ C = 100\ \text{pF},\ R = 9.6\ \text{k}\Omega$$

The current source applies a step of 10 mA at $t = 0$. Show that the voltage across the capacitor is given by the following expression:

$$v(t) = 96\{1 - \exp(-0.75\tau)[\cos\tau - (7/24)\sin\tau]\}$$

in which τ is the time in microseconds.

Plot this function for the interval $0 < \tau < 4$, and hence obtain values for the following quantities:

  (i)   the time taken to reach 10% of the final value;
  (ii) the time taken to reach 90% of the final value;
  (iii) the maximum value reached by the voltage.

The **rise time** is often defined as the time between 10% and 90% of the final response. Calculate this rise time. Calculate also the magnitude of the overshoot in the response, expressed as a percentage of the final value.

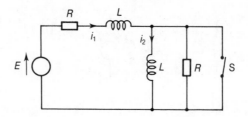

Figure 9.17

9.12 The natural response of a parallel $L-C-R$ circuit is described by the expression

$$\exp(-\alpha t)[A\cos(\beta t) + B\sin(\beta t)]$$

in which $-\alpha \pm j\beta$ are the roots of the equation

$$LCm^2 + \frac{L}{R}m + 1 = 0$$

Derive expressions for α, β in terms of the resonant frequency and Q factor of the circuit.

In an experiment such a circuit was excited by an impulse. After the impulse had died away, the resulting voltage was a damped sine wave of frequency 120 kHz. The amplitude of the sine wave dropped from 5 V to 2.5 V in 31 cycles. Estimate the resonant frequency and Q factor. Estimate also the bandwidth of the circuit.

9.13 An inductor is equivalent to an inductance of 2.7 H in series with 2.1 kΩ resistance, all in parallel with 0.926 nF capacitance. At a time when the inductor is carrying a steady current of 0.1 A the supply is interrupted. Obtain an expression for the voltage across the inductor after the interruption. Estimate the maximum voltage (positive or negative) reached.

9.14 Set up a scheme for numerical solution of Problem 9.6. Using a time step of 0.2 s, follow the intial phase to estimate the time of opening of the switch.

9.15 For the circuit shown in Figure 9.17 set up equations valid after the switch is opened which express di_1/dt and di_2/dt in terms of i_1 and i_2.

If the means are available, try a numerical solution for the case $R = 1$, $L = 1$, $E = 1$. (In determining the initial conditions, assume the inductors to have a very small series resistance.)

For comparison, the analytic solution may be shown to be

$$i_1(t) = 1 + (1/\sqrt{5})[\exp(-\alpha_2 t) - \exp(-\alpha_1 t)]$$

$$i_2(t) = 1 - (1/\sqrt{5})[(\sqrt{5} - 1)\exp(-\alpha_2 t) + (\sqrt{5} + 1)\exp(-\alpha_1 t)]$$

in which $\alpha_1 = (3 - \sqrt{5})/2$, $\alpha_2 = (3 + \sqrt{5})/2$.

10 ♦ *Transient Analysis*
2: Application of the Laplace transform

10.1 Introduction	10.5 Transform analysis
10.2 Definition of the	directly from the circuit
Laplace transform	diagram
10.3 Transient analysis	10.6 Summary
10.4 Some more transforms	Problems

In this chapter is explained the use of the method of the Laplace transform to determine the response of a network to transient excitation. It is shown that the method leads to a formulation which closely parallels that already found for phasor analysis.

10.1 ♦ Introduction

In the last chapter the determination of the response to transient excitation was presented in the form of direct solution of the differential equation derived through analysis of the network. In this chapter it is shown how such problems can be solved by using the Laplace transform technique directly on the branch equations of a network, leading to a formulation of transient problems which is closely parallel to formulation in the case of steady-state AC. The next section describes the mathematics of the Laplace transform process and subsequent sections illustrate the application by treating problems already solved by the direct methods of Chapter 9.

10.2 ♦ Definition of the Laplace transform

In this section the process of taking a Laplace transform will be described and illustrated by working out some useful transforms. It is assumed that all variables, such as voltage and current, are zero before a definite time, conveniently taken as the origin. Consider a function of time $v(t)$, identically zero for $t < 0$. The Laplace transform of $v(t)$ is a function $V(s)$ defined by the integral

$$V(s) = \int_0^\infty v(t) \exp(-st) \, dt \tag{10.1}$$

In this integral the variable s, which may be complex, is to be regarded as quite arbitrary, save that its real part is positive and sufficiently large to make the integral converge. The integral can be evaluated by some means for any specific value of s, and so defines the corresponding value of $V(s)$. We consider some particular cases.

10.2.1 ◆ The exponential function

This is defined as

$$\begin{aligned} v(t) &= 0 & t &< 0 \\ &= \exp(-\alpha t) & t &> 0 \end{aligned} \tag{10.2}$$

Using Equation 10.1, we have

$$\begin{aligned} V(s) &= \int_0^\infty \exp[-s(\alpha + t)] \, dt \\ &= [-(s + \alpha)^{-1} \exp(-st)]_0^\infty \\ &= \frac{1}{s + \alpha} \end{aligned}$$

The two equations

$$\begin{aligned} v(t) &= \exp(-\alpha t) \\ V(s) &= (s + \alpha)^{-1} \end{aligned} \tag{10.3}$$

are said to form a Laplace transform pair. This is often denoted by the symbolism

$$V(s) = \mathcal{L}[v(t)]$$

10.2.2 ◆ Cosine and sine functions

The easiest way to determine these transforms is to use Equation 10.3 with an imaginary value for α:

$$\begin{aligned} \mathcal{L}[\exp(j\beta t)] &= \frac{1}{s - j\beta} \\ &= \frac{s + j\beta}{s^2 + \beta^2} \end{aligned}$$

Hence

$$\mathcal{L}[\cos(\beta t)] = \frac{s}{s^2 + \beta^2}$$

and

$$\mathcal{L}[\sin(\beta t)] = \frac{\beta}{s^2 + \beta^2}$$

10.2.3 ◆ Damped oscillation

The transforms of the functions $\exp(-\alpha t)\cos(\beta t)$ and $\exp(-\alpha t \sin(\beta t)$ can be found by the same technique of using Equation 10.3 with α replaced by $\alpha - \mathrm{j}\beta$:

$$\mathcal{L}[\exp(-\alpha t \exp(\mathrm{j}\beta t)] = \frac{1}{s + \alpha - \mathrm{j}\beta}$$

$$= \frac{s + \alpha + \mathrm{j}\beta}{(s + \alpha)^2 + \beta^2}$$

Hence

$$\mathcal{L}[\exp(-\alpha t)\cos(\beta t)] = \frac{s + \alpha}{(s + \alpha)^2 + \beta^2}$$

$$\mathcal{L}[\exp(-\alpha t)\sin(\beta t)] = \frac{\beta}{(s + \alpha)^2 + \beta^2}$$

10.2.4 ◆ Step function

The unit step can be regarded as the special case of equation 10.3 with $\alpha = 0$. Denoting the step function by $H(t)$ we have

$$\mathcal{L}[H(t)] = \frac{1}{s}$$

10.2.5 ◆ Differential of a function

A very important result is obtained by considering the transform of the differential of a function, such as $\mathrm{d}v(t)/\mathrm{d}t$. From Equation 10.1

$$\mathcal{L}(\mathrm{d}v/\mathrm{d}t) = \int_0^\infty \frac{\mathrm{d}v}{\mathrm{d}t} \exp(-st)\,\mathrm{d}t$$

Integrating by parts we find

$$\mathcal{L}(\mathrm{d}v/\mathrm{d}t) = [\exp(-st)v(t)]_0^\infty - \int_0^\infty (-s)\exp(-st)v(t)\,\mathrm{d}t$$

$$= -v(0) + sV(s) \tag{10.4}$$

The results of the last few sections can conveniently be assembled for future reference in tabular form (Table 10.1).

Table 10.1

$v(t)$	$V(s)$
$\exp(-\alpha t)$	$\dfrac{1}{s + \alpha}$
$\cos(\beta t)$	$\dfrac{s}{s^2 + \beta^2}$
$\sin(\beta t)$	$\dfrac{\beta}{s^2 + \beta^2}$
$\exp(-\alpha t)\cos(\beta t)$	$\dfrac{s + \alpha}{(s + \alpha)^2 + \beta^2}$
$\exp(-\alpha t)\sin(\beta t)$	$\dfrac{\beta}{(s + \alpha^2) + \beta^2}$
Step $\quad H(t)$	$\dfrac{1}{s}$
$\dfrac{dv}{dt}$	$-v(0) + sV(0)$

The above analysis shows how the Laplace transform of a given function can be determined. It is plausible and it can be shown rigorously that there is a one-to-one correspondence between a time function and its Laplace transform. It will be shown in the next section how a circuit problem can be solved in terms of Laplace transforms.

10.3 ◆ Transient analysis

In this section the Laplace transform method will be applied to determining the transient response of the series $R-C$ and series $L-C-R$ circuits considered in Chapter 9.

10.3.1 ◆ The series $R-C$ circuit

The circuit, previously shown as Figure 9.1, appears as Figure 10.1. Using the notation in the diagram we have

$$e(t) = v + v_R$$
$$v_R = iR \qquad\qquad\qquad (10.5)$$
$$i = C\frac{dv}{dt}$$

In each equation the two sides are equal for every value of time, so that the Laplace transforms of the two sides will be equal. Hence, denoting the Laplace transforms by capital letters, we have

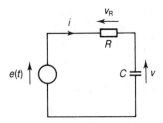

Figure 10.1 Series $R-C$ circuit: variables for transient analysis.

$$E(s) = V(s) + V_R(s)$$

$$V_R(s) = I(s)R \tag{10.6}$$

For the last equation we use Equation 10.4 to find

$$I(s) = C[-v(0) + sV(s)] \tag{10.7}$$

We note that apart from the term containing $v(0)$ the equations have exactly the form of an AC phasor analysis with $j\omega$ replaced everywhere by s. We can eliminate the quantities $I(s)$ and $V_R(s)$ between the equations to obtain

$$E(s) = -RCv(0) + (RCs + 1)V(s)$$

or

$$V(s) = \frac{E(s)}{RCs + 1} + \frac{RCv(0)}{RCs + 1} \tag{10.8}$$

We have in this way obtained an expression for the Laplace transform of the voltage across the capacitor, in which the transform $E(s)$ is known once the excitation waveform $e(t)$ is specified. To see how the time function $v(t)$ is derived, consider the case when $v(0) = 0$ and $e(t)$ is a step function of height $\hat{E}$. Using the entry in Table 10.1 we see $E(s) = \hat{E}/s$, so that

$$V(s) = \frac{\hat{E}}{s} \frac{1}{RCs + 1}$$

To interpret this we perform an expansion into partial fractions. Writing $\alpha = 1/RC$

$$V(s) = \hat{E}\frac{1}{RC} \frac{1}{s(s + \alpha)}$$

$$= \hat{E}\frac{1}{RC} \left[\frac{1}{\alpha s} - \frac{1}{\alpha(s + \alpha)} \right]$$

$$= \hat{E}\left(\frac{1}{s} - \frac{1}{s + \alpha} \right)$$

The time functions corresponding to the two terms in parentheses can be recognized from entries in Table 10.1. Hence

$$v(t) = \hat{E}[1 - \exp(-t/RC)]$$

This is the result expressed by Equation 9.6.

Consider next the case of sinusiodal excitation of this circuit, studied in Section 9.6, in which

$$\begin{aligned} v(t) &= 0 &\quad t < 0 \\ &= \hat{E}\cos(\omega t) &\quad t > 0 \end{aligned}$$

Referring to Table 10.1, in this case

$$E(s) = \hat{E}\frac{s}{s^2 + \omega^2}$$

Using as before $\alpha = 1/RC$, Equation 10.5 becomes

$$V(s) = \hat{E}\frac{s}{s^2 + \omega^2}\frac{\alpha}{s + \alpha}$$

This expression has a partial fraction expansion in the form

$$V(s) = \hat{E}\left(\frac{As + B}{s^2 + \omega^2} + \frac{D}{s + \alpha}\right)$$

in which

$$A = -D = \frac{\alpha}{\omega^2}B, \qquad B = \frac{\alpha\omega^2}{\alpha^2 + \omega^2}$$

On referring to Table 10.1 the expression for $v(t)$ is found to be

$$\begin{aligned} v(t) &= \hat{E}\frac{1}{\alpha^2 + \omega^2}\left[\alpha^2\cos(\omega t) + \alpha\omega\sin(\omega t) - \alpha^2\exp(-\alpha t)\right] \\ &= \hat{E}\frac{1}{1 + (RC\omega)^2}\left[\cos(\omega t) + RC\omega\sin(\omega t) - \exp(-t/RC)\right] \end{aligned}$$

Substituting, as in Section 9.6, $\omega RC = \tan\theta$ we find

$$\begin{aligned} v &= \hat{E}\frac{1}{[1 + (RC\omega^2)]^{1/2}} \\ &\quad \times \left[\cos\theta\cos(\omega t) + \sin\theta\sin(\omega t) - \cos\theta\exp(-t/RC)\right] \\ &= \hat{E}\frac{1}{[1 + (RC\omega)^2]^{1/2}}\left[\cos(\omega t - \theta) - \cos\theta\exp(-t/RC)\right] \end{aligned}$$

This is the result obtained in Section 9.6.

Most of the steps used in applying the Laplace transform technique to circuit problems have been illustrated by this study of the series R–C circuit. In the next section this is shown by working out the response of the L–C–R circuit to a step input, a problem solved in Section 9.7.

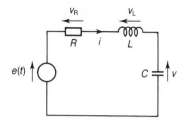

Figure 10.2 Series *L–C–R* circuit: variables for transient analysis.

10.3.2 ◆ The series *L–C–R* circuit

The circuit and notation are shown in Figure 10.2, which previously appeared as Figure 9.11. The relevant equations were grouped into Equations 9.17, here repeated:

$$e = v_L + v_R + v$$
$$v_L = L\frac{di}{dt}$$
$$i = C\frac{dv}{dt}$$
$$v_R = iR$$

$$(10.9)$$

As before, we take the Laplace transform of each of these equations, using Table 10.1 for the differentials. We find

$$E(s) = V_L(s) + V_R(s) + V(s)$$
$$V_L(s) = -Li(0) + LsI(s)$$
$$I(s) = -Cv(0) + CsV(s)$$
$$V_R(s) = I(s)R$$

$$(10.10)$$

Once again it is seen that, excluding the initial conditions, these equations are exactly the phasor equations with $j\omega$ replaced by s. An expression for $V(s)$ can be obtained:

$$E(s) = -Li(0) + (Ls + R)I(s) + V(s)$$
$$= -Li(0) - C(Ls + R)v(0) + (LCs^2 + RCs + 1)V(s)$$

$$(10.11)$$

We consider, as in Section 9.7.2, the case of an initially quiescent circuit, in which case

$$V(s) = \frac{E(s)}{LCs^2 + RCs + 1}$$

Assuming that it is required to find the response to a step we put $E = \hat{E}/s$, in which case

$$V(s) = \hat{E}\frac{1}{s}\frac{1}{LCs^2 + RCs + 1}$$

The way in which the time function $v(t)$ is found depends on whether or not the quadratic can be factorized in real form. If $(RC)^2 > 4LC$, the expression for $V(s)$ can be written

$$V(s) = \hat{E}\frac{1}{LC}\frac{1}{s}\frac{1}{(s + \alpha_1)(s + \alpha_2)}$$

in which α_1, α_2 are given by Equations 9.21. A straightforward partial fraction expansion can be found in the form

$$V(s) = \hat{E}\left[\frac{A}{s} + \frac{B}{s + \alpha_1} + \frac{D}{s + \alpha_2}\right]$$

giving

$$v(t) = \hat{E}[A + B\exp(-\alpha_1 t) + D\exp(-\alpha_2 t)]$$

If we take the case of Section 9.7.2 the quadratic does not have real factors. We then write

$$V(s) = \hat{E}\frac{1}{LC}\frac{1}{s}\frac{1}{(s + \alpha)^2 + \beta^2}$$

in which α, β are given by Equations 9.23. The partial fraction expansion then takes the form

$$V(s) = \hat{E}\frac{1}{LC}\left[\frac{A}{s} + \frac{Bs + D}{(s + \alpha)^2 + \beta^2}\right]$$

It is found that

$$A = 1/(\alpha^2 + \beta^2) = LC, \ B = -A, \ D = -2\alpha A$$

Hence

$$V(s) = \hat{E}\left[\frac{1}{s} - \frac{(s + \alpha) + \alpha}{(s + \alpha)^2 + \beta^2}\right]$$

Referring to Table 10.1 we find

$$v(t) = \hat{E}\left[1 - \exp(-\alpha t)\cos(\beta t) - (\alpha/\beta)\exp(-\alpha t)\sin(\beta t)\right]$$

This will be found to agree with Equation 9.29.

10.3.3 ◆ Initial conditions

In the previous examples only the case of initial quiescence has been considered. From the forms of Equations 10.8 and 10.11 it can be seen that non-zero values for initial voltages across capacitors or currents through

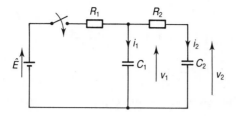

Figure 10.3 Circuit for Example 10.1.

inductors merely lead to extra source terms, effectively modifying $E(s)$. The interpretation is done in the same way. For example, the term in $v(0)$ in Equation 10.8 leads immediately to a term

$$v(t) = v(0) \exp(-t/RC)$$

This is a natural response term, in accord with results in Chapter 9.

EXAMPLE 10.1 In the circuit of Figure 10.3, $R_1 = R_2 = 1\,\text{k}\Omega$, $C_1 = C_2 = 10\,\mu\text{F}$. The capacitors are initially uncharged. Obtain an expression for the voltage across C_2 after the switch is closed.

In this circuit the closure of the switch is equivalent to a step of height E being applied. Using the notation in the diagram, the relevant equations are

$$E = v_1 + R_1(i_1 + i_2)$$

$$v_1 = v_2 + R_2 i_2$$

$$i_1 = C_1 \frac{\text{d}}{\text{d}t} v_1$$

$$i_2 = C_2 \frac{\text{d}}{\text{d}t} v_2$$

$$(10.12)$$

Taking Laplace transforms, remembering that E represents a step and using upper-case letters to denote the transforms, we find

$$\hat{E}\frac{1}{s} = V_1 + R_1(I_1 + I_2)$$

$$V_1 = V_2 + R_2 I_2$$

$$I_1 = -C_1 v_1(0) + s C_1 V_1$$

$$I_2 = -C_2 v_2(0) + s C_2 V_2$$

$$(10.13)$$

In this problem the quantities $v_1(0)$, $v_2(0)$ are zero. Eliminating I_1, I_2 we find

$$\hat{E}\frac{1}{s} = V_1 + s R_1 C_1 V_1 + s R_1 C_2 V_2$$

$$V_1 = V_2 + s R_2 + s R_2 C_2 V_2$$

Hence

$$\hat{E}\frac{1}{s} = V_2 \left[(1 + R_1C_1s)(1 + R_2C_2s) + R_1C_2s \right]$$

$$= R_1R_2C_1C_2 \left[s^2 + s \left(\frac{1}{R_2C_1} + \frac{1}{R_1C_1} + \frac{1}{R_2C_2} \right) + \frac{1}{R_1R_2C_1C_2} \right] V_2$$

Substituting the given values, this expression becomes

$$\hat{E}\frac{1}{s} = 10^{-4}(s^2 + 3 \times 10^2 s + 10^4)V_2$$

The quadratic can be factorized into the form

$$(s + \alpha_1)(s + \alpha_2)$$

in which

$$\alpha_1 = \tfrac{1}{2}(3 - \sqrt{5})10^2 = 38.2$$

$$\alpha_2 = \tfrac{1}{2}(3 + \sqrt{5})10^2 = 261.8$$

Hence

$$V_2(s) = \hat{E} \times 10^4 \frac{1}{s(s + \alpha_1)(s + \alpha_2)}$$

$$= \hat{E} \times 10^4 \left[\frac{1}{\alpha_1\alpha_2 s} - \frac{1}{\alpha_1(\alpha_2 - \alpha_1)} \frac{1}{s + \alpha_1} \right.$$

$$\left. + \frac{1}{\alpha_2(\alpha_2 - \alpha_1)} \frac{1}{(s + \alpha_2)} \right]$$

Referring to Table 10.1 we find

$$v_2(t) = \hat{E}[1 - 1.171 \exp(-38.2t) + 0.171 \exp(-261.8t)]$$

10.4 ◆ Some more transforms

Although the transforms used in the majority of circuit problems are included in Table 10.1, the table can usefully be extended to cover some other cases. These will be considered in this section.

10.4.1 ◆ Delay

A common operation is to delay the application of a waveform. If $v(t)$ represents a waveform identically zero for $t < 0$, its application at a time τ is equivalent to the excitation $v(t - \tau)$, zero for $t < \tau$. We have

$$\mathcal{L}[v(t - \tau)] = \int_0^\infty v(t - \tau) \exp(-st) \, dt$$

Since $v(t - \tau)$ is zero for $t < \tau$,

$$\mathcal{L}[v(t - \tau)] = \int_\tau^\infty v(t - \tau) \exp(-st) \, dt$$

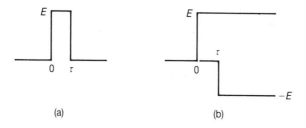

Figure 10.4 Pulse input: (a) waveform; (b) decomposition into steps.

Changing the variable of integration to $t' = t - \tau$, this becomes

$$\mathcal{L}[v(t - \tau)] = \int_0^\infty v(t') \exp[-s(t' + \tau)] dt'$$

$$= \exp(-s\tau) \int_0^\infty v(t') \exp(-st') dt'$$

$$= \exp(-s\tau) V(s) \qquad (10.14)$$

The transform of a waveform delayed by τ is seen to be the transform of the original multiplied by $\exp(-s\tau)$.

10.4.2 ◆ Exponential decay

Another transform easy to evaluate is that of $\exp(-\alpha t)v(t)$. We have

$$\mathcal{L}[\exp(-\alpha t)v(t)] = \int_0^\infty \exp(-\alpha t)v(t) \exp(-st) dt$$

$$= \int_0^\infty v(t) \exp[-(s + \alpha)t] dt$$

$$= V(s + \alpha) \qquad (10.15)$$

The entries in Table 10.1 for $\cos(\omega t)$ and $\exp(-\alpha t)\cos(\omega t)$ are special cases of this result.

10.4.3 ◆ The impulse function

The concept of the impulse function, $\delta(t)$, was introduced in Section 9.5: it was postulated as the limit of the waveform shown in Figure 10.4(a) as the pulse length τ tends to zero with the amplitude increasing to keep $E\tau = 1$. This limiting waveform will be shown to have a simple Laplace transform. The waveform can be regarded as the superposition of two step waveforms, as in Figure 10.4(b). The transform of a step appears in Table 10.1, and that for the delayed step then follows by application of Equation 10.14. The transform of the rectangular pulse is therefore given by

$$\frac{1}{\tau}\left[\frac{1}{s} - \frac{\exp(-s\tau)}{s}\right] = \frac{1}{s\tau}[1 - \exp(-s\tau)]$$

As $s\tau$ becomes smaller, the exponential becomes approximately equal to $1 - s\tau$, so that the expression in square brackets is approximately equal to $s\tau$. Hence

$$\mathcal{L}[\delta(t)] = 1 \tag{10.16}$$

It is a result of using the definition of Equation 10.1 that the right-hand side of Equation 10.16 has the dimension of time. The time function and its transform do not have the same dimensions: for example, the transform of a voltage has dimensions of voltage × time.

10.4.4 ♦ Ramp function

This is defined by

$$v(t) = 0 \qquad t < 0$$
$$\quad\ = t \qquad t > 0$$

We have

$$\mathcal{L}[v(t)] = \int_0^\infty t \exp(-st)\,dt$$

Integrating by parts

$$\mathcal{L}[v(t)] = \int_0^\infty t \frac{d}{dt}[-\frac{1}{s}\exp(-st)]\,dt$$

$$= \int_0^\infty \frac{1}{s}\exp(-st)\,dt$$

$$= \frac{1}{s^2} \tag{10.17}$$

By applying equation 10.15 we see

$$\mathcal{L}[t\exp(-\alpha t)] = \frac{1}{(s + \alpha)^2} \tag{10.18}$$

A more general version of this result is

$$\mathcal{L}[t^n \exp(-\alpha t)] = \frac{n!}{(s + \alpha)^{n+1}}$$

Not only does this result give the transform of the time function but also it enables us to interpret a repeated factor when it arises in a partial fraction expansion.

These further results are amalgamated with those in Table 10.1 in Table 10.2.

Table 10.2

	Function	Transform
Impulse	$\delta(t)$	1
Unit step	1	s^{-1}
Ramp	t	s^{-2}
	t^n	$n!s^{-n-1}$
Exponential	$\exp(-\alpha t)$	$(s + \alpha)^{-1}$
	$t^n \exp(-\alpha t)$	$n!(s + \alpha)^{-n-1}$
	$\exp(-\alpha + j\beta t)$	$(s + \alpha - j\beta)^{-1}$
Cosine	$\cos(\beta t)$	$s/(s^2 + \beta^2)$
Damped cosine	$\exp(-\alpha t)\cos(\beta t)$	$(s + \alpha)/[(s + \alpha)^2 + \beta^2]$
Sine	$\sin(\beta t)$	$\beta/(s^2 + \beta^2)$
Damped sine	$\exp(-\alpha t)\sin(\beta t)$	$\beta/[(s + \alpha)^2 + \beta^2]$
	$v(t)$	$V(s)$
Delay $0 < \tau$	0	$\exp(-s\tau)V(s)$
$\quad\quad t > \tau$	$v(t - \tau)$	
	$\exp(-\alpha t)v(t)$	$V(s + \alpha)$
	dv/dt	$-v(0) + sV(s)$

EXAMPLE 10.2 Determine the impulse response of the series R–C circuit in Section 10.3.1.

The relevant equations have been derived in Section 10.3.1, Equations 10.5. The transformed equations appear as Equations 10.6 and 10.7:

$$E(s) = V(s) + V_R(s)$$

$$V_R(s) = I(s)R$$

$$I(s) = -Cv(0) + sCV(s)$$

The impulse response was defined in Section 9.5. It is the response when the voltage source $e(t)$ is the unit impulse, $\delta(t)$, with no initial charge on the capacitor. Putting $E(s) = 1$, the equations give

$$(1 + sCR)V(s) = 1$$

$$V(s) = \frac{1}{1 + RCs} = \frac{1}{RC}\frac{1}{(s + 1/RC)}$$

Hence, using the table,

$$v(t) = \frac{1}{RC}\exp\left(\frac{-t}{RC}\right)$$

This agrees with the result expressed in Section 9.5, Equation 9.12.

EXAMPLE 10.3 Determine the step response of the series L–C–R circuit in Section 10.3.2 when $R^2 = 4L/C$.

Equation 10.11 expresses the result of the analysis for any values of L, C, R. Taking $v(0), i(0)$ to be zero we have

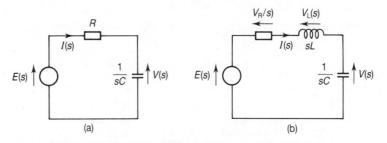

Figure 10.5 Laplace transform variables: (a) s-domain equivalent for R–C circuit (Figure 10.1); (b) s-domain equivalent for L–C–R circuit (Figure 10.2).

$$V(s) = \hat{E}\,\frac{1}{s}\,\frac{1}{LCs^2 + RCs + 1}$$

In the present case this reduces to

$$V(s) = \hat{E}\,\frac{1}{s}\,\frac{1}{LC}\,\frac{1}{(s + \alpha)^2}$$

where $\alpha = R/2L = 2/RC$. The partial fraction expansion is found to be

$$V(s) = \hat{E}\left[\frac{1}{s} - \frac{1}{s + \alpha} - \frac{\alpha}{(s + \alpha)^2}\right]$$

Using the appropriate entries in Table 10.2 we find

$$v(t) = \hat{E}[1 - \exp(-\alpha t) - \alpha t \exp(-\alpha t)]$$

10.5 ◆ Transform analysis directly from the circuit diagram

In the illustrative examples in Sections 10.3.1 and 10.3.2, the Laplace transform of the required quantity has been obtained from the transformed version of the network equations. It was commented in both the examples that the transformed equations are similar in form to the equations which would have been obtained in a phasor analysis with the substitution $s = j\omega$. In fact, the transformed equations can conveniently be set up directly from a circuit diagram in the same way that phasor equations can be set up. We consider first the case of an initially quiescent network: no inductor carries a current and no capacitor is charged at $t = 0$.

10.5.1 ◆ Initial quiescence

We take the example of the series R–C circuit (Section 10.3.1). Consider the diagram in Figure 10.5(a), in which the transforms E, V and I are assumed to obey Kirchhoff's laws in the same way as phasor representations do. Assuming also Ohm's law for the resistor, Equations 10.6 follow immediately. Equation 10.7 with $v(0)$ zero reads

$$I(s) = sCV(s)$$

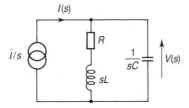

Figure 10.6 Circuit for Example 10.4.

which would be obtained if the capacitor were attributed an 'impedance' $1/sC$. In a similar fashion Equations 10.10 follow in this case of initial quiescence from the circuit of Figure 10.5(b), with the inductor attributed 'impedance' sL. Such circuits may be called s-domain equivalent circuits. To illustrate the process we consider an example.

> **EXAMPLE 10.4** For the circuit of Figure 10.6 find the waveform of the voltage across the capacitor when the current generator provides a current $\hat{I}$ applied to the quiescent network at $t = 0$.
>
> Treating the load on the generator as an admittance sC in parallel with an impedance $R + sL$, we have
> $$V(s) = ZI(s)$$
> where
> $$\frac{1}{Z} = sC + \frac{1}{sL + R}$$
> Hence
> $$V(s) = \hat{I}\frac{1}{s}\frac{sL + R}{s^2 LC + RCs + 1}$$
> $$= \hat{I}\frac{1}{sC}\frac{s + 2\alpha}{s^2 + 2\alpha s + \omega^2}$$
> in which $2\alpha = R/L$, $\omega^2 = 1/LC$. This expression can be processed in the standard way to find the waveform. For simplicity, take $C = 1$, $L = 1$, $R = \sqrt{2}$ for which values $\alpha = 1/\sqrt{2}$, $\omega = 1$.
> $$V(s) = \hat{I}\frac{1}{s}\frac{s + \sqrt{2}}{(s + 1\sqrt{2})^2 + 1/2}$$
> $$= \hat{I}\left[\frac{\sqrt{2}}{s} - \sqrt{2}\frac{s + 1/\sqrt{2}}{(s + 1/\sqrt{2})^2 + 1/2}\right]$$
> Hence
> $$v(t) = \sqrt{2}\hat{I}[1 - \exp(-t/\sqrt{2})\cos(t/\sqrt{2})]$$

10.5.2 ◆ Initial conditions

An assumption inherent in the Laplace transform method is that all network variables are identically zero before $t = 0$. However, the initial presence of charges and currents can be allowed for by introducing sources

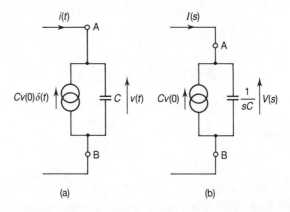

Figure 10.7 Allowance for initial voltage on a capacitor: (a) voltage established by current impulse; (b) s-domain equivalent circuit.

which act to bring about the conditions required just after $t = 0$. Consider first the case of a capacitor initially charged to $v(0)$. The relevant equation has appeared the discussion of the series R–C circuit (Section 10.3.1, Equation 10.7):

$$I(s) = -Cv(0) + sCV(s) \tag{10.19}$$

If we assume an initially uncharged capacitor, this equation is the transform of

$$i(t) = -Cv(0)\delta(t) + C\frac{\mathrm{d}v}{\mathrm{d}t}$$

This corresponds to the circuit shown in Figure 10.7(a), showing a capacitor in parallel with a source delivering a current impulse. This source deposits charge on the capacitor to bring the voltage to the required level. The s-domain equivalent circuit is shown in Figure 10.7(b). An alternative arrangement can be found by rearranging Equation 10.19 in the form

$$V(s) = v(0)\frac{1}{s} + \frac{1}{sC}I(s) \tag{10.20}$$

This represents a step voltage source in series with a capacitor, as shown in Figure 10.8(a); the s-domain equivalent circuit is shown in Figure 10.8(b). It is to be noted that the terminals of the capacitor in the physical circuit correspond to the terminals A, B in the diagrams: these arrangements are all equivalent circuits.

EXAMPLE 10.5 Determine the response of the series R–C circuit (Section 10.3.1) to a voltage step of height $\hat{E}$ when the capacitor voltage is initially $\hat{E}/2$.

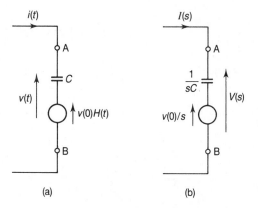

Figure 10.8 Allowance for initial voltage on a capacitor: (a) voltage established by step voltage source; (b) *s*-domain equivalent circuit.

The modified circuit is shown in Figure 10.9(a) and the *s*-domain circuit in Figure 10.9(b). From the diagram we have

$$\frac{\hat{E}/s - V(s)}{R} + \frac{1}{2}\hat{E}C = sCV(s)$$

whence

$$V(s) = \hat{E}\frac{1}{s(1 + sCR)} + \frac{1}{2}\hat{E}\frac{CR}{1 + sCR}$$

The first term is that already inverted in Section 10.3.1, and the second corresponds to $(1/2)\hat{E}\exp(-t/RC)$. Therefore

$$v(t) = \hat{E}[1 - \exp(-t/RC)] + \tfrac{1}{2}\hat{E}\exp(-t/RC)$$

The correctness of this expression can be verified from earlier results. Using the alternative equivalent circuit of Figure 10.8(b) gives rise to Figure 10.9(c). The voltage across the physical capacitor is represented by $V(s)$ taken between A and B. We have

$$I(s) = \left(\frac{\hat{E}}{s} - \frac{1}{2}\hat{E}\frac{1}{s}\right)\frac{1}{R + 1/sC}$$

$$V(s) = I\frac{1}{sC} + \frac{1}{2}\hat{E}\frac{1}{s}$$

From these equations we find, in agreement with the previous result,

$$V(s) = \hat{E}\frac{1}{s(1 + sCR)} + \frac{1}{2}\hat{E}\frac{CR}{1 + sCR}$$

Treatment of the cases when a current is flowing initially in an inductor follows similar lines. For this case, Equation 10.10 contains the relevant equation

$$V(s) = -Li(0) + sLI(s) \tag{10.21}$$

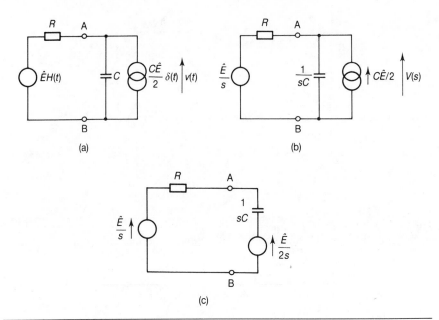

(a)

(b)

(c)

Figure 10.9 Example 10.5: (a) physical circuit using equivalence of Figure 10.7(b); (b) *s*-domain circuit; (c) *s*-domain circuit using equivalence of Figure 10.8.

This can be put in the alternative form

$$I(s) = V(s)/sL + i(0)/s \qquad\qquad (10.22)$$

These correspond to the circuit configurations shown in Figures 10.10 and 10.11.

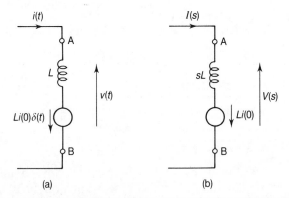

(a)

(b)

Figure 10.10 Allowance for initial current in inductor: (a) current established by voltage impulse; (b) *s*-domain circuit.

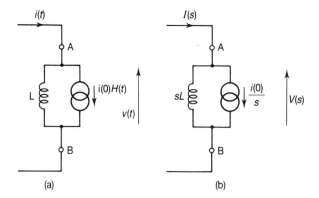

Figure 10.11 Allowance for initial current in inductor: (a) current established by step current source; (b) *s*-domain circuit.

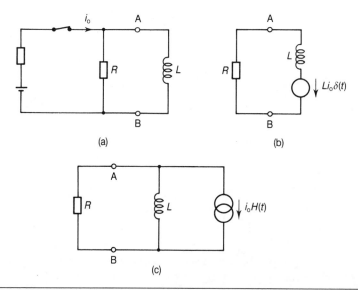

Figure 10.12 Comparison of alternative ways of establishing initial current in inductor (Section 10.5.2): (a) circuit; (b) *s*-domain circuit using equivalence of Figure 10.10; (c) *s*-domain circuit using equivalence of Figure 10.11.

As an example, consider the circuit shown in Figure 10.12(a) and the situation ensuing when the switch is opened. Prior to the opening of the switch the inductor carries a current of i_0, determined by the source voltage and resistance. The two possible circuit equivalents are shown in Figures 10.12(b) and 10.12(c). Consideration of these circuits show that in each case the resulting current is correctly given as $i_0 \exp(-Rt/L)$. In the circuit of Figure 10.12(b) the impulse generator serves to establish the

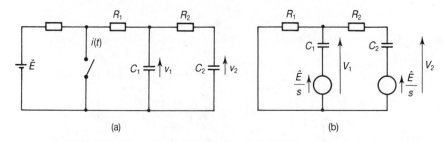

Figure 10.13 Example 10.6: (a) circuit; (b) *s*-domain circuit.

correct initial current in the inductance. In the circuit of Figure 10.12(c) the initial current is delivered through the resistance by the step generator and subsequently diverted through the inductance: the current $i(t)$ is then formed as the sum of the generator current and that through the inductance. Both examples illustrate Thévenin's theorem: the equivalent circuit yields the correct terminal relationship but differs from the actual within the transformed circuit.

EXAMPLE 10.6 The circuit shown in Figure 10.13(a) is allowed to reach a steady state. The switch then is closed. Determine the subsequent behaviour of the current $i(t)$.

In the steady state the voltages across both C_1 and C_2 will have reached $\hat{E}$. Using the equivalent circuit of Figure 10.8 the transformed circuit appears as in Figure 10.13(b). Drawing up a current balance at the node we find

$$\left(\frac{\hat{E}}{s} - V_1\right)sC_1 + \left(\frac{\hat{E}}{s} - V_1\right)\left(R_2 + \frac{1}{sC_2}\right) = \frac{V_1}{R_1} = I(s)$$

Solving for V_1 we find

$$I(s) = \frac{V_1}{R_1} = \hat{E}\frac{C_1 + C_2 + sR_1C_1C_2}{1 + s(R_1C_1 + R_2C_2 + R_1C_2) + s^2R_1R_2C_1C_2}$$

[It will be noticed that the right-hand side has dimensions of voltage × capacity, equal to current × time: this is correct for the Laplace transform of a current.] Following factorization of the denominator, the right-hand side can be expressed in partial fraction form and the inverse found. To simplify presentation, consider the case $R_1 = R_2 = 1\,\Omega$, $C_1 = C_2 = 1\,\text{F}$. The expression for $I(s)$ then becomes

$$I(s) = 2\hat{E}\frac{1 + \frac{1}{2}s}{1 + 3s + s^2}$$

The denominator factorizes into

$$1 + 3s + s^2 = (s + \alpha_1)(s + \alpha_2)$$

where

$$\alpha_1 = (3 - \sqrt{5})/2 = 0.382$$

$$\alpha_2 = (3 + \sqrt{5})/2 = 2.628$$

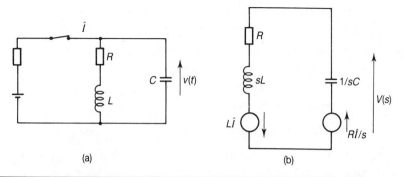

Figure 10.14 Example 10.7: (a) circuit; (b) *s*-domain circuit.

Performing the expansion we find

$$I(s) = \hat{E}\frac{2}{\sqrt{5}}\left(\frac{0.809}{s + \alpha_1} + \frac{0.309}{s + \alpha_2}\right)$$

Hence

$$i(t) = \hat{E}\frac{2}{\sqrt{5}}[0.809\exp(-0.382t) + 0.309\exp(-2.628t)]$$

As a check, consider $i(0)$. Since both v_1 and v_2 are then equal to $\hat{E}$, the current in R_2 will initially be zero and the current in R_1 will therefore be $\hat{E}/R_1$, or in the numerical case, $\hat{E}$. This is in agreement with the value of

$$\hat{E}\frac{2}{\sqrt{5}}(0.809 + 0.309)$$

EXAMPLE 10.7 The circuit of Figure 10.14(a) is allowed to reach a steady state with the switch closed. The current flowing in the circuit is $\hat{I}$. The switch is then opened. Obtain an expression showing how the voltage across the capacitor subsequently varies with time.

In the steady state the current through the inductor is $\hat{I}$, and the voltage across the capacitor is $R\hat{I}$. It is convenient to incorporate these values as series generators, giving the *s*-domain equivalent circuit shown in Figure 10.14(b). We then have

$$\left(R + sL + \frac{1}{sC}\right)I(s) = \hat{L}I + \hat{R}I\frac{1}{s}$$

The voltage across the capacitor is given by

$$V(s) = R\hat{I}\frac{1}{s} - \frac{1}{sC}I(s)$$

$$= R\hat{I}\frac{1}{s} - \hat{I}\frac{L + R/s}{LCs^2 + RCs + 1}$$

$$= \hat{I}\frac{L(RCs - 1) + R^2C}{LCs^2 + RCs + 1}$$

To simplify the algebra, we write

$$\omega^2 = 1/LC, \ \alpha = R/2L, \ \beta^2 = \omega^2 - \alpha^2$$

We then find that the expression can be rearranged to give

$$V(s) = R\hat{I}\frac{s + \alpha + (\alpha - \omega^2/2\alpha)}{(s + \alpha)^2 + \beta^2}$$

Referring to Table 10.2, we find

$$v(t) = R\hat{I}\exp(-\alpha t)\left[\cos(\beta t) + \frac{1}{\beta}\left(\alpha - \frac{\omega^2}{2\alpha}\right)\sin(\beta t)\right]$$

It can be checked that the correct initial conditions follow from this expression: $v(0) = R\hat{I}$, $-C[\mathrm{d}v/\mathrm{d}t]_{t=0} = \hat{I}$.

10.6 ◆ Summary

The examples in the previous sections have shown the way in which analysis of networks can be carried out when they are subject to excitation by any arbitrary waveform beginning at $t = 0$, and with any particular conditions pre-existing at that instant. The latter are accounted for by adding extra sources to the network, and subsequent analysis follows exactly the same path as would a phasor analysis but with the impedances of inductance and capacitance replaced by sL and $1/sC$ respectively. Thus the algebraic process is identical in the two cases except for the replacement of jω by s. One very important consequence of this replacement is that all expressions are real functions of the variable s. This is so useful that a nominally phasor analysis is often conducted in terms of s until a final replacement by jω needs to be made: the same analysis suffices to deal with both phasor analysis and with transient excitation under conditions of initial quiescence. This condition of initial quiescence is by far the most important: problems arising in practice which involve a non-quiescent initial state are usually rather specialized, one-off cases. Viewed in this way, the Laplace transform provides a unifying principle in the treatment of networks. (For table of Laplace transforms see Table 10.2, p. 263.)

PROBLEMS

The following problems should be done using the Laplace transform method: some of them have already appeared as problems in Chapter 9. The relative difficulty of the methods should be considered.

10.1 A series R–C circuit is excited by a source $e(t)$ defined by

$$t < 0 \qquad e(t) = 0$$
$$t > 0 \qquad e(t) = At$$

Obtain an expression for the voltage across the capacitor.

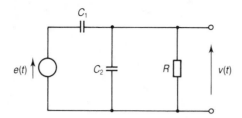

Figure 10.15

10.2 In the circuit of Figure 10.15 the source is defined by

$$t < 0 \qquad e(t) = 0$$

$$t > 0 \qquad e(t) = \hat{E} \sin(\omega t)$$

Obtain an expression for $v(t)$. Check that after a long time has elapsed this expression reduces to the result predicted by steady-state AC theory.

10.3 In the circuit of Figure 10.16 a step of height E is applied at $t = 0$. Assuming initial quiescence, show that the Laplace transform of $v_2(t)$ is given by

$$V_2(s) = E \frac{R_2}{L_2} \frac{1}{(s + \alpha_1)(s + \alpha_2)}$$

in which α_1, α_2 are the roots of the equation

$$L_1 L_2 s^2 + (R_1 L_1 + R_1 L_2 + R_2 L_1)s + R_1 R_2 = 0$$

Obtain an expression for $v_2(t)$ in the case

$$L_1 = L_2 = 1, \ R_1 = 1, \ R_2 = 2$$

By inspection of the circuit, it is to be expected that for small values of time, $v_2(t) \approx ER_2 t/L_2$. Apply this check to the expression obtained.

10.4 In the circuit shown in Figure 10.17, with the switch in position 1, the current from the source finally reaches a value I. In this state the switch is operated. Obtain expressions for the current $i_1(t)$ and for the voltage across the resistance R. (This problem appeared as Problem 9.7.)

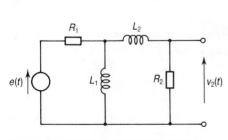

Figure 10.16

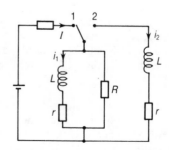

Figure 10.17

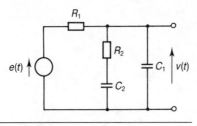

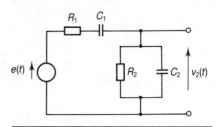

Figure 10.18   **Figure 10.19**

10.5   In the circuit of Figure 10.18 the components have the values

$$R_1 = 10, \ R_2 = 40, \ C_1 = 1/12, \ C_2 = 1/100$$

The source $e(t)$ is defined by

$$t < 0 \qquad e(t) = 0$$
$$t > 0 \qquad e(t) = \hat{E} \cos(t)$$

Assuming initial quiescence, show that the voltage $v_2(t)$ is given by the expression

$$v_2(t) = 0.03\hat{E}[18\cos(t) + 16\sin(t) - 3\exp(-3t) - 15\exp(-t)]$$

10.6   In the circuit of Figure 10.19, the source $e(t)$ is a step of height E. Obtain an expression for the Laplace transform of the voltage $v_2(t)$. Assume initial quiescence.

In a particular case the component values are given by

$$R_1 = R_2 = 1, \ C_1 = C_2 = 1$$

Show that

$$v_2(t) = E\frac{1}{\sqrt{5}}\left[\exp(-\alpha_1 t) - \exp(-\alpha_2 t)\right]$$

in which

$$\alpha_1 = (3 - \sqrt{5})/2, \ \alpha_2 = (3 + \sqrt{5})/2$$

Show that for small values of t

$$v_2(t) \approx Et$$

Check this by inspection of the circuit.

[Use the expansion $\exp(x) = 1 + x/1! + x^2/2! + \dots$]

10.7   The circuit shown in Figure 10.20 is allowed to reach a steady state with the switch closed. Obtain expressions for the currents $i_1(t)$, $i_2(t)$ after the switch is opened. In determining initial conditions, assume that the inductors have a very small series resistance. (This problem appeared as a numerical exercise in Problem 9.15.)

10.8   In an L–C–R circuit $L = 10$ H, $C = 5$ μF, $R = 3$ kΩ. These components and a switch, initially open, are connected in series and the capacitor charged to 50 V. The switch is then closed. Obtain expressions for the voltage across the capacitor and for the current in the circuit. (This problem appeared as Problem 9.9.)

10.9   The inductance, capacitance and resistance of Problem 10.8 are connected in a loop. The terminals of the capacitor are connected through a switch to a DC

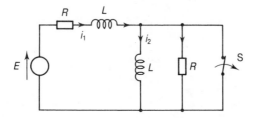

Figure 10.20

power supply. After a long time the voltage across the capacitor reaches a steady value of 50 V. The switch is then opened. Obtain an expression for the voltage across the capacitor. (This problem appeared as Problem 9.10.)

10.10   In the circuit of Figure 10.21

$$L = 6.4 \text{ mH}, \; C = 100 \text{ pF}, \; R = 9.6 \text{ k}\Omega$$

The current source applies a step of 10 mA at $t = 0$. Show that the voltage across the capacitor is given by the following expression:

$$v(t) = 96\{1 - \exp(-0.75\tau)[\cos \tau - (7/24)\sin \tau]\}$$

(This problem appeared as part of Problem 9.11.)

10.11 Show for the circuit of Figure 10.22 that, if the circuit is initially quiescent and $V(s)$, $E(s)$ denote the Laplace transforms of $v(t)$, $e(t)$ respectively,

$$\frac{V(s)}{E(s)} = \frac{R}{(R + sL)(LCs^2 + RCs + 2)}$$

Hence obtain an expression for $v(t)$ when $e(t)$ is a step of height E for the case $L = 1, C = 2, R = 1$.

10.12   One stage of an amplifier has the equivalent circuit shown in Figure 10.23. The amplifier contains n such stages in cascade. Show that the response to an impulse of magnitude K (V s) at the input to the first stage is given by

$$v_{n+1}(t) = (-1)^n K \frac{g}{C} (gR)^{n-1} \frac{(\alpha t)^{n-1}}{(n-1)!} \exp(-\alpha t)$$

in which $\alpha = 1/RC$.

10.13   An amplifier consists of three stages of the type appearing in Figure 10.23. Obtain an expression for the response when the input is a step of height E.

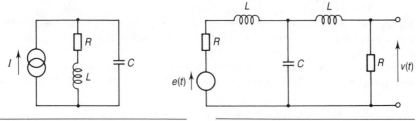

Figure 10.21   **Figure 10.22**

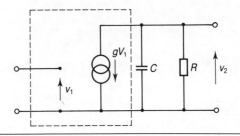

Figure 10.23

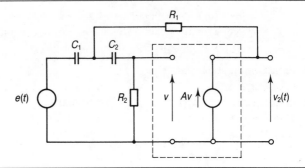

Figure 10.24

Taking for convenience values $gR = 1$, $\alpha = 1$, plot this response to determine the times taken to reach 10% and 90% of the final value. Hence determine the rise time. What would this rise time be in the case $R = 5$ kΩ, $C = 15$ pF?

10.14 A series L–C–R circuit of high Q factor and resonant angular frequency ω_0 $(= 1(LC)^{1/2})$ is excited by a series voltage source $\hat{E} \sin(\omega_0 t)$ applied to the quiescent circuit at $t = 0$. Show that the current in the circuit is given by

$$i(t) = \frac{\hat{E}}{R}[\sin(\omega_0 t) - (\omega_0/\beta)\exp(-\omega_0 t/2Q)\sin(\beta t)]$$

in which

$$\beta = \omega_0(1 - 1/4Q^2)^{1/2}$$

For high Q circuits this expression represents a sine wave whose amplitude is roughly given by

$$\frac{\hat{E}}{R}[1 - \exp(-\omega_0 t/2Q)]$$

Making use of this approximate expression estimate, in terms of Q, the number of cycles taken for the amplitude to reach 90% of its final value.

10.15 The circuit shown in Figure 10.24 contains a voltage-controlled dependent voltage source. Obtain an expression for the transfer function $V_2(s)/E(s)$. In a particular case

$$R_1 = R_2 = 1 \text{ k}\Omega, \ C_1 = C_2 = 1 \ \mu\text{F}, \ A = 0.5$$

Derive an expression for the response when $e(t)$ is a step of height E.

11 ◆ *The Signal Spectrum*: *Fourier Series*

11.1 Introduction	11.5 Computational aids: the
11.2 Periodic waveforms	discrete Fourier transform
11.3 Application to transient	11.6 Non-recurrent signals:
problems	the Fourier integral
11.4 Signal spectrum and	11.7 Summary
channel frequency	Bibliography
response	Problems

The idea that a non-sinusoidal waveform is equivalent to an assembly of sine waves each of different frequency and phase has been encountered in earlier chapters. In the present chapter this idea is explored quantitatively: it is shown that such representation is possible both for periodic and for one-off waveforms.

11.1 ◆ Introduction

The idea that any signal can be regarded as equivalent to an assembly of sine waves, each of the appropriate magnitude and phase, has been referred to in the discussions of earlier chapters. This equivalent assembly is termed the spectrum of the signal. In the present chapter the concept is explored further and quantified in certain important cases.

The calculation of the spectrum of a particular signal waveform may appear from what follows to be dauntingly mathematical, although in fact only simple mathematical operations are involved. To offset this aspect of the subject it may help to point out that a spectrum can be, and regularly is, determined experimentally. To do this an instrument called a spectrum analyser is used, the basis of which can be briefly described as follows. It has been seen that certain circuits, such as the inductance–capacitance resonant circuits studied in Chapter 5, respond primarily to a small band of frequencies around a centre frequency. Suitable design can make this range very small, perhaps only a few hertz with a centre frequency of many

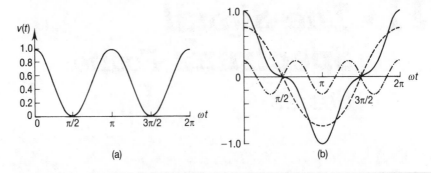

Figure 11.1 The decomposition of periodic, non-sinusoidal waveforms into harmonic components: (a) $\cos^2(\omega t)$; (b) $\cos^3(\omega t)$ (——, $\cos^3(\omega t)$; - - -, $0.75\cos(\omega t)$; — · —, $0.25\cos(3\omega t)$.

megahertz or gigahertz. In a spectrum analyser the signal is passed through such a narrow-band filter, the output of which gives a measure of the magnitude of the frequency component equal to the centre frequency. If this centre frequency is varied the spectrum can be investigated over a wide range of frequencies.

We begin by considering the case of signals in which the waveform consists of a regular repetition of a particular shape. Such waveforms are said to be periodic.

11.2 ◆ Periodic waveforms

Before investigating spectral representation of an arbitrary periodic waveform we will consider some illustrative examples derived directly from sinusoidal functions, the waveforms represented by an integer power of a cosine function.

11.2.1 ◆ The waveform $\cos^n(\omega t)$

Since $\cos\theta$ is periodic in 2π any function of $\cos\theta$ is also periodic with the same period. The waveform $\cos^n \omega t$ is therefore periodic in $T = 2\pi/\omega$. The waveforms for $n = 2$ and $n = 3$ are shown in Figure 11.1. We recall the well-known trigonometrical formulae

$$\cos^2\theta = \tfrac{1}{2}(1 + \cos 2\theta)$$

$$\cos^3\theta = \tfrac{1}{4}(3\cos\theta + \cos 3\theta)$$

These enable us to write

$$\cos^2\omega t = 0.5 + 0.5\cos 2\omega t \tag{11.1}$$

$$\cos^3\omega t = 0.75\cos\omega t + 0.25\cos 3\omega t \tag{11.2}$$

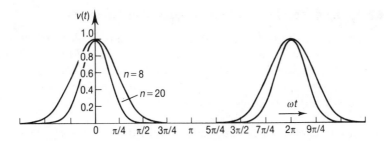

Figure 11.2 Graphs of $\cos^n (\omega t)$ for $n = 8,20$.

These equations show that in both cases the waveforms are equivalent to a linear sum of simple sinusoids of different frequencies. The frequencies involved are integer multiples of the frequency $1/T$. This is termed the fundamental frequency, and the integer multiples are referred to as the first, second, third, ..., harmonics of the fundamental $1/T$. In the case of Equation 11.1, the waveform has a zero frequency (or DC) component together with a second harmonic; in the case of Equation 11.2 the waveform has a fundamental component together with a third harmonic. The absence of a fundamental component in the first case shows that $\cos^2 \omega t$ is periodic in π/ω rather than $2\pi/\omega$. This is apparent from the graph in Figure 11.1(a).

As the value of n increases, the waveform represented by $\cos^n \omega t$ becomes more peaky as shown in Figure 11.2. In the case $n = 8$ we have

$$v(t) = \cos^8 \omega t$$
$$= \tfrac{1}{16}(1 + \cos 2\omega t)^4$$
$$= \tfrac{1}{16}(1 + 4\cos 2\omega t + 6\cos^2 2\omega t + 4\cos^3 2\omega t + \cos^4 2\omega t)$$

The relevant formulae are

$$\cos^2 \theta = \tfrac{1}{2}(1 + \cos 2\theta)$$
$$\cos^3 \theta = \tfrac{1}{4}(3\cos \theta + \cos 3\theta)$$
$$\cos^4 \theta = \tfrac{1}{4}(1 + \cos 2\theta)^2$$
$$= \tfrac{1}{8}(3 + 4\cos 2\theta + \cos 4\theta)$$

Hence we find for v(t) the expression

$$\tfrac{1}{128}(35 + 56\cos 2\omega t + 28\cos 4\omega t + 8\cos 6\omega t + \cos 8\omega t) \qquad \textbf{(11.3)}$$

Once again we find that $v(t)$ can be expressed as a sum of harmonic terms. The sharper spike associated with this waveform is accompanied by higher harmonics: as the value of n is increased to give a sharper pulse waveform so does the order of the highest harmonic, which will be equal to n.

11.2.2 ♦ The general periodic waveform: Fourier series

The examples of the last two sections indicate that similar expansions may be possible for any periodic waveform. Let $v(t)$ be such a waveform with period T. The expansion would be expected to take the form

$$v(t) = \sum_{n=0}^{\infty} [a_n \cos(n\omega t) + b_n \sin(n\omega t)] \qquad (11.4)$$

in which $\omega = 2\pi/T$. The sum has been written in a form involving an infinite number of terms: it is possible, as in the previous examples, that the series may terminate. In general very rapid rates of change or jumps in amplitude might be expected to lead to the presence of very high harmonics. If the series representation of Equation 11.4 is assumed to be possible then a way of calculating values for the coefficients can be devised. The method depends on the fact that the average over a period of either of the expressions $\cos(n\omega t)\cos(m\omega t)$ or $\sin(n\omega t)\sin(m\omega t)$ is zero unless $m = n$, and that the average of $\cos(n\omega t)\sin(m\omega t)$ is always zero. When $n = m$, the averages are $1/2$. The mathematical proof of these statements will be found in Appendix C. We proceed as follows: first multiply both sides of Equation 11.4 by $\cos(m\omega t)$, m being considered to be a particular integer, and integrate over one complete period. The result of this procedure may be written

$$\int_0^T v(t) \cos(m\omega t)\, dt$$

$$= \sum_{n=0}^{\infty} \left[a_n \int_0^T \cos(n\omega t)\cos(m\omega t)\, dt + b_n \int_0^T \sin(n\omega t)\cos(m\omega t)\, dt \right]$$

The results quoted above and proved in Appendix C show that all the integrals on the right-hand side are zero except for that multiplying a_m. Hence

$$\int_0^T v(t) \cos(m\omega t)\, dt = a_m \int_0^T \cos^2(m\omega t)\, dt$$

The integral on the right-hand side has the value $T/2$ unless $m = 0$, when the value is T. We therefore have the following expressions from which to calculate the coefficients a_m:

$$a_m = \frac{2}{T} \int_0^T v(t) \cos(m\omega t)\, dt \quad m \neq 0 \qquad (11.5)$$

$$a_0 = \frac{1}{T} \int_0^T v(t)\, dt \qquad (11.6)$$

It is to be noted that a_0 is the average, or DC, value of $v(t)$. Expressions for the coefficients b_n are found in a similar way: this time both sides of Equation 11.4 are multiplied by $\sin(m\omega t)$, followed by integration over a period. We find

$$\int_0^T v(t) \sin(m\omega t) \, dt$$

$$= \sum_{n=0}^{\infty} \left[a_n \int_0^T \cos(n\omega t) \sin(m\omega t) \, dt + b_n \int_0^T \sin(n\omega t) \sin(m\omega t) \, dt \right]$$

In this case only the integral multiplying b_m is non-zero, whence

$$\int_0^T v(t) \sin(m\omega t) \, dt = b_m \int_0^T \sin^2(m\omega t) \, dt$$

The expression for calculating the coefficients b_m is therefore

$$b_m = \frac{2}{T} \int_0^T v(t) \sin(m\omega t) \, dt \tag{11.7}$$

The expressions of Equations 11.5, 11.6 and 11.7 therefore suffice to determine all the coefficients in Equation 11.4 for any specified waveform $v(t)$. There may of course be instances in which analytic expressions cannot be obtained and when numerical evaluation will be necessary. That the series incorporating the values of the coefficients so determined actually adds up to the correct value for $v(t)$ at all instants of time can be proved mathematically, but the proof is outside the scope of this book. The series so obtained is known as the **Fourier series** for the function $v(t)$. Some examples will now be given.

EXAMPLE 11.1 Determine the Fourier series for the waveform $v(t)$ of period T for which

$$v(t) = 0 \qquad 0 < t < \tfrac{1}{2}T$$
$$v(t) = 1 \qquad \tfrac{1}{2}T < t < T$$

The waveform is shown in Figure 11.3(a). With this form for $v(t)$ the integrals in Equations 11.5, 11.6 and 11.7 break into two parts, one between 0 and $T/2$, which will be zero since $v(t)$ is zero in the range, the other between $T/2$ and T. Consider first Equation 11.6 which gives

$$a_0 = \frac{1}{T} \int_{T/2}^T 1 \, dt = \tfrac{1}{2}$$

Equation 11.5 gives

$$a_m = \frac{2}{T} \int_{T/2}^T \cos(m\omega t) \, dt$$

$$= \frac{2}{m\omega T} [\sin(m\omega t)]_{T/2}^T$$

Since $\omega = 2\pi/T$, $\sin(m\omega T)$ is zero at both end points, so that we find

$$a_m = 0 \qquad m \neq 0$$

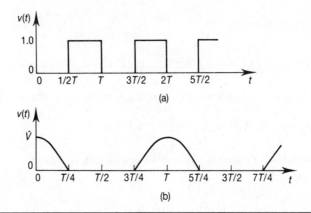

Figure 11.3 Waveform for (a) Example 11.1 and (b) Example 11.2.

Finally, from Equation 11.7 we have

$$b_m = \frac{2}{T}\int_{T/2}^{T} \sin(m\omega t)\,\mathrm{d}t$$

$$= \frac{2}{m\omega T}[-\cos(m\omega t)]_{T/2}^{T}$$

$$= \frac{1}{m\pi}[\cos(m\pi) - \cos(2m\pi)]$$

Hence

$$b_m = 0 \qquad m \text{ even}$$

$$= -2/m\pi \qquad m \text{ odd}$$

We can express this result more simply by writing

$$m = 2p + 1$$

and allowing p to take values 0, 1, 2, ... The Fourier series for $v(t)$ can then be written

$$v(t) = \frac{1}{2} - \frac{2}{\pi}\sum_{p=0}^{\infty}\frac{1}{2p+1}\sin\left[(2p+1)\frac{2\pi t}{T}\right] \qquad \textbf{(11.8)}$$

A certain amount of work could have been avoided had it been noticed that a_0 is in fact the average of $v(t)$ over the period and clearly equal to 1/2. Further, the function

$$v_1(t) = v(t) - a_0$$

is an odd function, $v_1(-t) = -v_1(t)$, and so cannot be represented by a series of cosines, which are even functions. The coefficients a_m must therefore be zero when $m \geqslant 1$.

The series of Equation 11.7 converges only very slowly, with individual terms decreasing as $1/m$. It may be shown that this behaviour is a direct consequence of the presence of discontinuities of amplitude in $v(t)$.

EXAMPLE 11.2 Determine the Fourier series for the waveform of period T for which

$$v(t) = \hat{V} \cos\left(\frac{2\pi t}{T}\right) \qquad |t| < \tfrac{1}{4}T$$

$$= 0 \qquad \tfrac{1}{4}T < |t| < \tfrac{1}{2}T$$

The waveform is shown in Figure 11.3(b), from which it will be seen that $v(t)$ is an even function about $t = 0$. It cannot be represented by a sine series, since these are odd functions, so that we deduce $b_m = 0$ for all values of m. This can be confirmed by evaluating the appropriate integral, as in the last example.

For the coefficient a_0 we have

$$a_0 = \frac{\hat{V}}{T} \int_{-T/4}^{T/4} \cos\left(\frac{2\pi t}{T}\right) dt$$

Changing the variable to $\theta = 2\pi t/T$, we find

$$a_0 = \frac{\hat{V}}{2\pi} \int_{-\pi/2}^{\pi/2} \cos\theta \, d\theta = \frac{\hat{V}}{\pi}$$

For non-zero values of m we have, using the same change of variable,

$$a_m = \frac{\hat{V}}{\pi} \int_{-\pi/2}^{\pi/2} \cos\theta \cos m\theta \, d\theta$$

This integral is evaluated in Appendix C, Section C.1. When m is an odd integer the integral vanishes, except for the case $m = 1$. We find

$$a_1 = \tfrac{1}{2}\hat{V}$$

For even integer values we write $m = 2p$, $p = 1, 2, 3, \ldots$. We find

$$a_{2p} = -\hat{V}\frac{2}{\pi} \frac{(-1)^p}{4p^2 - 1}$$

The required series is therefore

$$v(t) = \frac{\hat{V}}{\pi}\left[1 + \frac{\pi}{2}\cos\left(\frac{2\pi t}{T}\right) - 2\sum_{p=1}^{\infty}\frac{(-1)^p}{4p^2 - 1}\cos\left(\frac{4p\pi t}{T}\right)\right] \qquad (11.9)$$

It will be noticed in this case that as m increases the terms decrease as m^{-2}, as opposed to m^{-1} in Example 11.1. This faster rate of decrease occurs because in the present case the waveform is not discontinuous in amplitude, but only in slope (at $t = \pm T/4$).

EXAMPLE 11.3 Determine the Fourier expansion of the trapezoidal waveform of period 1s defined by

$$-0.5 < t < -0.3 \qquad v(t) = 0$$

$$-0.3 < t < -0.1 \qquad v(t) = 5(t + 0.3)$$

$$-0.1 < t < 0.1 \qquad v(t) = 1$$

$$0.1 < t < 0.3 \qquad v(t) = 5(0.3 - t)$$

$$0.3 < t < 0.5 \qquad v(t) = 0$$

Since the waveform is symmetrical about $t = 0$ and is of period 1, it has an expansion in the form

$$v(t) = \sum_{n=0}^{\infty} a_n \cos (2\pi nt)$$

By inspection we see

$$a_0 = \langle v(t) \rangle = 0.4$$

For other values of n, making use of symmetry, we have

$$a_n = 2 \times 2 \int_0^{0.5} v(t) \cos (2\pi nt) \, dt$$

$$= 4 \int_0^{0.1} \cos (2\pi nt) \, dt + 4 \int_{0.1}^{0.3} 5(0.3 - t) \cos (2\pi nt) \, dt$$

The second integral contains the term

$$\int_{0.1}^{0.3} t \cos (2\pi nt) \, dt$$

To evaluate this we integrate by parts to obtain

$$\left[t \, \frac{\sin (2\pi nt)}{2\pi n} \right]_{0.1}^{0.3} - \int_{0.1}^{0.3} \frac{\sin (2\pi nt)}{2\pi n} \, dt$$

$$= \left[t \, \frac{\sin (2\pi nt)}{2\pi n} \right]_{0.1}^{0.3} - \left[-\frac{\cos (2\pi nt)}{(2\pi n)^2} \right]_{0.1}^{0.3}$$

$$= \frac{1}{2\pi n} [0.3 \sin (0.6\pi n) - 0.1 \sin (0.2\pi n)]$$

$$+ \left(\frac{1}{2\pi n} \right)^2 [\cos (0.6\pi n) - \cos (0.2\pi n)]$$

Performing the remaining integrals and collecting terms, we find that the terms in $1/n$ cancel as they should, since the waveform has no discontinuity of amplitude, leaving

$$a_n = 5 \left(\frac{1}{\pi n} \right)^2 [\cos (0.2\pi n) - \cos (0.6\pi n)]$$

Numerical values obtained from this formula will be found tabulated in the solution to Example 11.5, Section 11.5.2.

11.2.3 ♦ Alternative forms for the Fourier series

It is sometimes useful to express the series of Equation 11.4 in a different form, combining the two trigonometrical functions as follows:

$$a_n \cos (n\omega t) + b_n \sin (n\omega t) = c_n \cos (n\omega t + \theta_n)$$

in which

$$c_n = (a_n^2 + b_n^2)^{1/2}$$

and θ_n is to be determined from

$$a_n = c_n \cos \theta_n, \ b_n = c_n \sin \theta_n$$

We may then write Equation 11.4 in the form

$$v(t) = \sum_{n=0}^{\infty} c_n \cos(n\omega t + \theta_n) \tag{11.10}$$

Alternatively we could use the form

$$v(t) = \sum_{n=0}^{\infty} c_n \sin(n\omega t + \phi_n) \tag{11.11}$$

in which

$$\phi_n = \theta_n - \pi/2$$

This latter form is used in the simulation program SPICE. Further information will be found in Section 15.14.

11.2.4 ◆ Complex form for the Fourier series

Another form which is useful is obtained by replacing the trigonometrical functions by complex exponentials. Using in Equation 11.10 the identity

$$\cos \theta = \tfrac{1}{2}[\exp(j\theta) + \exp(-j\theta)]$$

we have

$$v(t) = c_0 + \tfrac{1}{2}\sum c_n[\exp(jn\omega t + \theta_n) + \exp(-jn\omega t - \theta_n)]$$

We define a set of coefficients C_n for any integer n in the range $-\infty < n < \infty$ as follows:

$$C_0 = c_0$$

$$\text{for } n \geqslant 1 \quad C_n = \tfrac{1}{2}c_n \exp(j\theta_n)$$

$$\text{for } n \leqslant -1 \quad C_n = C_{-n}^*$$

With these definitions, Equation 11.10 can then be written

$$v(t) = \sum_{n=-\infty}^{\infty} C_n \exp(jn\omega t) \tag{11.12}$$

Expressions for the coefficients C_n can be found for any particular $v(t)$ by use of Equations 11.5, 11.6 and 11.7, but are more easily determined directly: multiply both sides of Equation 11.12 by $\exp(-jm\omega t)$ and integrate over one period. We find

$$\int_0^T v(t)\exp(-jm\omega t)\,dt = \sum_{n=-\infty}^{\infty} C_n \int_0^T \exp[j(n-m)\omega t]\,dt$$

The function $\exp[j(n-m)\omega t]$ is periodic in $T = 2\pi/\omega$ so that the integrals on the right-hand side all vanish except for the case $n = m$. The equation therefore reduces to

$$\int_0^T v(t)\exp(-jm\omega t)\,dt = C_m \int_0^T 1\,dt = C_m T$$

from which

$$C_m = \frac{1}{T}\int_0^T v(t)\exp(-jm\omega t)\,dt \tag{11.13}$$

This single expression has the advantage of being correct for all values of m, including zero.

11.3 ◆ Application to transient problems

When the excitation applied to a linear network has a waveform which is periodic in time the response also becomes periodic after the initial transients have died away. One way of determining this periodic response uses the Fourier series representation of the exciting waveform. The basis of this approach has already been encountered in Section 4.17, in which it was shown how the response to an excitation made up of several sinusoidal waveforms of different frequencies could be found. Consider the case of a linear network for which an output is related to the sinusoidal input of angular frequency ω by an amplification of $A(\omega)$ and a phase lead of $\phi(\omega)$, it being assumed that the functions $A(\omega)$ and $\phi(\omega)$ are known for all values of ω. Thus, an input

$$v_i(t) = \hat{V}\cos(\omega t)$$

produces an output

$$v_o(t) = A(\omega)\hat{V}\cos[\omega t + \phi(\omega)] \tag{11.14}$$

Consider an excitation which can be described as the superposition of many sinusoidal components by the series form of Equation 11.11:

$$v_i(t) = \sum_{n=0}^{\infty} c_n\cos(n\omega t + \theta_n)$$

The response to this waveform will be, for a linear network, the same as the sum of the responses obtained by applying each sinusoidal component separately. We therefore have

$$v_o(t) = \sum_{n=0}^{\infty} c_n A(n\omega)\cos[n\omega t + \theta_n + \phi(n\omega)] \tag{11.15}$$

An analytic form for the sum of the series, if it could be obtained, would give the response at all times throughout the cycle of the waveform. The

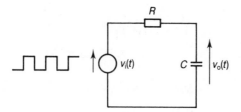

Figure 11.4 Circuit for Example 11.4.

problem is to obtain the sum: an analytic form can rarely be found and the summation of a series consisting of many terms of varying signs by numerical means is a significant computational task. However, the general form of solution represented by Equation 11.15 will be seen to be of great importance.

EXAMPLE 11.4 The square wave described in Example 11.1 forms the input to the resistance–capacitance circuit shown in Figure 11.4. The time constant (RC) of the circuit is equal to the period T of the square wave. Obtain an expression for the voltage across the capacitor in the form of a Fourier series.

We first consider the voltage across the capacitor when the input is sinusoidal. Let the phasor representation of the input of angular frequency ω be $\mathbf{V_i}$. Then, following the methods of Chapter 5, the output will be represented by the phasor $\mathbf{V_o}$ given by

$$\mathbf{V_o} = \frac{1}{1 + j\omega RC}\mathbf{V_i}$$

Hence, the function $A(\omega)$ of Equation 11.14 is given by

$$A(\omega) = \frac{1}{[1 + (\omega T)^2]^{1/2}}$$

and the phase $\phi(\omega)$ by

$$\phi(\omega) = -\tan^{-1}(\omega T)$$

The series for the excitation, Equation 11.8, gives

$$v_i(t) = \frac{1}{2} - \frac{2}{\pi} \sum_{p=0}^{\infty} \frac{1}{2p + 1}\sin[2p + 1)\Omega t]$$

where $\Omega = 2\pi/T$. The response is therefore

$$v_0(t) = \frac{1}{2} - \frac{2}{\pi} \sum_{p=0}^{\infty} \frac{1}{2p + 1}A_p\sin[(2p + 1)\Omega t + \phi_p]$$

where

$$A_p = [1 + (2p + 1)^2(\Omega T)^2]^{-1/2}$$

and

$$\phi_p = -\tan^{-1}[(2p + 1)\Omega T]$$

At the angular frequencies Ω, 3Ω, 5Ω, for which ωT has the values 2π, 6π, 10π, the values of A and ϕ are as follows:

ω	$A(\omega)$	$\phi(\omega)$ (deg)
Ω	0.1572	−81.0
3Ω	0.0530	−87.0
5Ω	0.0318	−88.2

The first four terms of the series for $v_0(t)$ become

$$v_0(t) = 0.5 - 0.100\sin(\omega t + \phi_0) - 0.018\sin(3\omega t + \phi_1)$$
$$- 0.006\sin(5\omega t + \phi_3) - \ldots$$

In this case it is possible to obtain an exact solution with which the series solution may be compared. This exact solution is derived as follows: the variation of the voltage across the capacitor over the period $0 < t < T$ can be regarded as the superposition of a natural response term $A\exp(-t/RC)$, resulting from the excitation prior to $t = 0$, together with the response to a unit step applied at $t = T/2$. Using the results of Equation 9.4, Section 9.2.3, and Equation 9.6, Section 9.4.1, we have

$$0 < t < 0.5T \qquad v_0(t) = A\exp(-t/T)$$
$$0.5T < t < T \qquad v_0(t) = A\exp(-t/T) + 1 - \exp[-(t - 0.5T)/T]$$

In order that the response be periodic we must have

$$v_0(T) = v_0(0)$$

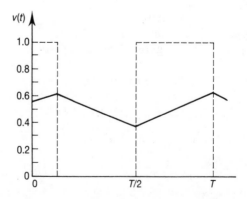

Figure 11.5 Calculated capacitor voltage waveform for Example 11.4.

Applying this condition yields an equation from which the constant A can be determined. We find

$$A = \frac{\exp(1) - \exp(0.5)}{\exp(1) - 1} = 0.622$$

The form of the resulting response is shown in Figure 11.5. It will be found that the first two terms of the series, a DC contribution of 0.5 V together with an AC component of 0.1 V peak amplitude, correspond to the true response to within a few per cent, although the sharp change of slope at $t = 0$, $0.5T$, T is not reproduced. Whether or not such differences are important depends on the use to which the solution is put. If the object of the present analysis were to estimate the magnitude of the fluctuations about the mean level and to choose a suitable value of the capacitor to reduce those fluctuations to a given low level, the analysis would be entirely adequate. For other purposes it might be necessary to sum many more terms of the series.

11.4 ♦ Signal spectrum and channel frequency response

The situation dealt with in Section 11.3 can be regarded in a more general way. It concerns basically the transmission of a signal through a network to a given output port. This is the constantly recurring matter of interest whenever signals are transmitted through a communication channel, the properties of which can be characterized by a channel frequency response. In this situation, the problem is not so much to find out what gross distortion takes place during transmission as to ensure that the channel will pass the signal substantially unaltered. This can be investigated by comparing the spectrum of the signal with the frequency response of the channel. The situation can be depicted in graphical form. The Fourier spectrum of the signal can be described by marking the magnitudes of the coefficients c_n of Equation 11.10 against a frequency axis, as in Figure 11.6(a). The frequency response of the network, $A(\omega)$, is defined in Equation 11.14. This can be depicted graphically as in Figure 11.6(b), using the same frequency axis as that in Figure 11.6(a). Equation 11.15, expressing the effect of the network on the signal, is obtained by multiplication of the signal spectrum by the frequency response, making due allowance for phase. This result of this process as far as the magnitudes are concerned is shown in Figure 11.6(c). If we are interested in transmitting the signal with little distortion there are two special cases to be noted: these are

(i) when each component of the spectrum is multiplied by the same constant, and

(ii) when the phase varies with frequency in a linear fashion.

In the first case, $A(\omega)$ is a constant, independent of frequency. This

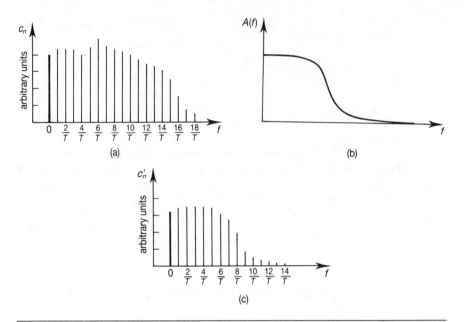

Figure 11.6 Signal spectrum and channel frequency response: (a) magnitude of components of frequency n/T; (b) magnitude of channel gain; (c) magnitude of frequency components after transmission.

corresponds to a multiplication of the signal by that constant. Most channels introduce attenuation, in which case the constant will be less than unity; to overcome this loss of signal amplifiers are introduced.

The second case corresponds to a delay of the signal, without change in shape:

$$v_o(t) = v_i(t - \tau)$$

The phase response which will bring about such delay can be seen as follows: consider the nth term of the Fourier series for $v_o(t)$, $c_n \cos(n\Omega t + \theta_n)$. The delay will make this term

$$c_n \cos(n\Omega t + \theta_n - n\Omega\tau)$$

This is equivalent to a phase response at an angular frequency ω given by

$$\phi = -\omega\tau$$

A channel having such a frequency characteristic corresponds to a simple delay. Although a delay without distortion does not alter the shape of the signal waveform and is usually permissible, it can lead to undesirable effects: for example, a delay of 0.5 s on a two-way telephone link renders conversation difficult. For most purposes, however, a channel required to be distortionless is designed so that the frequency response is of the form

$$H(j\omega) \approx A_o \exp(-j\omega\tau) \tag{11.16}$$

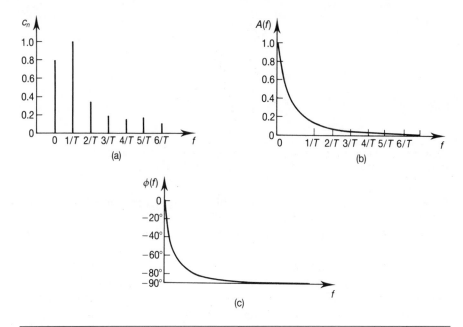

Figure 11.7 Spectral interpretation of Example 11.3: (a) spectral components of signal; (b) gain–frequency response of circuit; (c) phase–frequency response of circuit.

over the range of frequencies for which the signal spectrum has significant magnitude.

Viewed from this standpoint, it can readily be seen that the network of Example 11.3 would be expected to seriously distort the square wave signal. In Figure 11.7 is shown, in the form adopted in Figure 11.6, the square wave spectrum, the magnitude of the network frequency response and the phase of the response. From these figures it is clear that $A(\omega)$ falls off far more rapidly than do the components of the signal spectrum, corresponding to the gross distortion found.

In practice signals have frequency components which are of significant magnitude only up to a certain frequency, characteristic of the signal, so that an approximation to the distortionless channel characterized by Equation 11.16 can be designed. Thus good quality transmission of sound requires that all frequencies between about 20 Hz and 20 000 Hz are reproduced; colour TV signals extend to about 5 MHz.

The frequency content of a signal is limited by the finite rate of change of the signal with time. A rough equivalent relating time of rise to maximum frequency can be established by considering the case for a sine wave, as illustrated in Figure 11.8. Thus a rise time of τ_r corresponds to a maximum frequency given by

$$f_c \approx 1/2\tau_r \qquad\qquad (11.17)$$

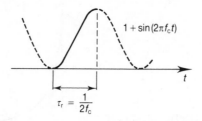

Figure 11.8 Diagram indicating relation between rise time and highest frequency component, Equation 11.17, Section 11.4.

It must be emphasized that this equivalence is only approximate: no practical signal has a truly sharp cut-off at a particular frequency, although it is often for design purposes convenient and adequate to assume this to be the case.

To summarize the general discussion of this section, a signal can be defined either by its waveform in time or by its frequency spectrum. In the first case it is said to be defined in the time domain, in the second case in the frequency domain. Determination of the waveform of a signal is a matter for experiment. Determination of the spectrum may involve calculation of the Fourier series but, as indicated in Section 11.1, may equally well be experimental, using a spectrum analyser.

11.5 ◆ Computational aids: the discrete Fourier transform

Evaluation of the coefficients a_m, b_m in the infinite series of Equation 11.4 by use of Equations 11.4, 11.5 and 11.6 is rarely straightforward. In practice, for waveforms of interest the coefficients are virtually zero above a certain frequency characteristic of the waveform, as discussed in Section 11.4. In this case Equation 11.4 becomes a finite series:

$$v(t) = \sum_{m=0}^{M} \left[a_m \cos\left(\frac{2\pi m t}{T}\right) + b_m \sin\left(\frac{2\pi m t}{T}\right) \right] \tag{11.18}$$

Since there is only a finite number of coefficients to be determined, an alternative to evaluating the integrals of Equations 11.4, 11.5 and 11.6 exists. As it stands, Equation 11.18 contains $2M + 1$ unknown coefficients. If values of $v(t)$ at $2M + 1$ different times are known then a set of simultaneous linear equations can be set up from which the unknown coefficients can be determined. These values of $v(t)$ can be acquired by calculation if a mathematical expression for $v(t)$ is available, or by sampling the waveform in real time by experiment. Consider the situation when N, conveniently taken to be an even number, equally spaced samples are taken over one period. The times of sampling will be taken to be

$$t = \frac{kT}{N}, \ k = 0, 1, 2, \ldots, N - 1$$

The value at $k = N$ will be the same as that at $k = 0$ on account of the assumed periodicity of the waveform. Denoting

$$v(kT/N) = v_k$$

and noting that for sample k the ratio t/T is equal to k/N, the set of equations takes the form

$$v_k = \sum_{m=0}^{M} \left[a_m \cos\left(\frac{2\pi mk}{N}\right) + b_m \sin\left(\frac{2\pi mk}{N}\right) \right], \ k = 0, 1, \ldots, N - 1$$

Since we have only N equations we can hope to determine only N coefficients. This number is reached if we take $M = N/2$ (N was assumed to be even). Although there are apparently $2M + 1$ coefficients to be determined, the coefficient b_M does not enter in, since the expression $\sin(2\pi mk/N)$ is zero for $m = N/2$. Noting that for this value of m the term $\cos(2\pi mk/N)$ reduces to $(-1)^k$, the set of equations takes the form

$$v_k = a_0 + a_M(-1)^k + \sum_{m=1}^{M-1} \left[a_m \cos\left(\frac{2\pi mk}{N}\right) + b_m \sin\left(\frac{2\pi mk}{N}\right) \right]$$

$$(11.19)$$

in which k can take the values $0, 1, \ldots, 2M - 1$. It has to be assumed that the number of samples, N, has been chosen to be sufficiently large to ensure that the higher coefficients are very small. The effective neglect of b_M, for example, will not then give rise to significant error.

It may be noted that an exact algebraic solution to these equations can be obtained in a form similar to that obtained in the case of the infinite Fourier series. The coefficients can be found from the expressions

$$a_0 = \frac{1}{N} \sum_{k=0}^{N-1} v_k$$

$$a_n = \frac{2}{N} \sum_{k=0}^{N-1} v_k \cos\left(\frac{2\pi kn}{N}\right) \qquad (11.20)$$

$$b_n = \frac{2}{N} \sum_{k=0}^{N-1} v_k \sin\left(\frac{2\pi kn}{N}\right)$$

These expressions define the discrete Fourier transform.

11.5.1 ◆ The Nyquist critical frequency

In the above analysis, the interval between samples is

$$\tau = \frac{T}{N} = \frac{T}{2M}$$

Associated with this interval is the special frequency known as the **Nyquist critical frequency** equal to

$$f_c = \frac{1}{2\tau}$$

This frequency is significant for the following reason: a sine wave of this frequency can take a zero value at each sampling point, as can be seen by considering $\sin(2\pi f_c t)$ at the times $k\tau = k/2f_c$. The presence of such a term could not be discovered from a set of samples taken with this interval. The principal conclusion is that in order to derive correctly the spectrum of a waveform by use of sampled values, a sampling interval must be chosen so that the corresponding Nyquist frequency is higher than the maximum frequency present in the waveform.

11.5.2 ◆ The fast Fourier transform

Straightforward solution of the set of Equations 11.19 involves in general a great deal of computation. It has been found, however, that if the number N is chosen to be an integer power of 2 then more efficient algorithms can be devised. These algorithms are referred to by the name of the **fast Fourier transform**. Further information can be found in the references given at the end of the chapter. Use of these transforms has made the interchange between waveforms in the time domain and spectra in the frequency domain relatively simple. In Section 11.1 a spectrum analyser was introduced as a narrow-band filter, accepting only the wanted frequency. It can now be seen that an alternative method of determining the spectrum of a waveform by experiment is to sample the waveform at a sufficiently rapid rate and then to use the fast Fourier transform algorithm. Using digital techniques, the sampled values can be transferred directly to the computer store for subsequent processing.

EXAMPLE 11.5   Obtain a finite expansion for the waveform of Example 11.3 in the form of Equation 11.19 in the cases $N = 8$ and $N = 16$.

   The coefficients in the series can be found by use of Equations 11.20 or, if available, a computer program realizing the fast Fourier transform. For the case $N = 8$, $\Delta t = 0.125$ so that the points are taken as follows:

k	t	v_k
0	0	1
1	0.125	0.875
2	0.25	0.25
3	0.375	0
4	0.5	0
5	0.625	0
6	0.75	0.25
7	0.875	0.875

For $N = 16$, $\Delta t = 0.0625$, the 16 points are

k	t	v_k	k	t	v_k
0	0	1	8	0.5	0
1	0.0625	1	9	0.5625	0
2	0.125	0.875	10	0.625	0
3	0.1875	0.5625	11	0.6875	0
4	0.25	0.25	12	0.75	0.25
5	0.3125	0	13	0.8125	0.5625
6	0.375	0	14	0.875	0.875
7	0.4375	0	15	0.9375	1

The values of the coefficients a_n resulting from processing these figures are given in the following table. As to be expected because of symmetry, the coefficients b_n are all zero. The coefficients from the infinite series obtained in Example 11.3 are included for comparison.

	$N = 8$	$N = 16$	*Example 11.3*
a_0	0.4063	0.3984	0.4
a_1	0.5594	0.5645	0.5664
a_2	0.1250	0.1398	0.1416
a_3	− 0.0594	− 0.0639	− 0.0629
a_4	− 0.0625	− 0.0313	− 0.0354
a_5		0.0046	0
a_6		− 0.0148	− 0.0157
a_7		− 0.0051	− 0.0116
a_8		0.0156	0.0070

As to be expected, the agreement is worse for the higher harmonics. However, for the 16-point series, the values are within 5% of that of the first harmonic and computation of the series shows a good approximation to the original trapezoidal waveform.

It may be noticed that the rise and fall times of the waveform, of the order of 0.2 s, when substituted in Equation 11.17 suggest significant harmonic content up to about 4 Hz. The fact that the waveform has discontinuities in slope make this an underestimate, but the table shows that the fall in value is rapid after the fourth harmonic.

11.6 ◆ Non-recurrent signals: the Fourier integral

In some systems the signal can be of short duration, not repeated or recurring only infrequently. An example of such a system is radar: typically, a pulse of perhaps 1 μs duration is emitted with a recurrence frequency of 500 Hz, successive pulses thus being 2 ms apart. We will consider this case in more detail. Clearly, the waveform is periodic and can therefore be represented by a Fourier series, which it is convenient to take in the exponential form of Equation 11.12:

$$v(t) = \sum_{n=-\infty}^{\infty} C_n \exp\left(j\frac{2\pi n t}{T}\right) \tag{11.21}$$

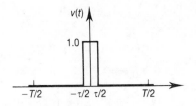

Figure 11.9 Repetitive pulse waveform for which $T \gg \tau$.

in which

$$C_n = \frac{1}{T} \int_{-T/2}^{T/2} v(t) \exp\left(-j\frac{2\pi n t}{T}\right) dt \tag{11.22}$$

(In this integral the limits do not matter provided that they cover a complete period.) The waveform will be taken in the form shown in Figure 11.9, with the period extending between $-T/2$ and $T/2$. In the example quoted earlier the pulse length τ would be 10^{-6} s and the period T 2×10^{-3} s. Substituting in Equation 11.21 we find

$$C_n = \frac{1}{T} \int_{-\tau/2}^{\tau/2} \exp\left(-j\frac{2\pi n t}{T}\right) dt$$

$$= \frac{1}{\pi n} \sin\left(\frac{n\pi\tau}{T}\right) \tag{11.23}$$

With the numerical values for τ and T this becomes

$$C_n = \frac{1}{\pi n} \sin\left(\frac{n\pi}{2} 10^{-3}\right)$$

Thus very many values of n will be significant and very many harmonics will have to be taken into account. It may also be noticed that the argument of the sine function changes only by a very small amount between successive values of n. We will rearrange Equation 11.23 using the following notation:

$$\frac{n}{T} = f_n$$

$$\frac{1}{T} = \Delta f$$

The frequency f_n is the frequency of the harmonic of order n, and Δf is the frequency between adjacent harmonics. Equation 11.23 then becomes

$$C_n = \Delta f \frac{1}{\pi f_n} \sin\left(\pi \tau f_n\right) \tag{11.24}$$

Using this expression in Equation 11.21 we find

$$v(t) = \sum_{n=-\infty}^{\infty} \Delta f \frac{1}{\pi f_n} \sin(\pi \tau f_n) \exp(j2\pi f_n t)$$

With the very small interval between successive values of the terms, this sum approximates closely to an integral:

$$v(t) = \int_{-\infty}^{\infty} \frac{1}{\pi f} \sin(\pi \tau f) \exp(j2\pi f t) \, df \tag{11.25}$$

The approximation becomes more correct as T grows larger, and Equation 11.25 is a correct expression in the case of a single pulse. In this limiting case Equation 11.22, leading to the expression $\sin(\pi \tau f)/\pi f$, takes the form

$$\frac{1}{\pi f} \sin(\pi \tau f) = \int_{-\infty}^{\infty} v(t) \exp(-j2\pi f t) \, dt \tag{11.26}$$

Equations 11.25 and 11.26 are said to be a **Fourier integral transform pair**. Evaluation of the integral in Equation 11.25, to check that it does in fact give the correct form for $v(t)$, is unfortunately not an elementary matter although it can be so checked by use of the appropriate mathematical technique. The arguments used to derive the pair of Equations 11.25 and 11.26 can be repeated without difficulty for any function of time which represents a signal of finite duration. The general Fourier integral transform pair take the form

$$v(t) = \int_{-\infty}^{\infty} V(f) \exp(j2\pi f t) \, df$$

$$V(f) = \int_{-\infty}^{\infty} v(t) \exp(-j2\pi f t) \, dt \tag{11.27}$$

The first of these integrals can be interpreted as expressing the strength of the frequency components in a small range df about f as $V(f)\,df$. For this reason $V(f)$ is called the **spectral density** of the time function $v(t)$. The spectral density for the radar pulse, giving by Equation 11.27 with the appropriate values for τ and T, is shown in Figure 11.10. Apart from the fact that the ordinate represents spectral density rather than strength of harmonic components, the diagram is similar to those of Figures 11.6 and 11.7 and can be used in a similar way: a circuit required to pass the pulse with reasonable preservation of shape needs to pass frequencies up to several times τ^{-1}. To ensure only a reasonable amplitude, rather than shape, passing frequencies up to τ^{-1} is found to be adequate.

EXAMPLE 11.6 Determine the spectral density associated with a single pulse having the form, shown in Figure 11.11(a),

$$v(t) = \cos(\pi t/2\tau) \qquad |t| < \tau$$
$$= 0 \qquad |t| > \tau$$

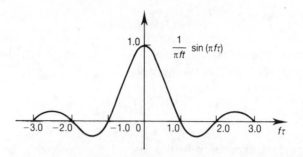

Figure 11.10 Fourier transform: spectral density of a single pulse with waveform of Figure 11.9.

For the spectral density we have, using Equations 11.27,

$$V(f) = \int_{-\infty}^{\infty} v(t) \exp(-j2\pi ft)\, dt$$

$$= \int_{-\tau}^{\tau} \cos\left(\frac{\pi t}{2\tau}\right)[\cos(2\pi ft) - j\sin(2\pi ft)]\, dt$$

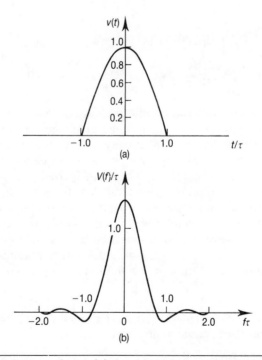

Figure 11.11 Fourier transform: (a) time domain waveform of single pulse; (b) spectral density.

The imaginary part of the integrand containing the sine term is an odd function of the variable f, and so does not contribute to the integral. Hence

$$V(f) = \int_{-\tau}^{\tau} \cos\left(\frac{\pi t}{2\tau}\right) \cos(2\pi ft)\,dt$$

Direct evaluation of the integral, or reference to tables of integrals, gives the result

$$V(f) = \frac{4\tau}{\pi} \cos(2\pi f\tau)/(1 - 16f^2\tau^2)$$

This expression is illustrated graphically in Figure 11.11(b).

11.7 ◆ Summary

It has been shown by means of examples how a signal waveform can be regarded as made up by superposition of a number of sinusoidal waveforms. These waveforms constitute the spectrum of the signal waveform. For periodic waveforms the spectrum contains only harmonic frequencies of the fundamental; for one-off transient signals all frequencies will be represented.

It has been shown also how the spectrum can be obtained by computation from a set of sample values.

The application of the simulation program SPICE to perform the Fourier analysis of waveforms is dealt with in Section 15.14.

Formulae
Fourier series
If $v(t)$ is periodic in period $T = 2\pi/\omega$ then

$$v(t) = \sum_{n=0}^{\infty} [a_n \cos(n\omega t) + b_n \sin(n\omega t)]$$

in which

$$a_0 = \frac{1}{T}\int_0^T v(t)\,dt$$

$$a_m = \frac{2}{T}\int_0^T v(t)\cos(m\omega t)\,dt, \ m \neq 0$$

$$b_m = \frac{2}{T}\int_0^T v(t)\sin(m\omega t)\,dt$$

If

$$c_n = (a_n^2 + b_n^2)^{1/2}$$

and θ_n is determined from

$$a_n = c_n \cos\theta_n, \ b_n = c_n \sin\theta_n$$

then

$$v(t) = \sum_{n=0}^{\infty} c_n \cos(n\omega t + \theta_n)$$

Alternatively

$$v(t) = \sum_{n=0}^{\infty} c_n \sin(n\omega t + \phi_n)$$

in which

$$\phi_n = \theta_n - \pi/2$$

If

$$C_m = \frac{1}{T} \int_0^T v(t) \exp(-jm\omega t) \, dt$$

then

$$v(t) = \sum_{n=-\infty}^{\infty} C_n \exp(jn\omega t)$$

The discrete Fourier transform

With $N = 2M$

$$v_k = a_o + a_M(-1)^k + \sum_{m=1}^{M-1} \left[a_m \cos\left(\frac{2\pi mk}{N}\right) + b_m \sin\left(\frac{2\pi mk}{N}\right) \right]$$

$$a_0 = \frac{1}{N} \sum_{k=0}^{N-1} v_k$$

$$a_n = \frac{2}{N} \sum_{k=0}^{N-1} v_k \cos\left(\frac{2\pi kn}{N}\right) \qquad n = 1, 2, \ldots M$$

$$b_n = \frac{2}{N} \sum_{k=0}^{N-1} v_k \sin\left(\frac{2\pi kn}{N}\right)$$

Fourier integral transform

$$v(t) = \int_{-\infty}^{\infty} V(f) \exp(j2\pi ft) \, df$$

$$V(f) = \int_{-\infty}^{\infty} v(t) \exp(-j2\pi ft) \, dt$$

Bibliography

Further information and listings of programs for the fast Fourier transform can be found from the following sources:

Lynn P. A. and Fuerst W. (1989). *Introductory Digital Signal Processing*. New York: Wiley

Press W. H., Flannery B. P., Teukolsky S. A. and Vettering W. T. (1986). *Numerical Recipes*. Cambridge: Cambridge University Press

For those using Turbo Pascal,

Turbo Pascal Numerical Methods Toolbox. Borland Ltd.

Each of these references makes the programs available on disc.

PROBLEMS

11.1 A periodic waveform has the form of a flat-topped rectangular pulse of magnitude $\hat{V}$, duration τ, repeated in time T. Defining $v(t)$ in the form

$$v(t) = \hat{V} \qquad 0 < |t| < \tau/2$$

$$v(t) = 0 \qquad \tau/2 < |t| < T/2$$

determine values for the coefficients a_n in the expansion

$$v(t) = \sum_{n=0}^{\infty} a_n \cos(2\pi nt/T)$$

11.2 By a change in time origin in the result of Problem of 11.1, obtain the Fourier expansion for the waveform

$$v(t) = \hat{V} \qquad 0 < t < \tau$$

$$= 0 \qquad \tau < t < T$$

11.3 Obtain a Fourier expansion for the periodic waveform of period T defined by

$$v(t) = \hat{V} |\sin(\pi t/T)|$$

11.4 The output of a rectifier circuit has the waveform of Problem 11.3. This output is connected to a resistive load, resistance R, through an inductance L. Obtain an expression, in the form of a Fourier series, for the current through the load. Obtain also an expression for the average voltage developed across the load.

11.5 In the situation described in Problem 11.4, $T = 10\,\text{ms}$, $R = 100\,\Omega$. The inductance L is to be chosen so that the peak-to-peak variation of the voltage across the load is less than 5% of the average voltage. By assuming that only the lowest harmonic in the Fourier series is significant, estimate a suitable value for L.

11.6 A voltage source generates a periodic waveform $e(t)$ for which the Fourier expansion is

$$e(t) = \sum_{n=0}^{\infty} E_n \cos(n\omega_0 t)$$

This voltage source is connected in series with inductance, capacitance and resistance. The series $L-C-R$ circuit so formed is tuned to the repetition frequency $\omega_0/2\pi$. Show that for a high Q circuit the voltage across the capacitor is given approximately by the series

$$E_0 + QE_1 \sin(\omega_0 t) - \sum_{n=2}^{\infty} E_n \frac{1}{n^2 - 1} \cos(n\omega_0 t)$$

11.7 In a particular case of the situation described in Problem 11.6, the waveform is that of Problem 11.1 with $\tau = 0.05$ μs, $T = 1$ μs and amplitude 10 V; the Q factor for the circuit is 10. Estimate the peak amplitudes of the first, second and third harmonics in the waveform across the capacitor. Express, as a percentage, the ratios of the second and third harmonics to that of the first.

11.8 Obtain a Fourier expansion for the waveform defined by

$$v(t) = \tfrac{1}{2}\hat{V}[1 + \cos(4\pi t/T)] \qquad |t| < T/4$$

$$= 0 \qquad\qquad\qquad T/4 < |t| < T/2$$

Calculate for values of n between 0 and 5 the ratio a_n/a_0.

11.9 In the waveform of Problem 11.8, $\hat{V} = 1$, $T = 1$. the waveform can be approximately represented by a finite series of the form

$$\sum_{n=0}^{4} a_n \cos(2\pi nt)$$

By using the discrete Fourier transform with $N = 8$ determine the coefficients a_n.

If computational means is available, repeat for a series including terms up to $n = 8$ $(N = 16)$.

11.10 A single pulse is of duration 2τ, during which interval the waveform is described by

$$v(t) = \tfrac{1}{2}[1 + \cos(\pi t/\tau)] \qquad -\tau < t < \tau$$

Using the Fourier integral transform show that the corresponding spectral density is given by

$$V(f) = \frac{1}{2\pi f} \frac{1}{1 - (2f\tau)^2} \sin(2\pi f\tau)$$

Make a plot of this expression for the case $\tau = 0.5$ μs. What bandwidth would be required to pass all frequency components up to the first zero?

11.11 A radio frequency pulse lasting for $2N$ cycles is described by

$$v(t) = \hat{V} \sin(2\pi t/T) \qquad -NT < t < NT$$

$$= 0 \qquad\qquad\qquad |t| > NT$$

Derive an expression for the spectral density. Illustrate by means of a sketch the variation of $|V(f)|$ with frequency. Show that as N increases the spectrum becomes closely confined to a small range about the frequency of the sine wave.

12 ◆ *Non-linear Circuits*

1: *Large-signal theory; rectifiers*

In this chapter some of the problems of analysing circuits containing non-linear components such as diodes are considered. It is shown how models of such devices make possible the use of linear theory to obtain useful approximations. The method is illustrated by application to rectifier circuits.

12.1 ◆ Introduction

Past chapters have all concerned linear circuits: circuits made of components for which, as far as magnitude is concerned, voltage and current are linearly related. Many components are linear over the whole range of values used in practice, perhaps deviating from linearity only at extreme values; some have by their nature markedly non-linear characteristics, notably electronic devices such as diodes and transistors. The inclusion of such devices in an otherwise linear network usually makes exact analysis difficult: results usually have to be obtained by numerical modelling and solution. However, in many cases it is not important to perform an exact analysis, and a degree of approximation may allow useful and worthwhile results to be more easily obtained. This can be for several reasons: it might be, for example, that an approximate analysis is sufficient to guide an experimental design process, or, as in the case of semiconductor devices, the devices themselves have such a wide spread of characteristics that exact prediction for one particular device is not particularly meaningful. In this situation approximations may be possible which allow the powerful tools of linear circuit analysis to be used, and it is with such approximations that this chapter and the succeeding one are concerned.

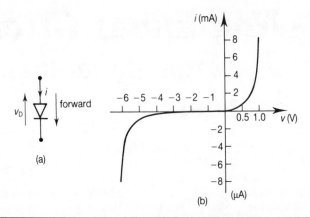

Figure 12.1 The semiconductor diode: (a) circuit symbol; (b) typical characteristic (note different forward and reverse scales).

Two classes of situations may be envisaged: large signal and small signal. When analysis of a circuit is to be carried out, the characteristics can often be modelled in terms of a number of linear characteristics. One example of this procedure is the analysis of the rectifier circuits used in converting AC to DC. This topic illustrates large-signal analysis, and is considered in the present chapter. If the excursions of voltage and current are small and take place in a region where device characteristics are not

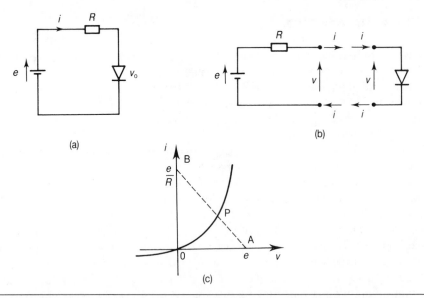

Figure 12.2 Diode circuits: (a) a simple circuit; (b) decomposition into linear and non-linear parts; (c) graphical solution using load line AB ($v = e - iR$).

changing too rapidly, the small region involved can be treated as linear over the local area. This technique forms the basis of the small-signal analysis often used in analysing transistor circuits: it is introduced in Chapter 13.

12.2 ◆ Diodes

A diode is a two-terminal device with the basic property that it offers less resistance to the passage of electric current in one direction than in the other. The preferential direction is known as the forward direction, as opposed to the reverse direction. The symbol used in circuit diagrams is shown in Figure 12.1(a), and in Figure 12.1(b) is shown a characteristic typical of a semiconductor diode (the difference in scales on the positive and negative axes should be noted.) A simple circuit in which a diode,

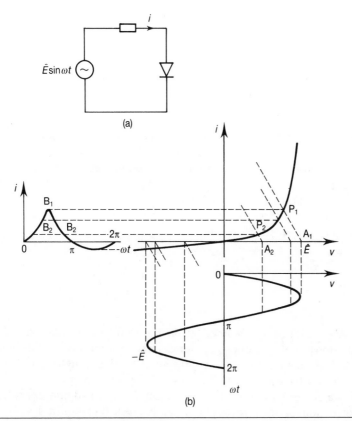

Figure 12.3 AC excitation: (a) diode circuit; (b) determination of current waveshape.

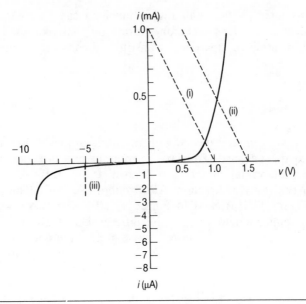

Figure 12.4 Characteristic for diode of Example 12.1.

resistor and battery appear in series is shown in Figure 12.2(a). Application of Ohm's law yields the equation

$$e = iR + v_D(i) \tag{12.1}$$

in which $v_D(i)$ is the voltage across the diode when carrying current i. If a mathematical expression is known for the function $v_D(i)$, Equation 12.1 can be solved for i by various numerical techniques. However, in general such an expression is not known, and in any case a graphical method is far simpler to use, and far more informative. The circuit can be regarded as composed of the two separate circuits shown in Figure 12.2(b). For the linear circuit on the left we have

$$i = \frac{e - v}{R} \tag{12.2}$$

For the diode

$$v = v_D(i) \tag{12.3}$$

The required solution is obtained when the same values of v and i satisfy both equations. These values can be derived graphically, as illustrated in Figure 12.2(c). Equation 12.2 is represented by the straight line AB in the figure: it is most easily constructed by joining the points $v = e$, $i = 0$ and $v = 0$, $i = e/R$. Equation 12.3 is represented by the diode characteristic. The intersection of these two graphs at P gives the required values. The point P is often referred to as the **working point**. The line AB characteristic of a linear circuit is of a type frequently found useful, and is called a **load line**.

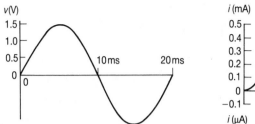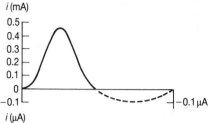

Figure 12.5 Waveform of diode current in Example 12.2.

The same technique can be used if in this circuit the battery is replaced by an AC source, as in Figure 12.3(a). In this case a succession of load lines have to be drawn, one for each instant of time as illustrated in Figure 12.3(b). A succession of working points is obtained, from which the waveforms of v and i can be plotted as shown in the diagram.

EXAMPLE 12.1 The resistance in the circuit of Figure 12.2(a) is 1 kΩ, and the diode has the characteristic shown in Figure 12.4. Find the working point when e has the value (i) 1 V, (ii) 1.5 V and (iii) −5 V.
   The relevant load lines are shown in Figure 12.4, giving

   (i)   0.84 V, 0.14 mA

   (ii) 1.03 V, 0.46 mA

   (iii) −5 V, −0.03 μA

In case (ii) the load line has been drawn by taking the second point $i = 1$ mA, $v = 1.0$ V, since the point $v = e/R$ is off the graph. For case (iii) the drop in the resistance by the reverse current is neglible on the scale of the graph so that the load line is virtually a line of constant voltage.

EXAMPLE 12.2 In the configuration of Example 12.1 the battery is replaced by an AC source of frequency 50 Hz and peak value 1.5 V. Determine the waveform of the current passing through the diode.
   By drawing a succession of load lines parallel to those shown in Figure 12.4 the waveform shown in Figure 12.5 is obtained. On the scale necessary to show the forward current the reverse current appears as zero: the dotted line shows it on an enlarged scale.

12.2.1 ◆ Modelling the diode characteristic

A number of possible circuit models for the diode characteristic are illustrated in Figures 12.6 and 12.7, in increasing order of complexity. The characteristic shown in Figure 12.6(a) is often taken as defining the 'ideal' diode: no voltage drop in the forward direction, no current in the reverse direction. [The term 'ideal' is used in many contexts, and can cause

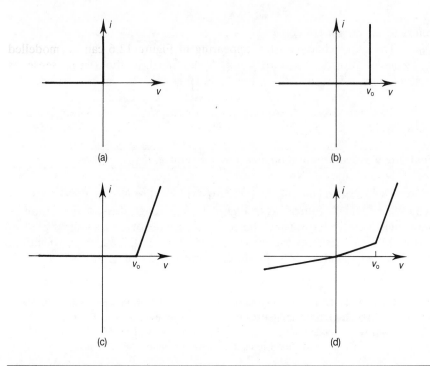

Figure 12.6 Linear approximations to diode characteristic: (a) the ideal diode; (b) inclusion of voltage drop; (c) inclusion of voltage drop and series resistance; (d) further complexity.

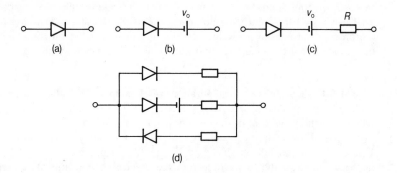

Figure 12.7 Circuit models for characteristics of Figure 12.6.

confusion. The above characteristic is ideal in that it simplifies calculations for the circuit designer. The same term ideal is used of a diode in a quite different sense by the semiconductor user, for whom it refers to the theoretical p–n junction diode having the characteristic

$$i = I_S[\exp(v/v_T) - 1]$$

in which the constant v_T has the value 0.025 mV. In the present text, ideal refers to the circuit context.]

The other characteristics appearing in Figure 12.6 can be modelled in terms of resistors, batteries and diodes ideal in the circuit sense as suggested by the circuits in Figure 12.7. Characteristics of the type shown in Figure 12.6 are known as **piecewise linear** approximations, since they are made up of a series of linear segments. When operation is within any one segment, linear theory applies. When operation crosses from one segment to the next, solutions have to be matched at the boundary. This can make analysis tedious, often so tedious that it has to be asked whether or not the complications entailed by using the better approximation are worth the benefits obtained, or whether a simpler approximation would provide results sufficiently accurate or informative for the purpose in hand. In the applications which follow the simplest approximation to diode behaviour will be used, the ideal characteristic of Figure 12.6(a). The diode symbol in the circuit diagrams will be assumed to signify a diode with this characteristic.

12.3 ◆ The 'battery charger'

A circuit often used with rechargeable cells is shown in Figure 12.8: the diode ensures that the charge transferred to the battery during one half-cycle is not subsequently drained away during the next half-cycle. The mode of operation is simple: the diode will begin to conduct as soon as the voltage of the AC source rises to the battery voltage. Since there is no drop across the diode, the difference between the source and battery voltages appears across the resistance. Inspection of the diagram of Figure 12.9(a) shows that conduction takes place over the period for which

$$E \cos(\omega t) \geqslant V_0$$

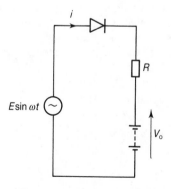

Figure 12.8 The battery charger circuit, Section 12.3.

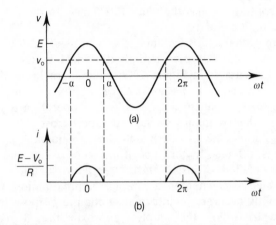

Figure 12.9 Waveforms for the battery charger circuit: (a) voltage waveforms, showing conduction angle 2α; (b) current waveform.

In the representative period $-\pi < \omega t < \pi$ conduction thus occurs when

$$-\alpha < \omega t < \alpha$$

in which

$$\alpha = \cos^{-1}(V_0/E)$$

The angle 2α is termed the **conduction angle**. The current waveform during this period is shown in Figure 12.9(b) and is given by

$$i(t) = \frac{1}{R}[E\cos(\omega t) - V_0], \qquad -\alpha < \omega t < \alpha$$

EXAMPLE 12.3 In the circuit of Figure 12.8 the peak voltage, E, of the source is 18 V, the battery voltage, V_0, is 12 V and the circuit resistance, R, is 10 Ω. Assuming that the diode has negligible voltage drop in the forward direction and passes negligible current in the reverse direction, estimate (i) the conduction angle, (ii) the peak charging current, (iii) the average charging current and (iv) the charge delivered to the battery in 1 h.

(i) $\alpha = \cos^{-1}(12/18) = 48.2°; \ 2\alpha = 96.4°$

(ii) $i_{pk} = (18 - 12)/10 = 0.6$ A

(iii) $\langle i \rangle = \dfrac{1}{2\pi} \displaystyle\int_{-\alpha}^{\alpha} \dfrac{E\cos(\omega t) - V_0}{R}\, d(\omega t)$

$= \dfrac{E}{2\pi R} \displaystyle\int_{-\alpha}^{\alpha} (\cos\theta - \cos\alpha)\, d\theta$

$= \dfrac{E}{\pi R} (\sin\alpha - \alpha\cos\alpha)$

Hence

$$\langle i \rangle = 0.105 \text{ A}$$

(iv) $Q = \langle i \rangle \times 3600 = 381 \text{ C}$

12.4 ◆ Rectifier circuits

While it is convenient to distribute electrical power in the form of alternating current, many devices require power to be delivered at a steady voltage for correct operation. As has been illustrated by the waveforms of Figures 12.5 and 12.9, the current through a diode fed from an AC source, although almost entirely in the forward direction, varies with time. To smooth out this variation it is necessary to provide a flow of charge from some reservoir during the time when the diode is not conducting. One way of providing such a reservoir is shown in the circuit of Figure 12.10, in which the reservoir is a capacitor placed in parallel with the 'load', R. When the diode is conducting the current flow not only supplies the required current through R but also charges up the capacitor. During the off period, current through R is provided by discharging the capacitor. In the absence of the capacitor, analysis for a general diode characteristic has been seen earlier to result in a non-linear algebraic equation, Equation 12.1. When the capacitor is incorporated, formal analysis leads to a differential equation in $i(t)$ containing a non-linear term:

$$i = \frac{v}{R} + C\frac{\mathrm{d}v}{\mathrm{d}t}$$

$$e - v = v_{\mathrm{D}}(i)$$

The presence of the differential prevents use of the graphical technique of the load line. Apart from numerical methods of solution, it is necessary to use an approximate characteristic which will permit a more straightforward mathematical solution to be obtained. In the following sections, the ideal case of Figure 12.6(a) is used.

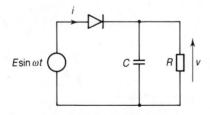

Figure 12.10 The half-wave rectifier circuit, Section 12.4.

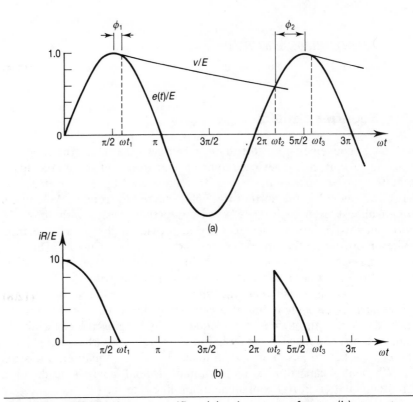

Figure 12.11 The half-wave rectifier: (a) voltage waveforms; (b) current waveform.

12.4.1 ◆ The half-wave, single-phase rectifier

This is the accepted name for the circuit of Figure 12.10. Consider the AC source to have the waveform $E \sin(\omega t)$ and to be connected to the circuit at an instant when the capacitor is uncharged. The ensuing sequence of events is illustrated in Figure 12.11(a). Conduction will start as soon as the applied voltage rises to zero. The diode then becomes effectively a short circuit so that the output voltage $v(t)$ is equal to the source voltage:

$$v(t) = e(t) = E \sin(\omega t)$$

The current through the diode, $i(t)$, is then given by

$$i = \frac{e}{R} + C\frac{de}{dt}$$

$$= \frac{E}{R}\cos(\omega t) + \omega C \cos(\omega t)$$

$$= \frac{E}{R}[1 + (\omega CR)^2]^{1/2} \cos(\omega t - \phi_1) \tag{12.4}$$

in which

$$\cos \phi_1 = \omega CR/[1 + (\omega CR)^2]^{1/2}$$
$$\sin \phi_1 = 1/[1 + (\omega CR)^2]^{1/2} \tag{12.5}$$

The current is given by the expression of Equation 12.4 as long as $i(t)$ is positive: as soon as it tries to reverse, the diode goes open circuit and the capacitor is left to discharge through the resistance. From Equation 12.4, conduction stops at the instant t_1 given by

$$\omega t_1 = \frac{\pi}{2} + \phi_1 \tag{12.6}$$

At this instant, the voltage $v(t_1)$ is equal to

$$v(t_1) = E \sin\left(\frac{\pi}{2} + \phi_1\right) = E \cos \phi_1 \tag{12.7}$$

The discharge of the R–C circuit is then described by

$$v(t) = E \cos \phi_1 \exp[-(t - t_1)/RC] \tag{12.8}$$

Conduction will recommence when $e(t)$ again rises to $v(t)$, at the instant t_2 shown on Figure 12.11(a). After this, the output voltage again follows the input voltage until once again the current falls to zero, at the instant t_3 on Figure 12.11(a). During this conduction period the current is again given by Equation 12.4. This being the case, the instant t_3 occurs at the same point in the cycle as did t_1. The instants t_1 and t_2 therefore determine the conduction period for each cycle. Figure 12.11 been calculated for the case when $\omega CR = 10$. In Figure 12.11(b) is shown the corresponding current waveform: the large value at the instant conduction recommences occurs because most of the current goes to the capacitor, and is dependent on dv/dt. Whereas t_1 is easy to determine, through the angle ϕ_1, Equation 12.6, the same is not true of t_2, or equivalently the angle ϕ_2 shown in the figure. Mathematically it is necessary to solve the equation

$$E \sin(\omega t) = E \cos \phi_1 \exp[-(t - t_1)/RC]$$

The simplest way is to use the graph of Figure 12.11(a) to find t_2.

It is clear that in order to reduce variation in the voltage across the capacitor, the capacitor has to be large. In the case of interest, therefore, the quantity ωCR is large and certain simplifying approximations become possible. Equation 12.5 gives

$$\phi_1 \approx \sin \phi_1 \approx 1/\omega CR$$

Hence conduction ceases very soon after the peak of the input voltage. Using this approximation in Equation 12.8, we find that in the off period

$$v(t) \approx E \exp[-(t - t_1)/RC]$$
$$\approx E[1 - (t - t_1)/RC] \tag{12.9}$$

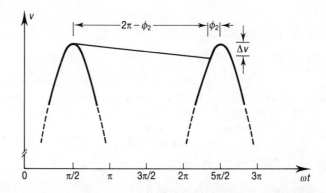

Figure 12.12 The half-wave rectifier: voltage waveform when $\omega CR \gg 1$.

The drop in voltage from its peak value of E until conduction recommences is therefore approximately given by

$$\Delta v \approx E(t_2 - t_1)/RC$$

$$< ET/RC \tag{12.10}$$

in which T is periodic time, $2\pi/\omega$. The fluctuation Δv is termed the peak-to-peak **ripple** voltage, and is an important parameter of a DC power supply. It is usually desired to reduce the ripple, so that the overestimate given in Equation 12.10, obtained by taking $t_2 = T$, is perfectly acceptable. However, to estimate the peak current, also an important parameter, it is necessary to obtain a better estimate of t_2. In Figure 12.12 is shown the form taken by $v(t)$ when $\omega CR \gg 1$. The angle ϕ_2 satisfies the equation

$$\Delta v = \frac{E}{\omega CR}(2\pi - \phi_2) = E(1 - \cos \phi_2)$$

Approximating $\cos \phi_2$ by $1 - \phi_2^2/2$ we obtain the equation

$$\frac{E}{\omega CR}(2\pi - \phi_2) = \frac{E}{2}\phi_2^2$$

Solving the quadratic and taking the positive root, we find

$$\phi_2 = \frac{1}{\omega CR}[(1 + 4\pi\omega CR)^{1/2} - 1]$$

Inserting the value $\omega CR = 10$ in this expression gives

$$\phi_2 = 1.03 \text{ rad, or } 58.8°$$

This compares favourably with the value of 54° derived from Figure 12.12. From Figure 12.11(a), $\omega t_2 = 5\pi/2 - \phi_2$, so that the maximum current is given by

$$i(t_2) = \frac{E}{R}(\cos \phi_2 + \omega CR \sin \phi_2)$$

For the value of $\omega CR = 10$ and with the value of ϕ_2 just found, this expression gives a peak current of $9.1E/R$, in good agreement with Figure 12.11(b).

When ωCR is large then we may approximate further to obtain

$$\phi_2 \approx 2(\pi/\omega CR)^{1/2} \tag{12.11}$$

$$i(t_2) \approx \frac{E}{R}[(1 - \phi_2^2/2) + \omega CR\phi_2]$$

$$\approx \frac{E}{R}2(\pi\omega CR)^{1/2} \tag{12.12}$$

When $\omega CR > 50$, which from Equation 12.10 corresponds to a ripple of less than 12%, the value derived from Equation 12.12 is less than 5% in error. This is of the order of error incurred through other approximations, such as the neglect of diode series resistance.

This analysis has been given in some detail in order to illustrate the method. The other models in Figure 12.7 can be used in the same way. That of Figure 12.7(b) is fairly simple to incorporate: it just reduces voltages by v_0 during the conducting period. Allowing a series resistor as in Figure 12.7(c) can be handled, but the analysis is considerably more complicated because the voltage $v(t)$ is no longer equal to the input voltage during conduction but has to be calculated by carrying out a transient solution. The only simple case is that when ωCR is so large that the output voltage $v(t)$ can be assumed to be virtually steady. This situation occurs when the circuit is used to rectify high-frequency signals, as in the diode detector used in ratio receivers. It is considered in Section 12.5.

EXAMPLE 12.4 An amplifier requires a current of 100 mA DC at approximately 25 V. A design suggests that a half-wave rectifier circuit is used which is fed from an 18 V RMS source of 50 Hz frequency, incorporating an available 2200 μF capacitor as reservoir. Assess the suitability of the design by estimating (i) the peak voltage across the capacitor, (ii) the peak-to-peak ripple voltage and (iii) the peak current through the diode.

The amplifier represents a resistance of $25/0.1 = 250$ Ω. Hence $\omega CR = 100\pi \times 2.2 \times 10^{-3} \times 250 = 173 \gg 1$.

(i) The peak voltage across the capacitor will be equal to the peak voltage of the source, $18\sqrt{2} = 25.5$ V.

(ii) Using Equation 12.10 we find

$$\Delta v = 25.4 \times 0.02/250 \times 2.2 \times 10^{-3} = 0.92 \text{ V}$$

(iii) For the peak current we apply Equation 12.12 which gives

$$i_{pk} \approx (25.4/250)[1 + (2\pi \times 100\pi \times 0.0022 \times 250)^{1/2}] \approx 3.4 \text{ A}$$

A design which gives a 1 V ripple on the supply line cannot be regarded in the present context as satisfactory. The predicted very high peak current would be modified appeciably by the diode series resistance not here taken in to account but, nevertheless, this type of circuit will in practice take high peak currents.

12.5 ◆ The diode detector

The circuit model for this case is shown in Figure 12.13. When the capacitor voltage is effectively constant, the problem is similar to the battery charger studied in Section 12.2.1, the only difference being that V_0 is no longer imposed but is to be determined. In the circuit of Figure 12.13 the diode current i divides into i_1 through the resistor and i_2 through the capacitor. In a steady state the average flow of charge must pass through the resistor, since a capacitor only stores charge. Hence, taking the average over one cycle, we must have

$$\langle i \rangle = i_1$$

We also have

$$i_1 = \frac{v}{R}$$

Hence, since v is assumed to be steady,

$$v = V_0 = R\langle i \rangle$$

Using the expression for the average current given in Example 12.3 (iii), we find

$$\langle i \rangle = \frac{E}{r\pi} \left(\sin \alpha - \alpha \cos \alpha \right)$$

in which $\cos \alpha = V_0/E$. Hence

$$V_0 = E \cos \alpha = \frac{ER}{\pi r} \left(\sin \alpha - \alpha \cos \alpha \right)$$

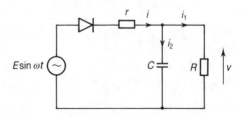

Figure 12.13 The diode-detector circuit, Section 12.5.

and

$$\tan \alpha - \alpha = \frac{\pi r}{R}$$

If, as usual, $R \gg r$ then α will be small and we may use the approximation

$$\tan \alpha = \alpha + \frac{\alpha^3}{3} + \ldots$$

to arrive at an expression for α:

$$\alpha \approx \left(\frac{3\pi r}{R}\right)^{1/3}$$

Once this is known the output voltage and peak current can be estimated.

12.6 ◆ Full-wave rectifier circuits

Full-wave circuits are so called because with them current is taken from the AC supply during both halves of the cycle, in contrast to the half-wave circuit just studied. A modification of the half-wave circuit is shown in Figure 12.14(a); the relevant waveforms are shown in Figure 12.14(b). In practice the two sources are derived from one AC supply through a centre-tapped transformer, as shown in Figure 12.14(c). The analysis of Section 12.3.1 applies except that the time over which the output current is supplied from the capacitor is now only half the period rather than the whole. This has the effect of reducing both the ripple voltage and the peak current by a factor of 2. A 'bridge' circuit which gives full-wave rectification is shown in Figure 12.15: it has precisely the same effect as the circuit of Figure 12.14 except that when conduction is taking place two diodes appear in series, rather than one. It also uses a single-winding transformer as opposed to the centre-tapped transformer of Figure 12.14(c), effecting a saving of size and weight.

It will be appreciated that circuits using reservoir capacitors to smooth the output necessarily take high peak currents. It might be said of the reservoir capacitor that it smooths by bypassing the fluctuating components of current rather than by removing them: an alternative is to use inductance smoothing, which has the effect of preventing the fluctuations developing.

12.6.1 ◆ Inductance smoothing

A circuit which reduces the current variation directly by use of inductance smoothing is shown in Figure 12.16(a). Assuming that one or the other diode is conducting at each instant of the cycle, the voltage waveform

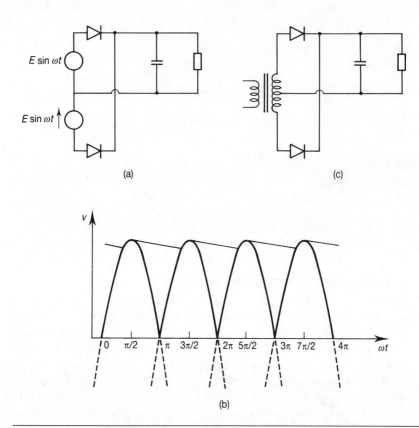

(a)

(c)

(b)

Figure 12.14 Full-wave rectifier with capacitance smoothing: (a) circuit for analysis; (b) voltage waveforms; (c) practical realization.

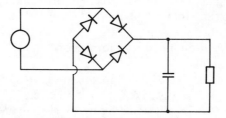

Figure 12.15 Full-wave rectifier using diode bridge.

between the points A and B is that shown in Figure 12.16(b). Since in a periodic state the average drop across an inductance is zero, the average voltage developed across the load R, must be equal to the average of that developed between A and B, which is equal to $2E/\pi$. To determine the ripple current is less simple: in principle it is necessary to determine the current flowing in an $L-R$ circuit under the periodic excitation of a

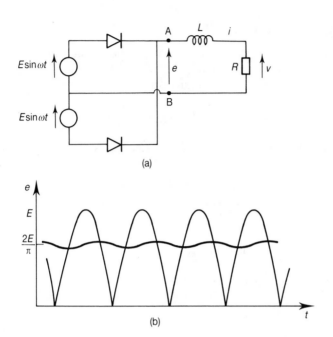

Figure 12.16 Full-wave rectifier with inductance smoothing: (a) circuit; (b) voltage waveforms.

half-sine waveform. This means finding a solution of the differential equation

$$L\frac{di}{dt} + Ri = E \sin(\omega t) \tag{12.13}$$

over the period $0 < \omega t < \pi$ for which, since the solution must be periodic, $i(0) = i(\pi/\omega)$. However, as with the case of the rectifier using a reservoir capacitor, an approximation can be obtained when the ripple is made small by use of a large enough inductance. Firstly, we note from Equation 12.13 that di/dt is zero when

$$Ri = E \sin(\omega t)$$

The values of $i(t)$ at the times satisfying this equation will be maximum or minimum. Since by supposition Ri is nearly constant at $2E/\pi$, this equation becomes

$$\sin(\omega t) \approx 2/\pi$$

Hence the times of maximum and minimum are given by

$$\omega t = \theta_c, \ \omega t = \pi - \theta_c$$

where

$$\theta_c = \sin^{-1}(2/\pi) = 0.699 \text{ rad} = 40°$$

Consideration of the mode of operation shows that the first value corresponds to the minimum value, the second to the maximum. Integrating Equation 12.13 between these limits we find

$$\int_{\theta_c/\omega}^{(\pi-\theta_c)/\omega} L\frac{di}{dt}dt = \int_{\theta_c/\omega}^{(\pi-\theta_c)/\omega} [E\sin(\omega t) - Ri]dt$$

Assuming as before that $Ri \approx 2E/\pi$, this becomes

$$L(i_{max} - i_{min}) = \frac{1}{\omega}\int_{\theta_c}^{\pi-\theta_c}\left(E\sin\theta - \frac{2E}{\pi}\right)d\theta$$

$$= \frac{2E}{\omega}\left[\frac{2\theta_c}{\pi} - 1 + \cos\theta_c\right]$$

Hence, inserting the value for θ_c,

$$\Delta v = R(i_{max} - i_{min}) \approx 0.41 E\frac{R}{\omega L}$$

In terms of the output voltage $v = 2E/\pi$

$$\frac{\Delta v}{v} = 0.64\frac{R}{\omega L}$$

The form of $v(t)$ calculated for $\omega L/R = 5$ is shown in Figure 12.16(b).

12.7 ◆ Conclusions

The examples of the foregoing sections indicate the methods by which rectifier circuits can be analysed. An exact analysis is not attempted but rather an analysis which will give a good indication of the operation of a practical circuit. It is important not to pursue the approximate analysis beyond the limits imposed by nature of the approximation made to the device characteristics. The type of analysis can be applied wherever the non-linear characteristic can be sensibly approximated to by one or more linear regions of operation.

PROBLEMS

12.1 A diode has the forward characteristic shown in Figure 12.17. It is connected into a circuit in the form of Figure 12.2(a). Determine the working point in the following cases:

(a) $e = 1.5$ V, $R = 300 \ \Omega$;
(b) $e = 1.0$ V, $R = 25 \ \Omega$;
(c) $e = 0.5$ V, $R = 50 \ \Omega$.

12.2 With the same diode and circuit as in Problem 12.1, the resistance R is 25 Ω and $e(t)$ is an AC source, given by

$$e(t) = 1.4 \sin(\omega t)$$

Plot the waveform of diode voltage and current over one positive half-cycle.

12.3 The characteristic shown in Figure 12.17 can be represented approximately by

$i = 0$ $0 < v < 0.25$

$\ = 0.145(v - 0.25)$ $v > 0.25$

Repeat Problem 12.1 using this approximate characteristic, and compare answers.

12.4 Repeat Problem 12.2 using the approximate characteristic defined in Problem 12.3.

12.5 In the Figure 12.18 is shown a circuit designed to produce a square waveform across the terminals AB. Make a careful sketch of the output waveform when the diodes (i) are ideal and (ii) have the forward characteristic shown in Figure 12.17 and carry negligible current in the reverse direction. Estimate in each case the

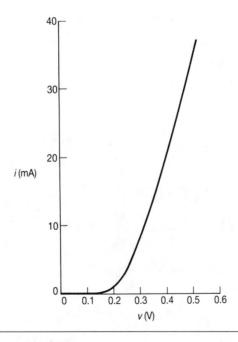

Figure 12.17

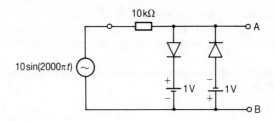

Figure 12.18

nominal amplitude of the square wave and the departure from constancy in each state.

12.6   In Figure 12.19 is shown an attempt to realize the circuit of Figure 12.18. Determine the output when the diodes have the characteristic shown in Figure 12.17.

12.7   A charger for 12 V car batteries has a circuit of the form shown in Figure 12.8. The AC source has an RMS value of 15 V with internal resistance 5 Ω. The diode characteristic is represented by the approximation

$$i = 0 \qquad 0 < v < 0.8$$
$$i = 0.5v \qquad v > 0.8$$

in which i is in amps, v in volts. The car battery has a resistance of 0.2 Ω. When the battery is charged to 12 V, estimate

  (i) the conduction angle,

 (ii) the peak current, and

(iii) the average charging current.

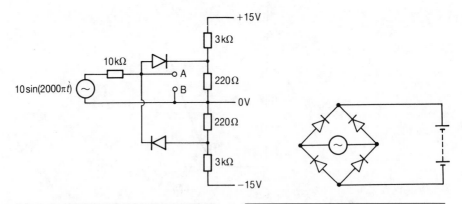

Figure 12.19   **Figure 12.20**

12.8 The diode rectifier in Problem 12.7 is replaced by a bridge assembly as shown in Figure 12.20. The AC source and diodes are the same as in Problem 12.7. Determine the new values of peak and average charging current.

12.9 A half-wave rectifier circuit with a reservoir capacitor is used to supply current to a load requiring 2 mA at 500 V, constant to within 0.1%. The circuit is supplied from a 2 kHz AC supply through a transformer. Assuming ideal diodes, suggest values for the AC voltage required from the transformer and for the reservoir capacitor.

12.10 In a full-wave circuit of the form shown in Figure 12.14 each half of the transformer delivers 140 V RMS at 50 Hz. The resistance of the load is 9 kΩ and the capacitor is 16 μF. Assuming ideal diodes, estimate the average output voltage and the peak-to-peak ripple.

12.11 In the circuit of Figure 12.21 the diode is ideal, the battery has negligible resistance and $\omega RC \gg 1$. Show that when $E > V$ the average current is given approximately by $(E - V)/R$. Show also, by considering the power balance, that as far as power is concerned the load on the source is equivalent to that of a resistance of value $R/[2(1 - V/E)]$.

12.12 In the circuit of Figure 12.22 the source delivers a train of pulses repeating at interval T. Each pulse is of height E and duration $\tau(< T)$. In the interval between pulses, the source looks like a short circuit. Assuming that the diode is ideal and the capacitor initially uncharged, make careful sketches illustrating the behaviour of the current from the source and of the voltages v_C and v_R over the time of the first three pulses.

At each positive-going edge the source has to deliver an impulse of current to recharge the capacitor. Considering the case $RC \gg T$, show that after the first pulse the charge transferred is equal to that passing through the resistor in one cycle.

12.13 In a circuit of the form of Figure 12.16(a)

$$\omega/2\pi = 50 \text{ Hz}, \ E = 75 \text{ V}, \ R = 50 \ \Omega, \ L = 0.8 \text{ H}$$

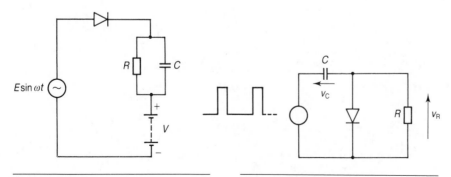

Figure 12.21   **Figure 12.22**

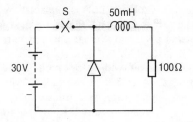

Figure 12.23

Assuming ideal diodes, calculate the average voltage across the resistor and estimate the peak-to-peak ripple.

12.14 In the circuit of Figure 12.16(a) the lower diode fails and becomes permanently open circuit. Show that conduction through the remaining diode is discontinuous, falling to zero over part of the cycle, and that during the conduction period the voltage across the resistor is given by the expression

$$v(t) = E\frac{R}{[R^2 + (\omega L)^2]^{1/2}}[\sin(\omega t - \theta) + \sin\theta\exp(-Rt/L)]$$

in which $\theta = \tan^{-1}(L\omega/R)$. For the case when the components have the values given in Problem 12.13, determine the conduction angle and make a careful sketch of the waveform.

12.15 In the circuit of Figure 12.23, 'S' represents an electronic switch which opens and closes regularly for equal periods of 50 μs. Estimate the average and peak-to-peak ripple of the voltage appearing across the resistor.

What is the function of the diode, and what would happen if it were to become an open circuit?

13 ◆ *Non-linear Circuits*
2: *Linearization*

In many devices using non-linear elements such as transistors, the transistors operate under steady voltages which are much larger than the small signals being processed. In such a situation it becomes possible to describe the operation of the transistor by a simple linear model. This technique of linearization is introduced in this chapter by application to diodes and transistors. Small-signal equivalent circuits are derived for important cases.

13.1 ◆ Introduction

As outlined in Section 12.1, it is possible to apply linear circuit theory to a network containing a non-linear device if the variations of current and voltage are sufficiently small. In this situation, in the region of interest the non-linear characteristic can be replaced by a linear characteristic. A simple example is the way that small excursions about a point on a curve may be considered as taking place on the tangent at the point. This technique is termed **linearization**, and is widely used in non-linear problems. It will be introduced by considering the diode circuit for which a graphical solution was obtained in Section 12.2, Figures 12.2 and 12.3.

13.2 ◆ Diode slope resistance

We consider the circuit of Figure 12.2(a), repeated in Figure 13.1(a). Determination of the working point depended on the solution of the pair of equations (cf. Equations 12.2 and 12.3 of Section 12.2)

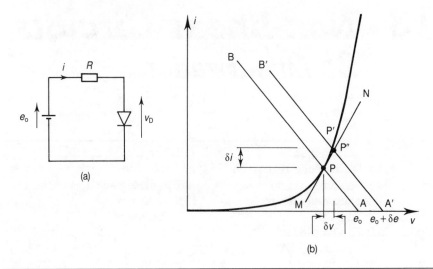

Figure 13.1 Small changes in a diode circuit: (a) circuit; (b) graphical treatment, illustrating 'slope' resistance.

$$i_P = (e_0 - v_P)/R \tag{13.1}$$

$$v_P = v_D(i_P) \tag{13.2}$$

in which v_P and i_P denote the working point, established by the graphical method. Consider now the situation when e_0 is changed by a small amount δe: the working point will move to an adjacent point P'. From the diagram in Figure 13.1(b) it can be seen that the new working point can be very closely determined from the intersection, P'', of the new load line A'B' with the tangent to the characteristic at P, MN in the diagram. The smaller the value of δe, the closer P'' is to P'. With the new value of e, $e_0 + \delta e$, let the coordinates of the new working point be $v_P + \delta v$, $i_P + \delta i$. We have

$$i_P + \delta i = (e_0 + \delta e - v_P - \delta v)/R \tag{13.3}$$

$$v_P + \delta v = v_D(i_P + \delta i) \tag{13.4}$$

Subtracting Equation 13.1 from Equation 13.3 we find

$$\delta i = (\delta e - \delta v)/R \tag{13.5}$$

This is the equation of the load line A'B' referred to a new origin at P. The right-hand side of Equation 13.4 can be expanded as a Maclaurin series in the form

$$v_D(i_P) + \delta i(\mathrm{d}v_D/\mathrm{d}i)_P + \frac{\delta i^2}{2!}(\mathrm{d}^2 v_D/\mathrm{d}i^2)_P + \ldots$$

If δi is sufficiently small, we may neglect terms in powers of δi beyond the first to obtain

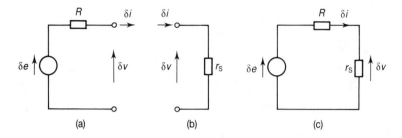

Figure 13.2 Equivalent circuits: (a) for source; (b) for diode; (c) combined.

$$v_P + \delta v = v_D(i_P) + \delta i(dv_D/di)_P \tag{13.6}$$

On subtracting Equation 13.4 we find

$$\delta v = \delta i(dv_D/di)_P \tag{13.7}$$

The differential coefficient $(dv_D/di)_P$ has the dimension of resistance. Its reciprocal has the dimension of conductance, and is the slope of the tangent MN in Figure 13.1(b). Denoting

$$r_s = (dv_D/di)_P \tag{13.8}$$

Equation 13.7 becomes

$$\delta v = r_s \, \delta i \tag{13.9}$$

The resistance r_s is the **slope resistance** of the diode at the working point P. Solving Equations 13.5 and 13.9 we find expressions for δv and δi:

$$\delta i = \frac{\delta e}{R + r_s} \tag{13.10}$$

$$\delta v = \delta e \frac{r_s}{R + r_s} \tag{13.11}$$

13.3 ◆ The small-signal equivalent circuit

Equation (13.5) can be given the circuit representation shown in Figure 13.2(a), and Equation 13.9 that in Figure 13.2(b). On combining these two circuits we find the circuit of Figure 13.2(c), from which the results expressed by Equations 13.10 and 13.11 can be obtained by inspection. Thus providing that we know the working point and the value of the slope resistance at that point (which can be estimated graphically) it becomes a straightforward matter to estimate the changes by means of the equivalent circuit of Figure 13.2(c). This procedure is most often useful with the configuration shown in Figure 13.3(a), in which the voltage applied to the circuit consists of a steady voltage, e_0, in series with a small variable

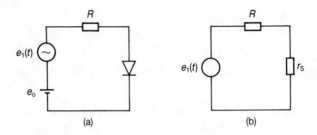

Figure 13.3 Diode circuit including small AC source.

source, which might be AC. Once the working point is known variations can be estimated by use of the small-signal equivalent circuit of Figure 13.3(b).

EXAMPLE 13.1 In the circuit of Figure 13.3(a) the diode has the characteristic shown in Figure 13.4, the resistance is 100 Ω and the battery voltage 0.5 V. The source $e_1(t)$ is sinusoidal, of frequency 50 Hz and RMS value 10 mV. Estimate (i) the working point, (ii) the diode slope resistance at this point and (iii) the RMS value of the fluctuation of the voltage across the diode. Give also an expression for the instantaneous voltage across the diode.

Drawing the 100 Ω load line on the characteristic we find the working point to be given by

(i) 0.24 V, 2.6 mA

Drawing the tangent at P we find it passes through the zero-current point at

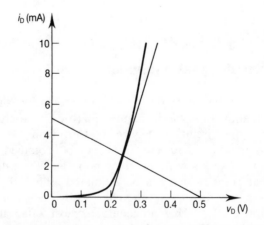

Figure 13.4 Diode characteristic for Example 13.1.

0.2 V and the point 0.355 V, 10 mA and hence corresponds to a drop of 0.155 V for 10 mA. Therefore

(ii) $r_s = (0.155/10) \times 10^3 = 15.5\,\Omega$

In the small-signal equivalent circuit of Figure 13.3(b) we therefore have

$$e_1(t) = 10 \times 10^{-3}\sqrt{2}\cos{(100\pi t)}$$

$$R = 100\,\Omega$$

$$r_s = 15.5\,\Omega$$

The voltage fluctuation is therefore given by

$$\frac{15.5}{100 + 15.5}10 \times 10^{-3}\sqrt{2}\cos{(100\pi t)}$$

The RMS value is given by

(iii) $10 \times 15.5/115.5 = 1.34$ mV

The complete expression for the diode voltage, in millivolts, is

$$v = 240 + 1.34\sqrt{2}\cos{(100\pi t)}$$

It may be verified from the characteristic in Figure 13.4 that with the 'signal' of 10 mV the deviation of the true point of operation from the tangent is negligibly small.

The great advantage of the small-signal equivalent circuit is that it enables us to replace, for the purpose of estimating small variations, a non-linear component by a linear one. To the resulting linear network we can apply the methods of linear theory. Clearly, if the variations about the working point are too large, errors will result, but in a very wide range of problems it is legitimate to use small-signal analysis.

Although it has been convenient to illustrate the small-signal technique by use of diode circuits, such applications are seldom encountered in practice: diodes are almost always in situations requiring large-signal analysis. Small-signal equivalent circuits are, however, of very great use in analysing the operation of transistor circuits.

13.4 ◆ Transistors

This book is concerned with the methods of network analysis and not the physics of transistors. For present purposes, a transistor will be taken as a device which has three terminals, the currents flowing into these terminals being related to the voltages between them in ways described by means of graphical characteristics. The small-signal equivalences associated with two types of transistors will be discussed, the field-effect transistor, Section 13.5, and the n–p–n bipolar transistor in Section 13.6.

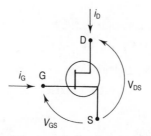

Figure 13.5 Junction field-effect transistor: circuit symbol and variables.

13.5 ◆ The field-effect transistor (FET)

The symbol and the electrical variables associated with the field-effect transistor are shown in Figure 13.5. The terminals G,D,S are referred to respectively as the gate, drain and source. Of the four variables i_G, i_D, v_{DS}, v_{GS}, the gate current i_G is usually small enough to be neglected. It is then convenient to take drain current i_D as a function of the two variables v_{DS}, v_{GS}:

$$i_D = f(v_{DS}, v_{GS}) \tag{13.12}$$

This functional relation is usually expressed graphically by measuring i_D as a function of v_{DS} for each of a sequence of constant values for v_{GS}. A set of such characteristics typical of those provided by manufacturers is shown in Figure 13.6. In a large class of transistor circuits, such as amplifiers, the circuitry establishes a working point for each transistor and when a signal is applied excursions from this point are small.

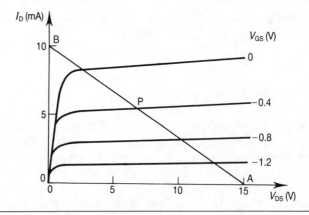

Figure 13.6 Output characteristic for FET.

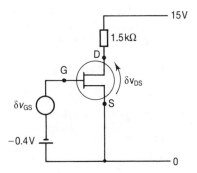

Figure 13.7 Circuit containing a field-effect transistor.

An example of a circuit which establishes a working point is that of the common-source amplifier, shown in basic form in Figure 13.7. In this circuit, the connection of the drain to the DC supply through the resistance R enforces a relation between i_D and v_{DS}:

$$v_{DS} = V_{DD} - Ri_D \tag{13.13}$$

For the particular values $V_{DD} = 15$ V, $R = 1.5$ kΩ, this relation is shown graphically in Figure 13.6 as the load line AB. If, for example, v_{GS} is then fixed at -0.4 V, the intersection of the load line with the -0.4 V characteristic determines the working point P. When a signal is superimposed on the steady-state conditions, the current and voltage vary by small amounts around the working point.

Two parameters determined from the characteristics can be used to estimate the changes in i_D which result from small changes in v_{DS} and v_{GS}. These parameters are as follows.

(i) The rate of change of v_{DS} with i_D, v_{GS} being held constant, gives the drain slope resistance, denoted by r_{DS}. This may be determined from the slope of the characteristic through the working point.

(ii) The rate of change of i_D with v_{GS} when v_{DS} is held constant, determines the mutual conductance, denoted by g_m.

Given values for these two parameters, we can express the change δi_D resulting from changes δv_{GS} and δv_{DS} in the form

$$\delta i_D = g_m \, \delta v_{GS} + \frac{1}{r_{ds}} \delta v_{DS} \tag{13.14}$$

[It may be observed that we can deduce Equation 13.14 directly from Equation 13.12 by differentiation, when g_m is seen to be the partial differential coefficient of i_D with respect to v_{GS}, v_{DS} being held constant, and $1/r_{ds}$ the partial differential coefficient with respect to v_{DS}, v_{GS} being held constant. This is of little direct use when no mathematical expression for the function $f(v_{GS}, v_{DS})$ is available.]

EXAMPLE 13.2 The transistor whose characteristics are given in Figure 13.6 is connected in the circuit of Figure 13.7. Determine the working point and estimate the drain slope resistance and the mutual conductance at that working point.

From the load line the working point is found to be

$$V_{GS} = -0.4 \text{ V}, \ V_{DS} = 6.8 \text{ V}, \ I_D = 5.5 \text{ mA}$$

The characteristic passing through the working point is approximately straight and goes through the points 15 V, 5.9 mA and 5 V, 5.4 mA. Hence

$$\frac{1}{r_{ds}} \approx \frac{5.9 - 5.4}{15 - 5} \times 10^{-3}$$

To estimate g_m we use the values of i_D for $v_{GS} = 0$ and -0.8 with $v_{DS} = 6.8$ V. These values are 8.5 mA and 3.05 mA respectively. Hence

$$g_m \approx \frac{8.5 - 3.05}{0.8} \times 10^{-3}$$

$$= 6.8 \times 10^{-3} \text{ S}$$

The unit for g_m is correctly the siemens, as given, but the value just calculated is often quoted as 6.8 mA V^{-1}.

These calculations may seem to be rather crude and inaccurate. However, a striving for great accuracy in such estimations is misplaced. This is primarily because the characteristics of individual transistors of the same type may vary quite considerably from one to the next, and characteristics such as those in Figure 13.6 represent an average. Transistor circuits can be designed to have precise and repeatable properties, but this is achieved through clever circuitry, not through acquiring identical transistors.

We can use Equation 13.14 to determine the effect of changing v_{GS} in the circuit of Figure 13.7 by a small amount. From Equation 13.13 changes in i_D and v_{DS} are, since V_{DD} is presumed constant, related by

$$\delta v_{DS} = -R \delta i_D \tag{13.15}$$

Combining this equation with Equation 13.14 we find

$$\delta v_{DS} \left(\frac{1}{R} + \frac{1}{r_{ds}} \right) = -g_m \ \delta v_{GS}$$

Hence

$$\delta v_{DS} = -g_m \frac{R r_{ds}}{R + r_{ds}} \delta v_{GS} \tag{13.16}$$

For the numerical case portrayed by the 1.5 kΩ load line in Figure 13.6, Equation 13.16 gives

$$\delta v_{DS} = -6.8 \times 10^{-3} \times \frac{1.5 \times 20}{1.5 + 20} 10^3 \ \delta v_{GS}$$

$$= -9.5 \ \delta v_{GS}$$

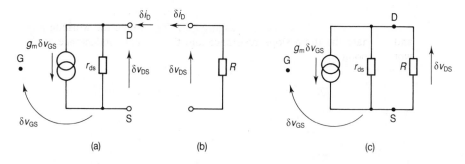

Figure 13.8 Small-signal equivalent circuits: (a) for FET; (b) for load; (c) combined.

13.5.1 ◆ The small-signal equivalent circuit

Equation 13.14 and Equation 13.15 just derived for the transistor are comparable with Equation 13.10 and Equation 13.11 derived for the diode circuit in Section 13.2, and like them can be given a circuit representation. The circuit representing Equation 13.14 is shown in Figure 13.8(a), that for Equation 13.15 in Figure 13.8(b), and that for the complete circuit in Figure 13.8(c). From the latter, Equation 13.16 can be written down by inspection. The circuit shown in Figure 13.8(a) is referred to as the **small-signal equivalent circuit** for the transistor. It is important to remember that the numerical values of g_m and r_{ds} have to be determined for the particular working point appropriate to the problem.

13.6 ◆ The n–p–n bipolar transistor

The symbol and electrical variables for this type of transistor are shown in Figure 13.9. The terminals B,C,E refer respectively to the base, collector and emitter. Unlike the gate current of the field-effect transistor, the base current is a significant variable in the operation of a bipolar transistor. The characteristics of bipolar transistors are presented in many forms: we shall

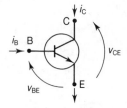

Figure 13.9 Bipolar transistor: circuit symbol and variables for an n–p–n bipolar junction transistor (BJT).

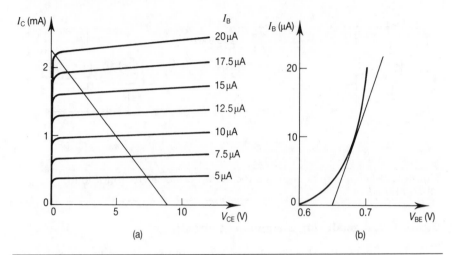

Figure 13.10 Characteristics for an n–p–n BJT: (a) output; (b) input.

assume that characteristics are available which express v_{BE} and i_C in terms of i_B and v_{CE}. Examples of such characteristics are shown in Figure 13.10. In Figure 13.10(a) are shown output characteristics, giving the variation of i_C with v_{CE} for various constant values of i_B. An input characteristic, which gives the variation of v_{BE} with i_B for constant v_{CE}, is shown in Figure 13.10(b). Although in principle a number of curves should appear in Figure 13.10(b), each for a different value of v_{CE}, it is found in practice that, firstly, in the normal mode of operation the i_B–v_{BE} curve is not very sensitive to the value of v_{CE} and, secondly, the differences that do exist are masked by the variations between different transistors of the same type. For these reasons only one curve appears in the figure.

In circuits involving bipolar transistors a working point is established in precisely the same way as previously considered for the field-effect transistor. Variations then take place about this working point.

> **EXAMPLE 13.3** The transistor in the circuit of Figure 13.11 has the characteristics shown in Figure 13.10. Determine the working point.
>
> As before, a load line can be superimposed on the output characteristics, Figure 13.10(a). This load line passes through the 9 V point on the zero-current axis and intercepts the zero-voltage axis at $9/3.9 = 2.3$ mA. The intersection with the 10 μA characteristic gives a working point at $i_C = 1$ mA, $v_{CE} = 5$ V. From the input characteristic, Figure 13.10(b), $V_{BE} = 0.684$ V.

Denoting excursions from the working point by δi_B, δi_C, δv_{BE}, δv_{CE} we can use the output characteristics of Figure 13.10(a) to derive an expression of the form

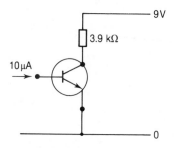

Figure 13.11 Circuit containing a bipolar transistor.

$$\delta i_C = h_{fe}\, \delta i_B + h_{oe}\, \delta v_{CE} \qquad (13.17)$$

in which h_{fe} is the rate of changes of i_C with i_B at constant v_{CE}, and h_{oe} is the rate of change of i_C with v_{CE} at constant i_B. From the input characteristics we can derive a similar expression for changes in v_{BE}:

$$\delta v_{BE} = h_{ie}\, \delta i_B + h_{re}\, \delta v_{CE} \qquad (13.18)$$

In this expression the parameter h_{ie} is the rate of change of v_{BE} with i_B at constant v_{CE}; h_{re} is the rate of change of i_B with v_{CE} at constant i_B. (From the above discussion of the input characteristics it can be concluded that the parameter h_{re} is small: as will be seen later, to take it as zero leads to a convenient simplification.) The coefficients h_{ie}, h_{re}, h_{fe}, h_{oe} are known as the hybrid parameters: the dimensions of these parameters are as follows:

h_{fe}, h_{re} dimensionless

h_{ie} resistance

h_{oe} conductance

EXAMPLE 13.4 For the transistor and circuit of Example 13.3, estimate values of the h parameters appropriate to the working point.

The 10 μA characteristic in Figure 13.10(a) is virtually a straight line which passes through the points 10 V, 1.04 mA and 2 V, 0.98 mA. Hence

$$h_{oe} \approx \frac{1.04 - 0.98}{10 - 2} \times 10^{-3} = 7.5\ \mu S$$

Keeping v_{CE} constant at 5 V, we find, from the same figure, for i_C and i_B the pair of values 1.32 mA, 12.5 μA and 0.7 mA, 7.5 μA. Hence

$$h_{fe} \approx \frac{1.32 - 0.7}{12.5 - 7.5} \times 10^3 = 124$$

Turning to the input characteristics, a tangent drawn at 10 μA, 684 mV passes through 0 μA, 650 mV and 20 μA, 720 mV. Hence

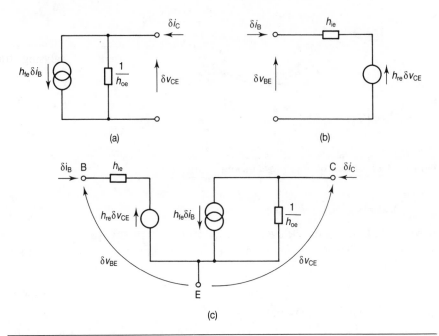

Figure 13.12 Small-signal equivalent circuits: (a) output side; (b) input side; (c) combined.

$$h_{ie} \approx \frac{720 - 650}{20 - 0} \times 10^3 = 3.5 \text{ k}\Omega$$

The single characteristic provided does not allow us to estimate h_{re}, which must be taken as zero.

13.6.1 ◆ The small-signal equivalent circuit

Equation 13.17 and Equation 13.18 have the circuit representations shown in Figures 13.12(a) and 13.12(b) respectively. Combined as in Figure 13.12(c) an equivalent circuit for the transistor is obtained. If it is permissible to neglect h_{re}, as it is for most silicon transistors, the circuit becomes that of Figure 13.13(a). This represents a simplification in that changes at the collector end do not affect conditions at the input, base, end. This, of course, must be the case if change of v_{CE} is considered not to affect the input characteristic. The circuit of Figure 13.13(a) can be rearranged to give the form shown in Figure 13.13(b), in which the dependent source has been made to depend on the change in base–emitter voltage, δv_{BE}, rather than the change in base current, δi_B. In this form the circuit is a slightly simplified version of the hybrid-π circuit much used in the analysis of transistor circuits, and the notation has been changed to conform. In terms of the h parameters the hybrid-π parameters are given by

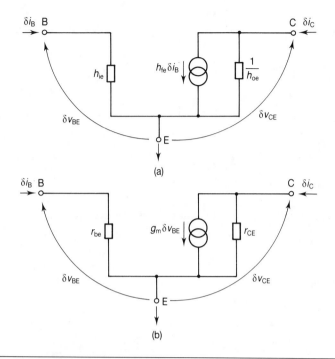

Figure 13.13 Comparison of small-signal equivalent circuits: (a) *h* parameters; (b) hybrid-π.

$$r_{be} = h_{ie}, \; g_m = h_{fe}/h_{ie}, \; r_{ce} = 1/h_{oe}$$

If it is not permissible to neglect h_{re}, the effect can be allowed for by placing an appropriate resistance between base and collector in the hybrid-π circuit. The hybrid-π circuit has the great advantage of having a configuration similar to that of the equivalent circuit for the field-effect transistor, Figure 13.8(a), the only difference being the presence of the resistance r_{be} across the input.

EXAMPLE 13.5 Give the small-signal equivalent circuit for the transistor circuit of Figure 13.11. With the aid of this circuit estimate the changes in v_{CE} and v_{BE} when i_B changes by 1 μA.

From the load line equation

$$v_{CE} = 9 - Ri_C \tag{13.19}$$

we deduce that

$$\delta v_{CE} = -R \, \delta i_C \tag{13.20}$$

which has the circuit representation shown in Figure 13.14(a). Combining this with the circuit of Figure 13.13(a), in which the parameter values have been found in Example 13.4, we arrive at the circuit shown in Figure

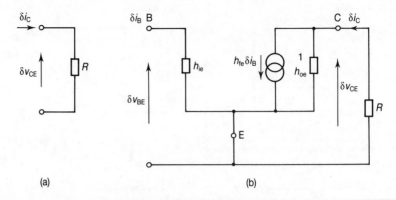

Figure 13.14 Small-signal equivalent circuits for Example 13.5.

13.14(b). Since h_{oe} represents a resistance of some 130 kΩ the resistance of the parallel combination in the output circuit can be taken as 3.9 kΩ. The change in v_{CE} is therefore related to the change in v_{BE} by

$$\delta v_{CE} = -h_{fe} R\ \delta i_B$$

$$= -124 \times 3900 \times 10^{-6}$$

$$= -0.48\ \text{V}$$

The change in v_{BE} is given by

$$\delta v_{BE} = h_{ie}\ \delta i_B$$

$$= 3500 \times 10^{-6}$$

$$\approx 3.5\ \text{mV}$$

13.7 ♦ A systematic approach

The small-signal equivalent circuits of Figure 13.8(c) and 13.14(b), derived from the transistor circuits of Figures 13.7 and 13.11 respectively, have been set up by considering separately that part of the circuit representing transistor action and that representing the response to small changes in the linear network in which the transistor is embedded. In both cases, the equivalent circuit for the latter component has been found by first deriving the load line Equations 13.13 and 13.19, from these deriving the relation for small changes and finally deducing a circuit representation. This long derivation is seen to be unnecessary once it is realized that the circuits of Figures 13.8(b) and 13.14(a) are merely those of the linear part of the network with the constant supply voltage placed equal to zero. This result can be derived by applying the principle of superposition, as illustrated in Figure 13.15. The result is correct for any linear network.

The series of steps by which a small-signal equivalent circuit can be derived will be illustrated by reference to the bipolar circuit of Figure

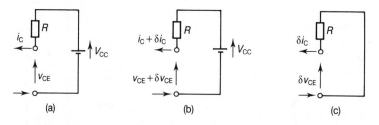

Figure 13.15 Example of derivation of the small-signal equivalent circuit for a linear network: (a) state with no signal: (b) small signal present; (c)superposition by subtraction of (a) from (b).

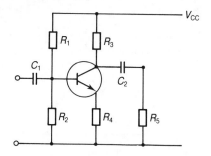

Figure 13.16 Transistor circuit.

13.16. The first step, Figure 13.17(a), is to extract the transistor, leaving the linear network unaltered but labelling the transistor terminals; in the second step, Figure 13.17(b), sources of constant voltage are reduced to zero by replacing each source by a short circuit (if a constant source of direct current were present it would be replaced by an open circuit); finally, into the circuit so derived is inserted the small-signal equivalent for the transistor, and any convenient rearrangement carried out, Figure 13.17(c). The labelling of the transistor terminals is useful because the nodes between which δv_{BE} and δv_{CE} appear are unambiguously established. To the resulting circuit the normal procedures of network analysis, including phasor analysis, can be applied.

13.7.1 ◆ Notation

It is clearly inconvenient to denote the small changes which represent the all-important signals by the prefix δ. This is avoided in normal usage by the convention that these quantities are denoted with lower-case subscripts, as opposed to the upper-case subscripts. These latter are used to denote the complete quantity, steady component plus signal. According to this convention we replace the variables hitherto used as follows:

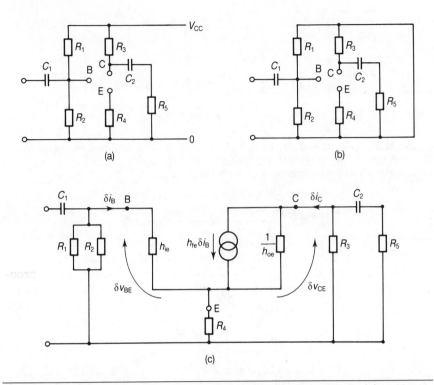

Figure 13.17 Steps in forming the small-signal equivalent circuit: (a) remove transistor, labelling terminals; (b) reduce fixed voltages to zero; (c) incorporate the small-signal equivalent circuit for the transistor to obtain complete circuit.

$$\delta i_G \rightarrow i_g \qquad \delta i_B \rightarrow i_b$$

$$\delta i_D \rightarrow i_d \qquad \delta i_C \rightarrow i_c$$

$$\delta v_{GS} \rightarrow v_{gs} \qquad \delta v_{BE} \rightarrow v_{be}$$

$$\delta v_{DS} \rightarrow v_{ds} \qquad \delta v_{CE} \rightarrow v_{ce}$$

13.7.2 ◆ Transistor circuits in Chapter 3

The circuits appearing as examples in Chapter 3 containing dependent current sources can now be seen as particular cases derived from the transistor circuit of Figure 13.16. Examples 3.6 and 3.7 concern the setting up of appropriate working points, it being assumed that the voltages applied to the transistor and the resulting currents are consistent with the transistor characteristics. Figure 3.13 represents the small-signal equivalent circuit of Figure 13.13(b), with a notation more appropriate to the context. Example 3.8, Figure 3.14, concerns a common-emitter amplifier; Example

3.9, Figure 3.15(a), concerns an emitter-follower circuit; Example 3.10, Figure 3.16, is an amplifier circuit incorporating an emitter resistance (which has the effect of stabilizing the collector current).

13.7.3 ◆ Transistor circuits in Chapter 5

Example 5.6 is derived from a problem encountered with the use of transistors as amplifiers at high frequencies. In Example 5.6, Figure 5.8a, the circuit is basically a common-source FET amplifier. The analysis illustrates how the presence of capacitance between gate and drain has the effect of greatly modifying the apparent impedance between gate and ground. The effect is known as the **Miller effect**. Since some capacitance is necessarily present in virtue of the construction of the transistor (the gate–drain or Miller capacitance, C_{gd}) the effect indicates a limit on the performance of an FET at high frequency. In the example the drain slope resistance r_{ds} has been implicitly combined with the resistance R.

Example 5.19, Figure 5.21(a), likewise illustrates one of the problems encountered in designing transistor circuits at high frequencies. Construction of the transistor inevitably means a short length of conductor joining the actual semiconductor material to the other components: this conductor when carrying current produces a magnetic field, which in circuit terms means inductance. As the example shows, this inductance can have serious effects on performance.

13.8 ◆ Summary

It has been shown how non-linear devices can be represented by linear circuits when signals constitute only small excursions about a DC operating point. The method of estimating the relevant parameters from the DC characteristics has been illustrated for diodes, field-effect transistors and bipolar transistors.

Small-signal parameters
Field-effect transistor

$$i_d = g_m v_{gs} + \frac{1}{r_{ds}} v_{ds}$$

Once a working point has been established, the mutual conductance g_m is given by the rate of change of I_D with respect to V_{DS} for constant V_{GS}; the drain slope resistance r_{ds} is given by the rate of change of V_{DS} with respect to I_D for constant V_{GS}.

Bipolar transistor

$$v_{be} = h_{ie} i_b + h_{re} v_{ce}$$
$$i_c = h_{fe} i_b + h_{oe} v_{ce}$$

With respect to a given working point, the hybrid parameters are defined as follows:

- h_{ie}: rate of change of V_{BE} with respect to I_B, V_{CE} constant;
- h_{re}: rate of change of V_{BE} with respect to V_{CE}, I_B constant;
- h_{fe}: rate of change of I_C with respect to I_B, V_{CE} constant;
- h_{oe}: rate of change of I_C with respect to V_{CE}, I_B constant

PROBLEMS

13.1 The forward characteristic of a diode is closely described by

$$I_D = 10^{-6}[\exp(20v_D) - 1]$$

in which I_D is in amps, V_D in volts. Plot this characteristic over the range $0\text{ V} < v_D < 0.5\text{ V}$.

The diode is connected in a circuit of the form of Figure 13.1(a) in which $R = 150\,\Omega$, $e_0 = 1.5\text{ V}$. Construct a suitable load line and hence determine the working point of the diode. Estimate the slope resistance of the diode at this point.

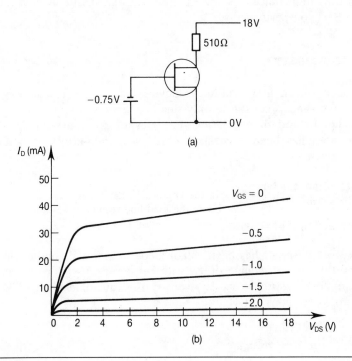

(a)

(b)

Figure 13.18

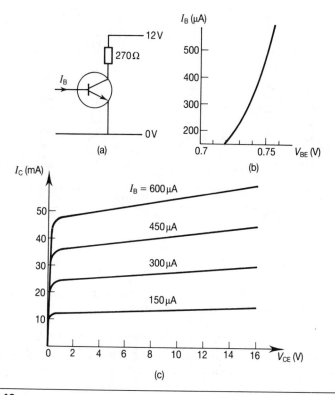

Figure 13.19

13.2 Set up the form of Equations 13.1 and 13.2 appropriate to Problem 13.1. By trial and error solve these equations to find the working point, and check against the previous solution. Determine the slope resistance by finding the value of di/dv at the working point.

13.3 The field-effect transistor in the circuit of Figure 13.18(a) has the characteristics shown in Figure 13.18(b). Using the load line technique, estimate the working point and the corresponding values of the parameters g_m, r_{ds}.

13.4 In the circuit of Figure 13.18(a) an AC voltage source of 0.5 V peak amplitude is placed in series in the gate connection. Derive a small-signal equivalent for the circuit. Hence estimate the peak value of the AC component of voltage across the 510 Ω resistor.

13.5 The bipolar transistor in the circuit shown in Figure 13.19(a) has the input and output characteristics shown in Figures 13.19(b) and 13.19(c). Determine the working point when the base current is fixed at 0.3 mA. Estimate the corresponding values of the parameters h_{ie}, h_{fe}, h_{oe}. (It may be assumed that h_{re} is zero.)

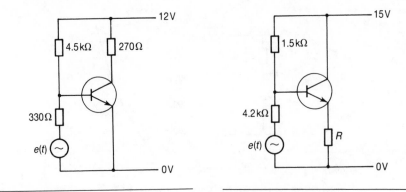

Figure 13.20   **Figure 13.21**

13.6 In Figure 13.20 is shown a modified version of the circuit of Figure 13.19(a), in which the AC source $e(t)$ delivers 15 mV peak.

   (i) Show that the base current I_B is still approximately 0.3 mA.

   (ii) Derive a small-signal equivalent circuit and hence estimate the peak value of the AC component of the collector–emitter voltage.

13.7 The bipolar transistor in the circuit shown in Figure 13.21 has the characteristics shown in Figure 13.19. In the circuit, $e(t)$ represents an AC voltage source of 1.5 V peak amplitude.

   (i) Determine the value of the resistance R which will give the transistor the same operating point as in Problem 13.5

   (ii) Derive a small-signal equivalent circuit and hence estimate the peak amplitude of the AC component in the voltage across R.

13.8 In the circuit of Figure 13.22, the field-effect transistor has the characteristics shown in Figure 13.18. In the figure, $e(t)$ represents an AC voltage source of 0.5 V peak amplitude.

   (i) Determine R_1, R_2 so that the transistor is operating under the same conditions as in Problem 13.3.

   (ii) Derive a small-signal equivalent circuit and hence estimate the peak value of the AC component of voltage across R_2.

13.9 It was mentioned in Section 13.5 that small-signal parameters such as g_m and r_{ds} could be obtained by differentiation if a suitable mathematical expression for the characteristics were available. One analytical model used for the field-effect transistor in the range $V_{DS} > (V_{GS} - V_T)$ takes the form

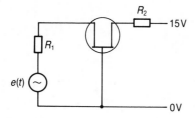

Figure 13.22

$$I_D = \beta(1 + \lambda V_{DS})(V_{GS} - V_T)^2$$

in which β, λ and V_T are constants of the device. Using this model, obtain expressions for

$$g_m = \partial I_D/\partial V_{DS}, \; r_{ds} = (\partial I_D/\partial V_{DS})^{-1}$$

13.10 The characteristics of Figure 13.18 have been calculated using the values

$$\beta = 5 \times 10^{-3} \text{ A V}^{-2}, \; \lambda = 0.02 \text{ V}^{-1}, \; V_T = -2.5 \text{ V}$$

Calculate the parameter values at the working point found in Problem 13.3.

14 ◆ *From Circuit to Circuit Diagram*

A circuit, either existing or to be made, is usually described by a circuit diagram. In this chapter errors which can be introduced in this apparently straightforward step are discussed: circuit elements in the diagram may not accurately describe physical components, and interconnections may introduce unwanted effects.

14.1 ◆ Introduction

The step from an actual circuit, perhaps part of some equipment, to its representation by a circuit diagram, or the converse step of realization in hardware of a designed circuit, may seem simple. This is not so, for a number of reasons. In the first place, the circuit diagram is made up of elements with the properties of inductance, resistance and capacitance: these elements have to be realized as hardware, nominally as the physical objects called inductors, resistors or capacitors. In previous chapters these terms have been used to denote ideal components, so that an inductor, for example, has the sole property of inductance. In fact, an inductor is a coil of wire which like any other electrical circuit requires the attributes of inductance, resistance and capacitance for complete description. So the realization of inductance in the circuit diagram by an inductor in the physical circuit introduces other, unwanted, electrical attributes. Similar comments can be made about resistors and capacitors. In this way the fidelity with which a circuit diagram can be realized is limited, and indeed restrictions can be placed on the type of circuit which it is useful to try to realize.

The other feature of a circuit diagram which seems simple to realize is the network of connections between the components. In going from circuit diagram to circuit, the immediate response is to replace these connections by wires. A wire, however, does not have the basic property of a connection in the diagram, that it carries current from one circuit element to another without any voltage difference being developed between the ends of the connection. At the very least, the wire will have resistance; further, the voltage difference between two adjacent wires will induce charges on the wires, an effect to be associated with capacitance, and the current carried by a wire will give rise to a magnetic field, bringing in inductance. Thus the action of the physical connection has to be described electrically by introducing additional resistance, capacitance and inductance into the circuit diagram. In many cases, the effects of these additional elements will be small; in other cases these 'stray' components may be of vital importance in arriving at a correct description of circuit behaviour. This is particularly true at high frequencies.

The aim of this chapter is to provide some indication of departures from the ideal in these various aspects.

14.2 ◆ Components

The electrical concepts of resistance, inductance and capacitance came through attempts to explain the behaviour of electrical circuits. The three properties are found to be sufficient in all cases: they are the circuit representations of the processes of, respectively, the dissipation of electrical energy into non-electrical forms, of the storage of energy in the associated magnetic field and of the storage of energy in the electrostatic field. The designer of components wishes to make a component which expresses only one of these processes, and not surprisingly finds the task impossible. A component is a physical circuit the behaviour of which is explained only by introducing all three processes, and in general must be characterized by a circuit containing all three ideal components. Whether or not the deviations from the ideal are important depends on the circumstances, and in particular the frequency of operation, but it is always important to be aware of the imperfections inherent in the components used to build an operational circuit. In the next sections the three types of components are briefly reviewed.

14.3 ◆ Resistors

The problem with making resistors can be seen with the aid of a simple calculation. Consider the resistance of a copper wire 0.1 mm in diameter and 1 m long. For copper the conductivity is $5.9 \times 10^7 \, S\,m^{-1}$, so

$$R = 1/(5.9 \times 10^7 \times \pi \times 10^8/4)$$

$$= 2.2 \, \Omega$$

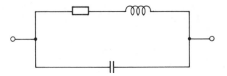

Figure 14.1 Resistors: a physical resistor may require added inductance and capacitance to give an accurate circuit description.

Even the metal alloys used for resistors, such as constantan, have conductivities only of the order of ten times less than copper, so that wire-wound resistors as low as a few tens of ohms necessarily involve long lengths of thin wire. To make such a resistor compact requires that the wire be coiled in some form, thus introducing significant inductance. Clearly, higher resistivity material is desirable, or a technique for using metal alloys in much thinner layers than possible in wire form. Apart from the realization of suitable values, perhaps $10\,\Omega$ to $10^7\,\Omega$, other aspects have to be considered:

- precision of manufacture;
- change of resistance with temperature;
- stability with time;
- power handling capacity;
- dependence of resistance on applied voltage.

Wire-wound resistors have certain advantages. They come out very well when judged by the list above, but they are bulky and, being necessarily coiled, they are associated with stray inductance and capacitance. Such a resistor has to be regarded as having an equivalent circuit of the form shown in Figure 14.1.

Metal resistors of more convenient size can be made in several ways. For example, a thin metallic film can be laid down on a ceramic rod, perhaps grooved into a helix to increase resistance, or metal foil can be used. Such resistors have some of the desirable properties of wire-wound resistors, but they can only handle low power and tend to be expensive.

Low quality discrete resistors are made of carbon composition, a mixture of carbon and a non-conducting filler giving a high resistivity mix. Such resistors are difficult to make to close tolerance values, have a bad temperature coefficient of resistance and are liable to change value with age. On the other hand they can be made to handle high power levels and are cheap. High stability versions improve on some of the properties. Stray reactances can usually be neglected, although care has to be exercised with high values at high frequencies.

Resistors in some integrated circuits are made by laying down thin films on a suitable substrate, ceramic or perhaps silicon. By correct choice

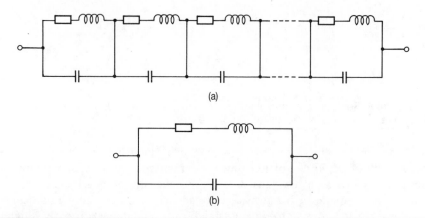

(a)

(b)

Figure 14.2 Inductors: a physical inductor will allow for resistance and capacitance as well as inductance for a full circuit description: (a) distributed components; (b) lumped approximation.

of film composition good, low-temperature-coefficient resistors can be fabricated. It is not easy to make such resistors to high precision, but by suitable design they can be adjusted using 'laser trimming'. A less precise but useful way is by adjusting the doping of a base layer of the semiconductor substrate.

It will be realized that in several ways a resistor is likely to depart from the ideal. Whether or not these departures are significant will depend on the circuit: the suitability of a particular type must be judged with the aid of the manufacturer's specification.

14.4 ◆ Inductors

Of the three basic components inductors are without doubt the furthest from the ideal. The basic concept of an inductor as a coil of wire shows that we have to allow for resistance in series with the inductance, and in addition there will be capacitance between adjacent turns, as shown in Figure 14.2(a). For most purposes this distributed capacitance can be replaced by a capacitance in parallel with the terminals, as shown in Figure 14.2(b). This equivalent circuit is of the same form as that for the resistor, but in the one the resistance will predominate and in the other the inductance. In order to give some idea of orders of magnitude, a solenoidally wound coil of some 50 mm diameter and 50 mm long with 100 turns of copper wire has an inductance of some 100 μH. This is a small value compared with the several henries that might be needed at power frequencies.

In order to increase inductance the coil may be wound on a material of high magnetic permeability. At low frequencies this might be a silicon

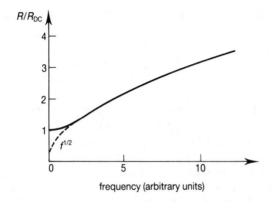

Figure 14.3 Form of the variation with frequency of the series resistance of an inductor.

iron alloy or a higher permeability alloy such as mu-metal. Such magnetic cores would be built up of thin sheet stampings. Another class of magnetic material is that of ferrites, which have the great advantage of high resistivity as well as high magnetic permeability (the relevance of the resistivity is discussed below).

14.4.1 ◆ Resistance in air-cored coils

It is convenient to commence by discussing coils wound without use of magnetic material, since the latter gives rise to an extra source of loss which must be represented by additional resistance. At low enough frequencies, the effective series resistance in the equivalent circuit of Figure 14.2(b) would be the resistance of the wire measured at DC. However, as the frequency is increased the resistance is found to increase in the manner indicated in Figure 14.3. For high frequencies the increase is proportional to $\sqrt{f}$. The resistance increases because at high frequencies the current becomes concentrated in a thin layer near the surface of the wire, rather than being uniformly distributed over the cross-section. This pattern arises because of eddy currents produced by the changing magnetic field within the wire. At a particular place within the wire the magnetic field will be due in part to the local current flow and in part to the current flow in the whole coil. The first effect is referred to as **skin effect**, the second as **proximity effect**. The magnitude of the effect can be gauged by comparing the diameter of the wire with a parameter known as the skin depth. This parameter can be expressed in terms of the conductivity, σ, and permeability, μ, of the wire by the formula

$$(\pi \mu \sigma f)^{-1/2}$$

For copper, when the frequency is expressed in megahertz, this reduces to

$$66f^{-1/2} \ \mu m$$

If the diameter of the wire is small compared with the skin depth then the current distribution over the wire will be nearly uniform, and the change in resistance small.

The proximity effect can be responsible for a seemingly paradoxical effect: of two coils of identical geometry and turns but wound with wires of different thickness, that wound with the thinner wire may well be found to have the lower resistance at high frequency.

Apart from resistance in the wire it is possible that the insulator on which the coil is wound is slightly conducting: if this is so, an additional resistance has to be added in parallel with the circuit of Figure 14.2(b). It would also be necessary to consider the capacitor in the equivalent circuit to be lossy, as discussed in a later section.

14.4.2 ◆ Coils with magnetic cores

As mentioned in Section 14.4.1, the presence of a magnetic core introduces extra sources of loss and hence increases the resistance. These losses arise from eddy currents in the core and from magnetic hysteresis.

Eddy currents are of importance chiefly in ferromagnetic alloys, which are relatively good conductors. The effect is reduced by making the cores of thin stampings each insulated from the other: for the high-performance materials used in small cores the stampings may be as thin as 0.1 mm. As far as the equivalent circuit is concerned, eddy currents in the core can be represented by a shunt resistance. At low frequencies this resistance may be taken as constant for a given core and winding.

Hysteresis losses occur because changing the magnetic flux density within a magnetic material requires work. This work goes in changing the orientation of the molecular magnetic dipoles, and appears as heat. The effect may be regarded as equivalent to an extra component of resistance in parallel with the circuit of Figure 14.2(b), but this resistance will vary with frequency and will further be dependent on the applied voltage, making the equivalent circuit non-linear.

Cores made of unlaminated ferrite material do not exhibit significant eddy current loss since the resistivity of ferrites is very high. Hysteresis losses are also low.

14.4.3 ◆ Inductance values

Unlike resistors and capacitors, inductors are not generally available in a range of standard values: they are usually made to conform to a particular design specification. In general, the parameter of interest is inductive reactance rather than inductance, so that large inductances are only required at low frequencies. This is as well, since inspection of the equivalent circuit of Figure 14.2(b) shows that every inductor has a self-resonant frequency, given in terms of the parameters in the figure by $1/2\pi(LC)^{1/2}$. Below this frequency the inductor presents an inductive reactance, above it presents a capacitive reactance. Further, since self-capacity originates as capacity between turns, increasing the number of turns also increases the self-capacity. Choice of the method of winding gives some control over the self-capacitance, but the maximum inductance which can be realized at a given frequency is strictly limited. For example, an inductor sold as an 'RF choke' for use at 500 kHz may be quite useless for this purpose at 10 MHz. A similar comment may be made concerning power transformers: a winding designed for operation at 50 Hz may be quite unsatisfactory at 2 kHz.

In general terms, inductors with ferromagnetic cores are restricted to power and audio frequencies. In the latter case, modern electronic circuitry tries to avoid the use of inductors altogether: they are bulky and expensive as well as being far from ideal. It may also be mentioned that the passage of a steady current, DC, through such an inductor may bring about magnetic saturation of the core and consequently reduce the effective inductance to alternating current.

At frequencies less than a few megahertz air-cored inductors tend to be too large; ferrite cores are frequently used up to 50 MHz. The type and size of core has to be chosen with the aid of the manufacturer's specification.

One of the problems with inductors is the difficulty of making a variable inductance. This can be done by altering the magnetic path: for example, some ferrite cores are mounted to include an adjustable air gap in the magnetic path. Another device is the insertion of a magnetic slug into an air-cored coil: such a slug is composed of a powdered magnetic material in a non-conducting binder.

As discrete components, inductors are used up to frequencies of several hundred megahertz. Above this, circuits are more profitably considered in terms of transmission lines. In regarding inductors as discrete components defined in terms of terminal voltage and current, it must not be forgotten that the magnetic field produced by the inductor may extend for an appreciable radius around it. Interactions between different parts of the circuit can be brought about by this field and these must be taken into consideration.

14.4.4 ♦ *Q* factor

As discussed in Section 5.4.2, the Q factor, defined in Equation 5.32 in the form $Q = \omega L/R$, is often taken as a figure of merit for an inductor. It is this quantity which when the inductor is incorporated in a resonant circuit primarily determines the overall Q factor, since it is in the inductor that most of the losses arise. Q factors of several hundreds can be achieved by good design, with both air-cored and ferrite-cored coils. At very high frequencies the usable value of inductance becomes small, since capacity cannot be reduced indefinitely, with the result that coils may shrink to one turn of small diameter having a low Q factor. It is in this region of frequency that the role of discrete components is taken over by transmission lines.

The simple definition of Q factor would indicate that the Q factor should increase linearly with frequency. In practice this is not the case since, as discussed in Section 14.4.1, the resistance increases with frequency. Although the self-capacitance of an inductor does not directly affect the value of the Q factor, it is a complication in the measurement of Q factor. Most instruments, such as the Q meter, effectively measure Q as the ratio of reactance to resistance. To ascertain the 'true' Q factor it becomes necessary to make allowance for the effect of the self-capacitance of the reactance.

14.5 ◆ Capacitors

A capacitor consists basically of two conducting sheets separated by a thin sheet of dielectric material: the problem in manufacture is to package the large areas required in a small enough space. Consider two electrodes 20 mm wide and 1 m in length, separated by a dielectric sheet 0.1 mm thick and of relative permittivity 2.0. A parallel-plate capacitor consisting of two plates, each of area A, separated by dielectric of thickness t and relative permittivity ε_r, has a capacitance given by

$$C = A\varepsilon_r\varepsilon_0/t$$

Hence the two strips have a capacitance of about 3.5 nF. At the other extreme we may consider a wire 0.5 mm in diameter 2 mm above an earthed base plate. For this case, the capacitance per metre length is given approximately by

$$C \approx \pi\varepsilon_0/\ln(4h/d)$$

in which d, h are respectively wire diameter and height above base plate. With the figures given we find a value of about $10\,\text{pF}\,\text{m}^{-1}$. This figure provides an estimate of the capacitance between a connecting wire and earth.

14.5.1 ◆ Dielectrics

Although small value capacitors can be made using air as the dielectric they are bulky, even if only to obtain the requisite mechanical stability between the plates. Most capacitors will incorporate a dielectric material so that the properties of the dielectric play an important part in determining the properties of the capacitor, in particular determining the associated losses. As discussed in Section 5.5.1, a real capacitor is characterized by a loss angle as well as by its capacitance. The loss angle, δ, and hence the power factor, $\sin \delta$, represent primarily properties of the dielectric although for large-value capacitors the resistance of the foil electrodes may also play a part. It is found that, for most good dielectrics, the loss angle is very nearly constant over a very large range of frequencies, typically several decades. For this reason the loss angle, or the power factor, is usually quoted as a figure of merit of a capacitor. The discussion of Section 5.5.1 has shown that the combination of a capacitance C in series with a resistance R has a loss angle equal to ωRC: this shows that a real capacitor with constant loss angle cannot be represented by a frequency-independent resistance in series with a capacitance. The way in which the loss in a capacitor affects the Q factor of the resonant circuit in which it is used has been illustrated in Section 5.5.1 and Example 5.14.

14.5.2 ♦ Electrolytic capacitors

As the calculations have made clear, large capacitors require large areas and very thin dielectric layers. Capacitors of values greater than a few microfarads are usually of the electrolytic type in which the dielectric layer is formed by electrolytic action over the very large surface of an etched metal film or of a porous metal. With capacitors of this type the applied voltage must be limited to quite low values and must be of the correct polarity. They are not precision components, and may vary over a period of time. They also pass a leakage current, so that a charge will leak away relatively quickly. The value of the capacitor may change with frequency, either because the dielectric constant may fall off with increasing frequency or because the method of construction may introduce a series inductance.

14.5.3 ♦ Solid dielectric capacitors

A wide range of dielectric materials have been used in the construction of capacitors including paper, mica, polystyrene, polycarbonate and ceramic. Of these mica is probably the best as a stable, low-loss dielectric, but capacitors using mica tend to be bulky and limited to values less than about 10 nF. Any type of capacitor has to be assessed on the basis of a number of factors such as

- stability with time,
- precision of manufacture,
- effect of temperature,
- dielectric loss angle,

- leakage,
- presence of inductive effects.

At high frequencies the inductance of connecting leads may be important.

14.6 ◆ Connections

In a circuit diagram the 'wires' connecting the various components and the nodes formed by the interconnections have no physical existence. In an actual construction these wires acquire such physical existence, and consequently have resistance and inductance, and capacitance to neighbouring wires. The resistance of long, thin leads may be important in precision circuitry; series inductance and stray capacitance become important at high frequencies and be the cause of serious malfunction. Inductance is the property of a complete circuit, but for thin wires the inductance is dominated by the magnetic field near the wire. To give an idea of order of magnitude, a wire 0.3 mm in diameter running 5 mm above a conducting plane contributes about $2\,\mu$H for a run of 1 m, or for a more realistic length, $20\,\text{nH}\,\text{cm}^{-1}$. At 100 MHz this length therefore contributes a series reactance of about 12 Ω. In Section 14.5 a value of $10\,\text{pF}\,\text{m}^{-1}$ was assigned to a connecting wire: 1 pF at 100 Hz gives a shunt reactance of about 1200 Ω. Thus at high frequencies these strays become significant parts of the circuit, and must be taken into account. In digital circuits operating with fast rise times, of the order of nanoseconds, stray inductance and capacitance form resonant circuits, giving rise to 'ringing' on signal lines when transients are excited in these resonant circuits.

14.7 ◆ Earth

The concept of a circuit earth permeates electrical circuits. The ideal earth is an impedanceless conducting body through which currents may flow without producing any voltage drop. Needless to say, such an earth is unrealizable. Originally, 'earth' took the form of a good electrical contact made with the physical ground by burying suitable electrodes. It is still used in this sense in electric power circuits, where for safety the external metal case of equipment is earthed to reduce the possibility of operators and others making accidental contact with dangerous voltages. It retains this sense also in some communication circuits in which ground currents flow: for example in the vicinity of medium-frequency radio transmitters. The ground is in fact a medium of uncertain properties in which current flow will cause voltage drops to be developed.

    The term 'earth' is also used in electronic equipment, when it refers to the conductor formed by linking all metal cases of apparatus together. This may or may not be connected to mains earth but the discussion of the last section makes it clear that such a connection is meaningless at high frequencies. The practice ensures that the stray capacitance associated with a unit is made to earth and not to another unit. It also helps to reduce pick-up by sensitive circuits from extraneous sources. In the construction of electronic circuits an earth-plane or ground-plane is frequently used: this provides locally a common electrode in which the flow of current produces, ideally, negligible voltage drop. Since this ideal is unrealizable, care must be taken in the design of good circuitry to take account of earth currents and their effect on circuit operation.

The possibility of interaction between parts of a circuit through magnetic fields has been mentioned above. Electrostatic interaction also takes place since capacitance exists between any two adjacent electrodes, be they components or connecting wires. The close presence of an earth plane reduces direct capacitance and thus the interaction. At the same time capacitances to earth are increased, but this is usually a preferable situation. Earth planes also reduce magnetic interaction by the action of eddy currents, but usually magnetic material will be needed for such screening.

14.8 ♦ Conclusions

The foregoing discussion makes it clear that care has to be taken both in the realization of a particular circuit in hardware and in the interpretation of the circuit diagram of a piece of equipment. To obtain predictable and repeatable results in the design of circuits it is essential that the various 'strays' are carefully considered.

15 ◆ *Circuit Simulation:*
SPICE

This chapter illustrates the use of the simulation program SPICE. Examples given in previous chapters are reworked using this program.

15.1 ◆ Introduction

The simulation program SPICE (Simulation Program with Integrated Circuit Emphasis) has been referred to briefly in earlier chapters. Although designed for the purpose of investigating large-scale integrated circuits, the program can be used on the much simpler examples of this book. In the present chapter it is shown how a circuit can be analysed using SPICE. As a check on the validity of the simulation, various examples analysed by algebraic methods in earlier chapters are reworked.

It is, perhaps, as well to make at the beginning of the chapter the point that ability to do numerical simulation is no substitute for the ability to perform algebraic analysis of simple circuits. It is only possible to take advantage of the power of numerical simulation when circuit concepts are properly understood. Understanding does not come through number crunching. For this reason reference to SPICE has not been made in end-of-chapter problems: some guidance is given at the end of this chapter as to which of those problems can be profitably solved using SPICE.

The original program was developed at the University of California for use on a punched card machine. Subsequently a number of differing versions have been developed for modern computers. The exposition of the present chapter is broadly applicable to any version, although work has been carried out using the MicroSim version PSpice.

15.2 ◆ PSpice

The program PSpice is available both in the professional version, designed for analysing the behaviour and operation of large scale integrated circuits, and also in a reduced memory version (variously known by such names as 'demo version', 'student version' or 'evaluation version'), which will run on an IBM PC. In this latter version the program can be freely copied for use on any number of machines. The examples in this chapter have all been worked using this reduced-memory version. It is presumed that PSpice is available in the professional or the reduced-memory version.

15.2.1 ◆ PROBE

Although not discussed in any detail, attention should be drawn to the PROBE graphics program which forms part of PSpice. Use of this program allows high-quality plots of the calculated variables to be displayed: for example, transient waveforms can be displayed in the form in which they would be observed with an oscilloscope. The program also contains a Fourier analysis package, which is mentioned in Section 15.14.

15.3 ◆ The program

The professional program is designed to model circuits containing a large number of semiconductor devices, and the main difference between the professional and student versions is that the latter is restricted to handling only some ten transistors. This restriction does not affect application to the type of circuit considered in this book, and the experience gained can be later applied to the full scale program.

The types of analyses which can be performed are

- DC analysis,
- AC analysis,
- transient analysis.

The program presumes that in any application the circuit will contain semiconductor devices, and accordingly performs automatically a DC analysis in order to determine working points of such devices. When such devices are present, the AC analysis is small signal, using appropriate linearized models corresponding to the determined working points. (The semiconductor devices, including diodes, bipolar transistors and field-effect transistors, are described by analytic models built into the program. The parameters required for these models have to be provided in input data.) Transient analysis is large signal, using the non-linear characteristics provided by the models.

The output of the program appears in text file saved on disk. The basic quantities determined are the voltages of selected nodes with respect to the common node, and currents through sources. This latter provision can be used to find the current in any connection by the simple process of inserting a zero-voltage source in that connection. The sign convention between voltage and current is that used for passive components: the current through the source is positive when it enters the node designated as the positive node. This is opposite to the customary convention.

Use of the program makes it possible to simulate the usual procedures carried out on a circuit by experiment. As discussed in earlier chapters, the simulated results will only be accurate if the modelling of all devices and of strays is correct. For professional use, the program provides for considerable refinement of the semiconductor models, but for the present purposes only the simple models incorporated in the analysis of earlier chapters will be used.

To use PSpice it is necessary to prepare a text file which contains

(i) a circuit description, and

(ii) control statements.

These control statements specify, for example, the type of analysis to be carried out, or the form in which output data is to be produced. The steps in the preparation of such a text file will be considered in the next sections, relevant to the type of DC or AC analysis of linear circuits forming the subject matter of Chapters 3–8. The cases of transient analysis and of incorporation of semiconductor devices will be considered in later sections.

It is convenient to note at this point the circuit quantities which may be obtained from the program. As indicated in Chapter 6, the program uses a modified nodal analysis method, so that nodal voltages and currents through voltage sources are the unknowns determined by the analysis. It is also possible in some cases to obtain quantities derived from these, such as voltage between nodes.

15.4 ♦ Circuit description

As outlined in Chapter 6, a circuit can be described in tabular form once the individual nodes have been numbered. Nodes are chosen to delineate each separate ideal component, and the common, or reference, node is numbered 0.

The fact that a DC analysis is automatically performed places some

Table 15.1

Factor		SPICE	
		Exponential form	*Symbol*
femto	10^{-15}	1E-15	F
pico	10^{-12}	1E-12	P
nano	10^{-9}	1E-9	N
micro	10^{-6}	1E-6	U
milli	10^{-3}	1E-3	M
kilo	10^{3}	1E3	K
mega	10^{6}	1E6	MEG
giga	10^{9}	1E9	G
tera	10^{12}	1E12	T

restrictions on the circuit connections. For a viable DC state to be established, it is necessary that in the circuit

  (i) each node has two connections,

 (ii) each node has a DC path to the common reference node, and

(iii) no loop of zero resistance exists.

In applying these conditions, a voltage source is considered to be of zero resistance, and a current source to be an open circuit. A physically realizable circuit virtually always satisfies these conditions. With an idealized circuit it may be necessary to add some very small or very large resistances before the conditions are satisfied.

In the text file, each component is entered as a separate line, with the generalized format

identifier   nodes   value

The first letter of the identifier (which can contain up to eight alphanumeric characters) is specific to the type of component. Values are to be given in the appropriate unit, either using exponential notation for powers of 10 or the scale factors set out in Table 15.1.

15.4.1 ◆ Passive components (Table 15.2)

The components in this category are resistance, capacitance, inductance and magnetically coupled inductors. The examples in Table 15.2 show how entries must be made.

15.4.2 ◆ Independent sources, DC and AC (Table 15.3)

The initial letter of the identifier is V for a voltage source, I for a current source. In addition to end nodes and value, the statement must also include the type of source, AC or DC. Since the sources have signs, the order of the nodes is important.

Table 15.2

Component	Circuit	Text file entry
Resistance (R)		R4 2 5 100
Capacitance (C)		C1 3 4 15N (C1 3 4 15E-9)
Inductance (L)		L3 1 0 2.5M (L3 1 0 2.5E-3)
Magnetically coupled pair (K)[a] $k = M/(L1\,L2)^{1/2}$		L1 1 2 100U L2 3 4 2.5M KCC L1 L2 0.7

[a]In this case the ordering of the nodes is important: interchange changes the sign of M.

Table 15.3

Component	Circuit	Text file entry
DC voltage source (V)		V5 3 0 DC 15 (V5 0 3 DC −15)
DC current source (I)		I3 2 1 DC 5 (I3 1 2 DC −5)
AC voltage source (V)		VIN 4 0 AC 5.4M 45 (VIN 0 4 AC 5.4M 225)
AC current source (I)		IS 2 4 AC 2 0 (IS 4 2 AC 2 180)

15.4.3 ♦ Dependent sources (Table 15.4)

Characterization of an ideal dependent source with voltage control requires two pairs of nodes and one value parameter:

identifier   output nodes   input (control) nodes   parameter

A current-controlled dependent source requires a current input, which in SPICE must be determined using the device of a zero-voltage source in the appropriate lead. The entry for a dependent current-controlled source takes the form

identifier   output nodes   control source   parameter

In Table 15.4 this source is identified as V0.

Table 15.4

Component	Circuit	Text file entry
Voltage-controlled voltage source (E)		E10 3 4 1 2 −6
Voltage-controlled current source (G)	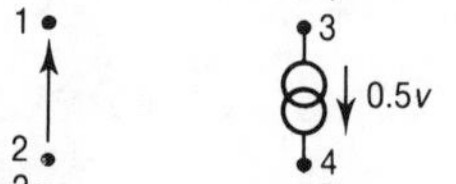	G2 3 4 1 2 0.5
Current-controlled voltage source (H)		H5 5 3 V0 50 V0 2 1
Current-controlled current source (F)		F3 5 3 V0 2 V0 2 0

In all of the entries in this table the order in which nodes are entered is important: interchange has the effect of changing the sign of the parameter.

15.5 ◆ Simple DC analysis

The introduction given in the last sections provides the basis for writing text files for the type of problem which appeared as worked examples in Chapter 3. Apart from lines describing the circuit, it is necessary to provide

(i) an initial title line, which identifies the output, and

(ii) the line

.END

to terminate the file.

EXAMPLE 15.1 As a first example, consider the problem of Example 3.2, Figure 3.7(a). With node numbering and component labels added the diagram is repeated in Figure 15.1. A suitable file is shown below, using the necessary extension for SPICE input files, .CIR.

File X15-1.CIR

```
Example 15.1
R1 1 2 3
R2 2 0 1
R3 2 3 3
R4 3 0 5
V1 1 0 DC 4
.END
```

On running the program, the results appearing in the output file X15-1.OUT give

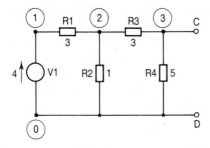

Figure 15.1 Circuit for Example 15.1.

$$V(1) = 4.0000, \ V(2) = 0.9143, \ V(3) = 0.5714$$

These agree with the previous results.

EXAMPLE 15.2 Rework Example 3.4, Figure 3.8(a). With added number-ing and labels, the circuit is shown in Figure 15.2. A suitable file appears below:

File X15-2.CIR

        Example 15.2
        R1 1 2 2
        R2 2 0 4
        R3 2 3 1
        R4 3 0 5
        V1 1 0 DC 5
        I2 3 0 DC −1
        .END

The output file X15-2.OUT gives

$$V(1) = 5.0000, \ V(2) = 3.6364, \ V(3) = 3.8636$$

in agreement with earlier results.

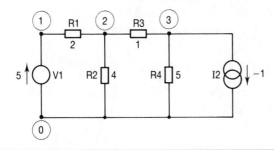

Figure 15.2 Circuit for Example 15.2.

15.6 ◆ Control statements for DC and AC analysis

The selection of control statements considered in this section concern DC and AC analysis and the format in which the output is presented. Each of the text file entries specifying the type of analysis has to be accompanied by another entry relating to output. It is therefore convenient to start by introducing the .PRINT command.

15.6.1 ◆ Output: .PRINT

The outputs which can be requested in any of the types of analysis are voltages between nodes and currents through voltage sources. For example, the general format for the voltage between nodes 1 and 2 is $V(1, 2)$; if one node is omitted, its value defaults to 0, the reference node. Thus $V(1)$ gives the voltage of node 1 to reference. To obtain the current through the voltage source V1, the appropriate entry is of the form $I(V1)$.

The line format is

.PRINT   DC/AC/TRAN   Output required (up to 8)

To obtain more than eight outputs a second .PRINT command must be used. The format for the output quantities given above is correct for DC analysis: the modifications for AC analysis will be given in a later section.

15.6.2 ◆ DC analysis: .DC

This command causes analysis of a circuit to be repeated for a sequence of values for one or two sources. Its main use is for calculating the DC characteristics of semiconductor devices, a use which will be illustrated in a later section. Another use will be illustrated by Example 15.3. The line format is of the form

.DC SCRNAME VSTART VSTOP VINCR

in which SCRNAME stands for any source label, VSTART is the initial voltage, VSTOP the final voltage and VINCR the step size. A second source name can be added, with initial and final voltages and step size. In the analysis the first source is swept through its range for each value of the second source. Thus

.DC V1 0 2 0.5 V2 0 20 5

will cause V1 to take the values 0, 0.5, 1.0, 1.5, 2.0 when V2 has each of the values 0, 5, 10, 15, 20.

EXAMPLE 15.3 For the circuit of Figure 15.1, find the Thévenin equivalent at the terminals C, D.

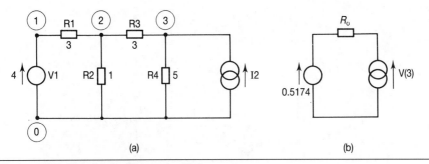

Figure 15.3 Circuits for Example 15.3.

One of the parameters required, the open-circuit voltage, has already been found in Example 15.1. To find the Thévenin resistance it is necessary to recalculate the voltage when a current is taken from the terminals. This can be simulated by adding to the circuit of Figure 15.1 a current source across C, D, as in Figure 15.3(a). A cursory inspection of the circuit indicates that the equivalent resistance is of the order of a few ohms, so that a current of 0.1 A should give a reasonable drop. The source I2 is therefore made to take values 0, 0.1. A suitable text file appears below.

File X15-3.CIR

```
Example 15.3
R1 1 2 3
R2 2 0 1
R3 2 3 3
R4 3 0 5
V1 1 0 DC 4
I2  3 0 DC 0
.DC I2 0 0.1 0.1
.PRINT DC V(3)
.END
```

The output file gives

I2	V(3)
0	0.5174
0.1	0.3571

The equivalent circuit of Figure 15.3(b) indicates that the drop on taking 0.1 A is $0.1R_0$. Hence

$$0.1R_0 = 0.5174 - 0.3571 = 0.1603$$

$$R_0 = 1.60 \ \Omega$$

Two final examples on DC analysis include dependent sources.

EXAMPLE 15.4 Rework Example 3.10, Figure 3.16. The circuit is shown in Figure 15.4. A text file is given below.

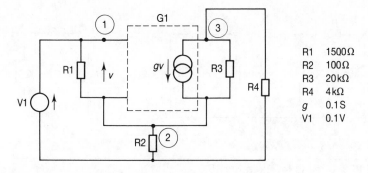

Figure 15.4 Circuit for Example 15.4.

File X15-4.CIR

Example 15.4
R1 1 2 1500
R2 2 0 100
R3 3 2 2E4
R4 3 0 4K
G1 3 2 1 2 0.1
V1 1 0 DC 0.1
.END

The output gives, in agreement with the earlier results,

V(1) = 0.1, V(2) = 0.0893, V(3) = −3.5444

EXAMPLE 15.5 Rework Example 3.11, Figure 3.17, repeated in Figure 15.5. Determine also the change in output voltage when a current of 200 mA is drained from the output of the amplifier.

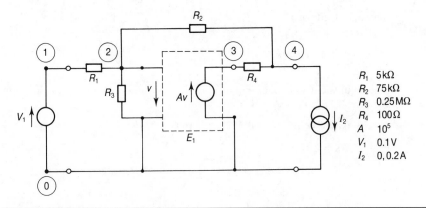

Figure 15.5 Circuit for Example 15.5.

The requirement to find the effect of an output current is met by incorporating a current source (sink) across the output terminals. A suitable text file is given below.

File X15-5.CIR

```
Example 15.5
R1 1 2 5K
R2 2 4 75K
R3 2 0 0.25 MEG
R4 3 4 100
E1 3 0 0 2 1E5
V1 1 0 DC 0.1
I2 4 0 DC 0
.DC I2 0 0.2 0.2
.PRINT DC V(4) I(V1)
.END
```

The output gives

I2	V(4)	I(V1)
0	−1.500	−2.000E-5
0.2	−1.503	−2.004E-5

Because of the sense in which the current through a source is read, the last column indicates a current into the amplifier of 0.02 mA, almost independent of the current drain on the output. Since the input voltage is 0.1 V, this corresponds to an input resistance of 5 kΩ. Because the output voltage is negative, a resistive load would sink a negative current, whereas in the simulation this current is positive. Under these conditions the output voltage will become more negative. The output resistance is approximately $3 \times 10^{-3}/0.2 = 15$ mΩ. These results confirm the conclusions drawn in Example 3.11.

15.6.3 ♦ AC analysis: .AC

In steady-state AC analysis all the sources in a circuit are presumed to be of the same frequency. The analysis can be carried out at a number of frequencies spaced between given end values on either linear or logarithmic scales. The command entry takes one of three forms

```
.AC LIN NP FSTART FSTOP
.AC DEC ND FSTART FSTOP
.AC OCT NO FSTART FSTOP
```

in which FSTART and FSTOP are respectively the initial and final frequencies of the sweep and NP, ND and NO relate to the number of points. (Frequency values have to be entered in full; exponential notation is not accepted. It is also to be noted that zero is not a permissible frequency.) Denoting FSTART by f1, FSTOP by f2 and by *n* the

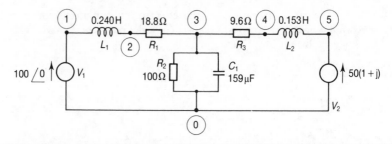

Figure 15.6 Circuit for Example 15.6.

appropriate integer NP, ND or NO, the frequencies for which calculation is performed are in each case given by

$$\left. \begin{array}{ll} \text{LIN} & \text{f1} + k(\text{f2} - \text{f1})/(n-1) \\ \text{DEC} & \text{f1} \times 10^{k/n} \\ \text{OCT} & \text{f1} \times 2^{k/n} \end{array} \right\} \quad k = 0, 1, 2, \ldots, n-1$$

As with DC analysis a .PRINT command is necessary to obtain the output. Since voltage and current have both magnitude and phase, several alternative formats for the output variables are provided. Taking as example the voltage between nodes 1 and 2, the alternatives are

VR(1,2)	real part
VI(1,2)	imaginary part
VM(1,2)	magnitude
VP(1,2)	phase(in degrees)
VDB(1,2)	20 log(magnitude)

For the current through a source the notation is similar: IR(V1), II(V1) and so on.

As an example, the following commands specify the voltage between nodes 1 and 2 in magnitude and phase at ten frequencies linearly spaced between 1 Hz and 1 kHz:

.AC LIN 10 1 1000
.PRINT AC VM(1,2) VP(1,2)

For logarithmic spacing between 10 Hz and 10 kHz with three points per decade:

.AC DEC 3 10 10000
.PRINT AC VM(1,2) VP(1,2)

In order to carry out single frequency calculations of the type in Chapter 4, the two frequencies f1 and f2 are set equal with $n = 1$.

EXAMPLE 15.6 Rework Example 4.7, Figure 4.18(a). The circuit is redrawn in Figure 15.6.

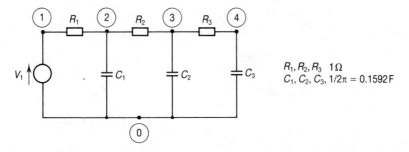

Figure 15.7 Circuit for Example 15.7.

A suitable text file appears below.

File X15-6.CIR

```
Example 15.6
R1 2 3 18.8
R2 3 0 100
R3 3 4 9.6
L1 1 2 0.240
L2 4 5 0.153
C1 3 0 159E-6
V1 1 0 AC 100 0
V2 5 0 AC 70.711 45
.AC LIN 1 50 50
.PRINT AC VM(3) VP(3) VR(3) VI(3)
.END
```

For comparison with the answer in Example 4.7 we use the real and imaginary parts to obtain

$$V(3) = -16.6 - j102.8$$

These agree with the previous results to the accuracy of that calculation.

EXAMPLE 15.7   In Figure 15.7 is shown a van der Pol network previously described in Chapter 5, Example 5.7. To simplify comparison with results easily obtained from Equations 5.13 and 5.14, values for R and C have been chosen to make f numerically equal to ωCR. The text file below calculates $V(4) = (1/M) \exp(-j\phi)$ for $f = 1, 2, 3, 4$.

File X15-7.CIR

```
Example 15.7
R1 1 2 1
R2 2 3 1
R3 3 4 1
C1 2 0 0.1592
C2 3 0 0.1592
```

```
C3 4 0 0.1592
V1 1 0 AC 1 0
.AC LIN 4 1 4
.PRINT AC VM(4) VP(4) VDB(4)
.END
```

The results of running the program, rounded off to three figures, are given below:

f	VM(4)	VP(4)	VDB(4)
1	0.156	−129	−16.1
2	0.051	−168	−25.8
3	0.022	+168	−33.1
4	0.011	+153	−39.0
ωCR	$1/M$	$-\phi$	$20 \log (1/M)$

It is left to the student to compare these results with those obtained by applying Equations 5.13 and 5.14. At a single frequency phase can only be determined in the range ± 180°: in order to compare with the graph of Figure 5.10(b) it is necessary to substract 360° from lines 3 and 4 of the table.

EXAMPLE 15.8 In Figure 15.8 is shown a linearized circuit model of a transistor amplifying stage supplied from a source of 1 kΩ internal imped- ance. This circuit was studied in Chapter 5, Example 5.5. The design purports to give a flat frequency response from 10 kHz to about 2 MHz. Check the design by simulating the circuit.

In this case a wide range of frequencies is involved, so a logarithmic scale is appropriate. For a preliminary investigation evaluation at two points per decade should suffice. In view of the theory of Example 5.6, the input impedance of the amplifier is monitored by recording V(2) and I(V1). A suitable text file follows.

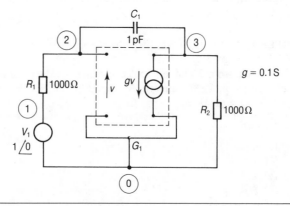

Figure 15.8 Circuit for Example 15.8.

File X15-8.CIR

    Example 15.8
    R1 1 2 1K
    R2 3 0 1K
    C1 2 3 1E-12
    G1 3 0 2 0 0.1
    V1 1 0 AC 1 0
    .AC DEC 2 10000 10000000
    .PRINT AC VM(3) VP(3) VM(2) VP(2) IM(V1) IP(V1)
    .END

The output data, rounded off for convenience, is given below.

Freq (MHz)	VM(4)	VP(4)	VM(2)	VP(2)	IM(V1)	IP(V1)
0.0100	100	180	1.00	−0.36	6.3E−6	−90
0.0316	100	179	1.00	−1.2	2.0E−5−	−91
0.100	100	176	1.00	−3.6	6.3E−5	−94
0.316	98	169	0.98	−11.3	2.0E−4	−102
1.000	84	147	0.84	−32.3	5.3E−4	−123
3.16	44	116	0.44	−62.6	8.9E−4	−154
10.0	15	99	0.15	−77.5	9.8E−4	−171

The values for VM(4) show that the gain at low frequencies is 100, falling off as frequency increases to 0.84 at 1 MHz and 0.44 at 3.16 MHz. Either more points or a graphical interpolation are needed to find the value at 2 MHz but the design seems about right. Bearing in mind the analysis of Example 5.6, it is of interest to note that the intrinsic gain of the amplifier, VM(4)/VM(2), is almost constant at 100: the loss of gain arises because the input impedance is falling as the frequency increases. To determine the behaviour of the input impedance requires manipulation of the figures: calculation of V(2)/I(V1) shows that it is almost purely capacitative over the range of frequencies in the table. This is in accord with theoretical expectations.

A final example in this section shows how the data for Bode plots can be calculated.

EXAMPLE 15.9 In Chapter 5, Example 5.16, an approximate method was used to make a Bode plot for the transfer function

$$H = 10 \frac{1 + j0.04f}{(1 + jf0.1)(1 + j0.01)}$$

Write a SPICE input file to provide the data for such a plot.

To use SPICE it is necessary to design a circuit (which need not be realizable) which will have the desired transfer function. Each term in the

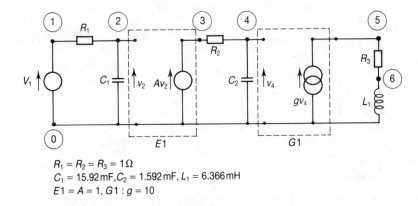

$R_1 = R_2 = R_3 = 1\,\Omega$
$C_1 = 15.92\,\text{mF}, C_2 = 1.592\,\text{mF}, L_1 = 6.366\,\text{mH}$
$E1 = A = 1, G1 : g = 10$

Figure 15.9 Circuit for Example 15.9.

transfer function taken in isolation can be simply realised: one way, using ideal 'buffer' elements, is shown in Figure 15.9. In that circuit we have

$$V(2) = V(1)\,\frac{1}{1 + j\omega R_1 C_1}$$

$$V(4) = A\,V(2)\,\frac{1}{1 + j\omega R_2 C_2}$$

$$V(5) = g\,V(4)\,(R_3 + j\omega L_1)$$

Hence

$$H = AgR_3\,\frac{1 + j\omega L_1/R_3}{(1 + j\omega R_1 C_1)\,(1 + j\omega R_2 C_2)}$$

Choosing the following values we obtain the required transfer function:

$$R_1 = R_2 = R_3 = 1\,\Omega$$

$$C_1 = 0.1/2\pi = 0.015\,92\ \text{F},\ C_2 = 0.01/2\pi = 0.001\,592\ \text{F}$$

$$L_1 = 0.04/2\pi = 6.366\ \text{mH}$$

A suitable text file circuit is given below.

File X15-9.CIR

```
Example 15.9
R1 1 2 1
R2 3 4 1
R3 5 6 1
C1 2 0 1.592E-2
C2 4 0 1.592E-3
L1 6 0 6.366E-3
E1 3 0 2 0 1
G1 5 0 4 0 −10
V1 1 0 AC 1 0
.AC DEC 3 0.1 100000
.PRINT AC VDB(5) VP(5)
.END
```

The exact points appearing in Figure 5.24 were obtained in this way.

15.7 ◆ Transient analysis

To perform a transient analysis several parameters must be specified. Firstly, to define a unique problem the initial conditions must be given: currents in inductors and voltages across capacitors. For a computational analysis it is necessary to give start and stop times and, most important, to set a step length appropriate to the problem. Provision is made for entering all these quantities. Initial conditions are added to the component entries; the remainder appear in the .TRAN control statement.

15.7.1 ◆ Initial conditions

The component entries described in Table 15.2 for the circuit description can have a final, optional element 'IC=' with which to specify initial conditions. Examples showing the complete entry are given in Table 15.5.

15.7.2 ◆ The control statement .TRAN

The control statement for transient analysis has the form

.TRAN TSTEP TSTOP ⟨ TSTART ⟨ TMAX ⟩⟩ ⟨ UIC ⟩

in which

TSTEP	printing interval	

TSTOP   stop time (computation starts at time zero)
The remaining items in angle brackets are optional:

TSTART   commencement of output (computation between zero and TSTART is not recorded for output)

TMAX   computational step length: if this is omitted the step lenghth defaults to either TSTEP or (TSTOP − TSTART)/50, whichever is the smaller

UIC   (use initial conditions) inclusion of this entry causes the program to use the initial conditions specified in the component entries; otherwise the initial conditions are those determined by the initial DC analysis.

Table 15.5

Component	Circuit	Text file entry
Inductance	① 0.5A → mm 0.25H ②	L1 1 2 0.25 IC = 0.5
Capacitance	③ 0.2 μF ④ + ◄——— −5V ———► −	C2 3 4 0.2U IC = 5

It will be noted that the order of entry of the nodes is important: reversal of order is equivalent to reversal of value in the initial condition.

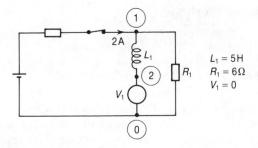

Figure 15.10 Circuit for Example 15.10.

In choosing suitable values for TSTEP and TMAX it is necessary to make some estimate of the rates of change likely to be encountered. For example, for a resonant circuit of period T the response goes from zero to maximum in $T/4$. Taking ten steps for this change indicates a computational step of about $T/40$. The progam can of course to rerun with a smaller step length to check the effect of step size.

To obtain output a .PRINT command is needed, specifying the analysis as TRAN:

.PRINT TRAN output variables

These matters will be illustrated by the following examples.

EXAMPLE 15.10 The analysis of Example 9.2 determined the decay of current in a parallel $L-R$ circuit, $L = 5\,\mathrm{H}$, $R = 6\,\Omega$, when the initial current in the inductor was 2 A. The circuit is redrawn in Figure 15.10, with a zero-voltage source included to allow an output of current. In this problem the time scale is given by the time constant L/R. This is of the order of 1 s, suggesting a computational step length of 0.1 s. In the following text file the initial current of 2 A appears in the component statement, and the entry UIC in the .TRAN command causes the program to take this condition. To avoid unnecessary output, the print step is taken as 0.2 s, and the duration is specified as 4 s.

File X15-10.CIR

```
Example 15.10
R1 1 0 6
L1 1 2 5 IC = 2
V1 2 0 DC 0
.TRAN 0.2 4 0 0.1 UIC
PRINT TRAN I(V1) V(1)
.END
```

The answers given in Example 9.2 can be verified.

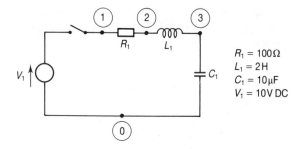

Figure 15.11 Circuit for Example 15.11.

EXAMPLE 15.11 In a series L–C–R circuit, $L = 2$ H, $C = 10 \ \mu F$, $R = 100 \ \Omega$. Determine the current response when the intially quiescent circuit is connected to a 10 V DC source.

We expect the response to generally have the form of a damped sine wave, the relevant parameters for which are resonant frequency and Q factor. We have

$$f_0 = 1/2\pi \ (LC)^{1/2} = 35.6 \ \text{Hz}$$

$$Q = 2\pi f_0 L/R = 4.47$$

$$1/f_0 = 28 \ \text{ms}$$

These values suggest that the response will take several cycles to decay. For the computational step length a value of 0.5 ms seems suitable. Initally at least, we look for the response over 100 ms, rather more than 4 cycles, with a print interval of 2 ms. The circuit is shown in Figure 15.11 and a text file appears below.

File X15-11.CIR

```
Example 15.11
R1 1 2 100
L1 2 3 2 IC = 0
C1 3 0 1E-5 IC = 0
V1 1 0 DC 10
.TRAN 2E-3 0.1 0 5E-4 UIC
.PRINT TRAN I(V1)
.END
```

The sudden application of the 10 V source to the quiescent circuit is equivalent to a step input, for which the response was determined in Section 9.7.2. The expression for the voltage across the capacitor was given in Equation 9.29:

$$v(t) = E\left\{1 - \frac{1}{\beta} \exp(-\alpha t) \ [\beta \cos(\beta t) + \alpha \sin(\beta t)]\right\}$$

The current in the circuit is given by

$$i(t) = C\frac{dv}{dt} = \frac{E}{L\beta} \exp(-\alpha t) \sin(\beta t)$$

In the following table results for about the first cycle are given for the output of the SPICE program and for calculation from the exact expression, allowing for the sign change implicit in I(V1).

t (ms)	I(V1) (mA)	
	SPICE	*Exact*
0	0	0
2	−9.18	9.20
4	−15.78	15.81
6	−18.80	18.82
8	−18.03	18.03
10	−13.96	13.94
12	−7.67	7.63
14	−0.54	0.49
16	6.01	−6.06
18	10.82	−10.86
20	13.15	−13.16
22	12.80	−12.78
24	10.09	−10.05
26	5.76	−5.69

While not perfect, agreement is good to about 1% of the maximum value.

15.7.3 ♦ Time-dependent waveforms

The examples of the last section have shown how the natural response and the step response can be calculated. SPICE makes provision for some other source waveforms relevant to transient problems.

The general format for a time-dependent source is similar to that for DC or AC sources. The entry for a source VT1 connected between nodes 2 and 0 takes the form

VT1 2 0 TYPE

where TYPE can have the possible entries

PULSE(...)
SIN(...)
EXP(...)
PWL(...)
SFFM(...)

The last type refers to a frequency-modulated source, which is outside the scope of this book. In the following entries TSTEP, TSTOP refer to the accompanying .TRAN statement.

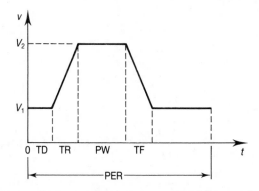

Figure 15.12 The time-dependent waveform PULSE (V1 V2 TD TR TF PW PER).

1. Pulse waveform: PULSE

The waveform described by this entry has the form of a repetitive trapezoidal pulse, Figure 15.12.

PULSE(V1 V2 TD TR TF PW PER)

parameter		*default value*
V1	initial value	
V2	pulse value	
TD	delay time	0
TR	rise time	TSTEP
TF	fall time	TSTEP
PW	pulse width	TSTOP
PER	period	TSTOP

2. Sinusoidal waveform: SIN

The waveform described by this entry is that of a damped sinewave, Figure 15.13.

SIN(VO VA FREQ TD THETA)

parameter	*default value*
VO offset	
VA amplitude	
FREQ frequency	1/TSTOP
TD delay time	0
THETA damping factor	0

The waveform

$$0 < t < t_d \qquad v(t) = V_0$$

$$t_d < t < T \qquad v(t) = V_0 + V_a \exp\left[-\alpha\left(t - t_d\right)\right] \sin\left(2\pi f t\right)$$

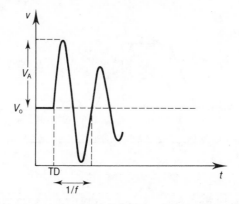

Figure 15.13 The time-dependent waveform SIN (VO VA FREQ TD THETA).

is specified by the entry

$$\text{SIN}(V_0 \; V_a \; f \; t_d \; \alpha)$$

3. Exponential pulse: EXP

The waveform described by this entry is basically a 'square' pulse with the exponential rising and falling edges, Figure 15.14.

EXP(V1 V2 TD1 TAU1 TD2 TAU2)

parameter	*default value*
V1 initial value	
V2 pulsed value	
TD1 rise delay time	0
TAU1 rise time constant	TSTEP
TD2 fall delay time	TD1 + TSTEP
TAU2 fall time constant	TSTEP

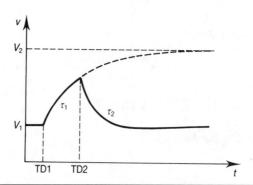

Figure 15.14 The time-dependent waveform EXP (V1 V2 TD1 TAU1 TD2 TAU2).

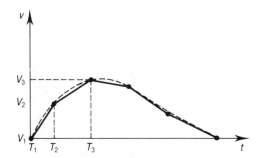

Figure 15.15 The time-dependent waveform PWL (T1 V1 T2 V2 . . .).

The waveform

$$0 < t < t_1 \qquad v(t) = V_1$$

$$t_1 < t < t_2 \qquad v(t) = V_1 + (V_2 - V_1)\{1 - \exp[-(t - t_1)/\tau_1]\}$$

$$t_2 < t < \text{TSTOP} \quad v(t) = V_1 + (V_2 - V_1)\{1 - \exp[-(t - t_1)/\tau_1]\}$$
$$- (V_2 - V_1)\{1 - \exp[-(t - t_2)/\tau_2]\}$$

is described by

$$\text{EXP}(V_1 \ V_2 \ t_1 \ \tau_1 \ t_2 \ \tau_2)$$

4. Piecewise linear: PWL

This entry allows a waveform to be simulated approximately by a series of straight line segments, as indicated in Figure 15.15.

$$\text{PWL}(T1 \ V1 \ T2 \ V2 \ldots)$$

The waveform takes the value V1 at T1, V2 at T2, and so on. Intermediate values are by linear interpolation. It is necessary that the times are entered in ascending order: T1 < T2 < . . ..

EXAMPLE 15.12 A source is connected to a dividing circuit consisting of two resistors in series, one of resistance 900 Ω, the other 100 Ω. The output is taken across the 100 Ω resistor, and should in all cases be of the same shape as the source but of 1/10 amplitude. Write SPICE files to verify this for the waveforms specified below. In these specifications, times are in seconds and frequencies in hertz.

(i) $0 < t < 1 \qquad v = 0$

$\quad 2 < t < 6 \qquad v = 10$

$\quad 8 < t < 10 \qquad v = 0$

between $t = 1$ and $t = 2$ the waveform rises linearly, and falls linearly between $t = 6$ and $t = 8$.

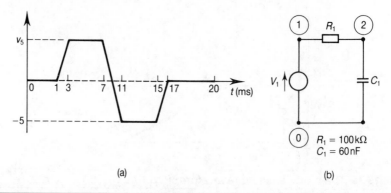

Figure 15.16 Example 15.13: (a) waveshape; (b) circuit.

(ii) $0 < t < 1$ $v = 0$
       $1 < t < 7$ $v = 10\{1 - \exp[-0.2(t - 1)]\}$
       $7 < t < 10$ $v = 10\{\exp[0.5(t - 7)] - \exp[-0.2(t - 1)]\}$

(iii) $0 < t < 1$ $v = 0$
       $1 < t < 10$ $v = 10\exp[-0.5(t - 1)]\sin[0.4\pi\,(t - 1)]$

(iv) The pulse of waveform (i) using the piecewise linear model.

The correct codes for the source specifications are

(i)   PULSE(0 10 1 1 2 4 10)

(ii) EXP(0 10 1 5 7 2)

(iii) SIN(0 10 0.2 1 0.5)

(iv) PWL(0 0 1 0 2 10 6 10 8 0 10 0)

A suitable .TRAN command is

.TRAN 0.5 10 0 0.1

EXAMPLE 15.13   The dipolar pulse shown in Figure 15.16(a) is applied to the series $R-C$ circuit, $R = 100$ kΩ, $C = 60$ nF, shown in Figure 15.16(b). Determine by means of a SPICE program the maximum and minimum values reached by the voltage across the capacitor.
     The pulse is most conveniently simulated as piecewise linear. A suitable text file appears below.

File X15-13.CIR

```
Example 15.13
R1 1 2 1E5
C1 2 0 6E-8
V1 1 0 PWL(0 0 1M 0 3M 5 7M 5 11M −5
+ 15M −5 17M 0 20M 0)
```

```
.TRAN 0.5M 20M 0 0.1M
.PRINT TRAN V(2)
.END
```

It is good practice to check against an analytical solution where possible. In this case such a solution can be found, but only by a tedious if straightforward method. It is here presumed that previous agreement is sufficiently good to give some confidence in the numerical solution. The output data give a maximum of 3.0 V at 8 ms, and a minimum of -2.2 V at 16 ms.

15.7.4 ◆ The control statement .PLOT

It is sometimes useful to supplement the .PRINT command with the .PLOT command, which has the same format:

.PLOT TRAN output variables

It results in a simple graphical display by the side of the tabular output. This command can be included in any of the above transient problems.

15.7.5 ◆ .PROBE

It is convenient to mention at this point that if the graphics program PROBE is to be used, it is necessary to add the line.

.PROBE

to the input textfile. This causes data to be saved on disc to an output file PROBE.DAT which is called by PROBE. Directions for proceeding are displayed in menu form.

15.8 ◆ Magnetic coupling

The form of entry for magnetically coupled inductors was given in Section 15.4.1, Table 15.2. The entry is used in the same way as for the other components. As mentioned before, in order to run a SPICE program it is necessary that all nodes have a DC path to the reference node: this sometimes requires modification of the circuit by adding a suitable link between primary and secondary. The incorporation of a coupled pair will be illustrated by two examples.

EXAMPLE 15.14 The circuit of Example 8.2 is redrawn in Figure 15.17. In Example 8.2 it was required to find the impedance at the primary. This is most easily done by feeding the primary from a current source of unit strength, when the voltage will be numerically equal to this impedance.

In this case the extra link between the two inductors does not affect

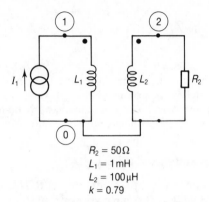

$R_2 = 50\,\Omega$
$L_1 = 1\,\text{mH}$
$L_2 = 100\,\mu\text{H}$
$k = 0.79$

Figure 15.17 Circuit for Example 15.14.

the required impedance: it could if necessary be a high-resistance link. A suitable text file appears below.

File X15-14.CIR

```
Example 15.14
R2 2 0 50
L1 1 0 1E-3
L2 2 0 1E-4
KCC L1 L2 0.79
I1 0 1 AC 1 0
.AC LIN 1 100000 100000
.PRINT AC VR(1) VI(1)
.END
```

The output gives

$$V(1) = 191.1 + j388.2$$

This value differs from that found in Example 8.2 only in the last decimal place.

EXAMPLE 15.15 The coupled inductors shown in the circuit of Figure 15.18 are each of inductance 1 mH and have a coupling coefficient of 0.5. The resistors are each of $2\,\text{k}\Omega$ and the capacitors of 10 nF. The current source is variable in frequency between 20 kHz and 100 kHz. Write a SPICE program to show the variation of secondary voltage with frequency.

File X15-15.CIR

```
Example 15.15
R1 1 0 2E3
R2 2 0 2E3
C1 1 0 10E-9
C2 2 0 10E-9
L1 1 0 1E-3
```

```
L2 2 0 1E-3
KCC L1 L2 0.5
I1 1 0 AC −1 0
.AC LIN 10 20000 100000
.PRINT AC VM(2) VP(2)
.END
```

The output from this problem shows the double-peaked frequency response which can appear in this sort of circuit when the coupling coefficient is sufficiently large. In this case the peaks occur at about 40 kHz and 70 kHz, that at 70 kHz being the slightly higher of the two.

15.9 ◆ Semiconductor devices

SPICE includes models for various semiconductor devices. The models are based on the theory of operation of the class of device, such as junction diodes or bipolar transistors. In order to specify a particular device appearing in a circuit, the model identifier is used to label the component together with a .MODEL statement which gives the values of a number of relevant parameters. These numerical values are found in the .LIB file: the file issued with the student version has a small representative sample of the total number available. Detailed consideration in the following sections is restricted to the type of problem encountered in Chapter 13. After a general introduction, examples will be given of circuits using diodes, field-effect transistors and bipolar transistors.

15.9.1 ◆ Device models

The devices in Table 15.6 are modelled in the SPICE program. The initial letter of the component label is shown in the second column of the table.

Within each category of device different types exist. The type appears in a .MODEL statement.

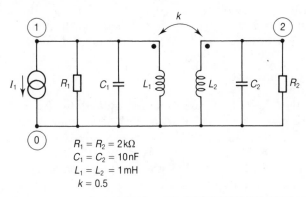

$R_1 = R_2 = 2\,\text{k}\Omega$
$C_1 = C_2 = 10\,\text{nF}$
$L_1 = L_2 = 1\,\text{mH}$
$k = 0.5$

Figure 15.18 Circuit for Example 15.15.

Table 15.6

Device	Identifier
Diode	D
Junction field-effect transistor (FET)	J
Bipolar transistor (BJT)	Q
MOS field-effect transistor (MOSFET)	M

Table 15.7

D(. . .)	Diode model
NJF(. . .)	n-channel JFET model
PJF(. . .)	p-channel JFET model
NPN(. . .)	n–p–n BJT model
PNP(. . .)	p–n–p BJT model
NMOS(. . .)	n-channel MOSFET model
PMOS(. . .)	p-channel MOSFET model

15.9.2 ◆ .MODEL statements

A .MODEL statement has the following form:

.MODEL component name type(parameters)

The types available are shown below in Table 15.7. The number of parameters appropriate to each type varies, depending on the analytical model and the degree of accuracy required of the simulation. Description of these parameters will be found in the SPICE *User's Guide*. For the limited purposes of this book only a minimal number will be specified in any example, sufficient to give DC characteristics typical of the device in question.

In the sections which follow only diode, JFET and BJT models will be used.

15.10 ◆ Diodes

This section will be illustrated by reference to the circuits associated with Sections 13.2 and 13.3. The characteristic used in numerical work in these sections referred to diode type IN752, which is included in the .LIB file provided with the Demo version. As a circuit element a diode is identified by an initial letter D to the component label, and the statement includes a type name. For example, an IN752 diode connected between nodes 1 and 2, forward direction from 1 and 2, would be described by an entry

D1 1 2 DIN752

This entry then has to be supplemented by a .MODEL statement giving values of the parameters particular to the IN752. Some of these parameters refer to DC characteristics, others to transient behaviour. In the following we shall only give the former, since only the DC characteristic is relevant to the problems to be considered: the remaining parameters will have the default values. Consulting the .LIB file we find the appropriate statement is

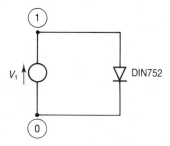

Figure 15.19 Circuit for Example 15.16.

.MODEL DIN752 D(IS = 0.5U RS = 6 BV = 5.20 IBV = 0.5U)

These parameters signify

IS	saturation current
RS	ohmic resistance
BV	reverse breakdown voltage
IBV	current at BV

EXAMPLE 15.16   Obtain the DC characteristic of the IN752 diode.

A simulation of a circuit configuration to determine the characteristic is shown in Figure 15.19. The text file is given below: a .DC analysis statement is included to allow V1 to take on a sequence of values.

File X15-16.CIR

```
Example 15.16. Diode IN752 characteristic
D1 1 0 DIN752
V1 1 0 DC 0
.MODEL DIN752 D(IS = 0.5U  RS = 6  BV = 5.20  IBV = 0.5U)
.DC V1 −6 2 0.2
.PRINT DC I(V1)
.END
```

For demonstration purposes, this file calls for current values in the range from −6 V to 2 V at intervals of 0.2 V. In fact, it may be preferable to use one file using a much coarser interval to give the reverse currents and another for the forward characteristic.

As mentioned earlier, in any circuit SPICE automatically determines the DC working point of any semiconductor device and the corresponding small-signal parameters: a small-signal AC analysis can then be carried out. This process is illustrated by the following reworking of Example 13.1.

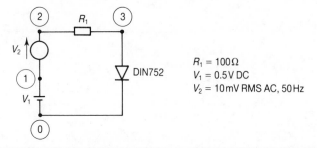

Figure 15.20 Circuit for Example 15.17.

EXAMPLE 15.17   Rework Example 12.1. A simulation of the circuit is shown in Figure 15.20. It is required to find the AC voltage across the diode.

A suitable text file appears below.

File X15-17.CIR

```
Example 15.17
R1 2 3 100
D1 3 0 DIN752
V1 1 0 DC 0.5
V2 2 1 AC 0.01 0
.MODEL DIN752 D(IS = 0.5U  RS = 6  BV = 5.20  IBV = 0.5U)
.AC LIN 1 50 50
.PRINT AC VM(3)
.END
```

Running the program gives the working point voltage of the diode as 0.2373 V; the AC voltage is given as 1.368 mV. These results are in agreement with those in Example 13.1.

15.11 ◆ Junction field-effect transistors

A circuit containing a field-effect transistor was considered in Section 13.5. The component entry for a junction field-effect transistor is illustrated by the entry for the circuit of Figure 15.21. This would be

J1 2 1 0 JF

The nodes appear in the order drain, gate, source and JF refers to the model type to be defined by a .MODEL statement. The .LIB file provided with the Demo version does not contain any model for such a transistor, but suitable characteristics are obtained with the following parameter values:

NJF(BETA = 2E-3 LAMBDA = 1E-2)

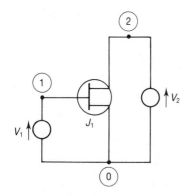

Figure 15.21 Circuit for Example 15.18.

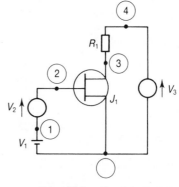

$R_1 = 1.5\,\text{k}\Omega$ $V_2 = 0.1\angle 0, 1\,\text{kHz}$
$V_1 = -0.4\,\text{V}$
$V_3 = 15\,\text{V}$

Figure 15.22 Circuit for Example 15.19.

EXAMPLE 15.18   Derive characteristics for the above model of a field-effect transistor.

A circuit is shown in Figure 15.21, and a suitable text file appears below.

File X15-18.CIR

```
Example 15.18
J1 2 1 0 JF
V1 1 0 DC 0
V2 2 0 DC 0
.MODEL JF NJF(BETA = 2E-3 LAMBDA = 1E-2)
.DC V2 0 20 2 V1 −2 0 0.4
.PRINT DC I(V2) V(1)
.END
```

The resulting characteristics were shown in Figure 13.6

In Section 13.5 the small-signal gain of the circuit of Figure 13.7 was calculated. This calculation is reworked in the next example.

EXAMPLE 15.19   Write a SPICE file to find the small-signal gain of the circuit of Figure 13.7.

A simulated circuit is shown in Figure 15.22, and a suitable text file appears below.

File X15-19.CIR

```
Example 15.19
R1 4 3 1.5K
J1 3 2 0 JF
```

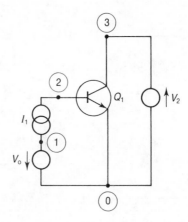

Figure 15.23 Circuit for Example 15.20.

```
V1 1 0 DC −0.4
V3 4 0 DC 15
V2 2 1 AC 0.1 0
.MODEL JF NJF(BETA = 2E-3 LAMBDA = 1E-2)
.AC LIN 1 1000 1000
.PRINT AC VM(3) VP(3)
.END
```

Running this program gives for the working point of the transistor

$$V_{GS} = -0.4 \text{ V}, V_{DS} = 6.8 \text{ V}$$

The AC output shows V(3) = 0.952∠180, giving a gain of −9.49.

15.12 ◆ Bipolar transistors

Treatment of circuits involving bipolar transistors follows a similar pattern: the identifier beginning the component label is Q; the accompanying .MODEL statement includes provision for both NPN and PNP transistors. A complete specification of a particular device requires values for 40 parameters: as with the diode, many of these refer to transient behaviour and all have default values. In the following examples the parameters given in the .LIB file will be used without further explanation. Using the circuit appearing in Figure 15.23 as example, the component entry would be

```
Q1 3 2 0 Q2N2222
```

The nodes appear in the order collector, base, emitter. This entry is supplemented by

```
.MODEL Q2N2222 NPN(IS = 3.108E-15 BF = 217
+ ISE = 190.7E-15 IKF = 1.296 NE = 1.541 BR = 6.18
+ IKR = 0 VAF = 131.5 RC = 1)
```

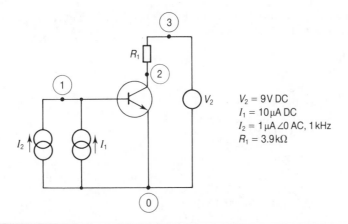

Figure 15.24 Circuit for Example 15.21.

(The use of the + sign to mark a continuing line will be noticed.) The incorporation of a bipolar transistor in a circuit is illustrated in the following two examples.

EXAMPLE 15.20 Obtain the DC characteristics of the 2N2222 transistor.

A simulation of a suitable circuit is shown in Figure 15.23. To fit in with the presentation of Figure 13.10(a), the base is fed at constant current. The zero-voltage source V0 is included only to get the base current printed in the output file: the current is determined by the .DC analysis statement.

File X15-20.CIR

```
Example 15.20. 2N2222 characteristics.
Q1 3 2 0 Q2N2222
I1 1 2 DC 0
V0 0 1 DC 0
V2 3 0 DC 0
.MODEL Q2N2222 NPN(IS = 3.108E-15 BF = 217
+ ISE = 190.7E-15 IKF = 1.296 NE = 1.541 BR = 6.18
+ IKR = 0 VAF = 131.5 RC = 1)
.DC V2 2 12 2 I1 5U 20U 2.5U
.PRINT DC V(2) I(V2) I(V0)
.END
```

This file calls for V2 to be varied between 2 V and 12 V at 2 V intervals. The region $0 < V2 < 0.5$ calls for separate investigation. The characteristics of Figure 13.10 were obtained using this program.

EXAMPLE 15.21 Rework the AC analysis of Example 13.4.

A simulation of the circuit is shown in Figure 15.24. In this circuit, the current source I1 provides the DC bias current and I2 the AC signal. A suitable text file appears below.

File X15-21.CIR

        Example 15.21
        R1 2 3 3.9K
        I1 0 1 DC 10E-6
        I2 0 1 AC 1E-6 0
        V2 3 0 DC 9
        Q1 2 1 0 Q2N2222
        .MODEL Q2N2222 NPN(IS = 3.108E-15 BF = 217
        + ISE = 190.7E-15 IKF = 1.296 NE = 1.541 BR = 6.18
        + IKR = 0 VAF = 131.5 RC = 1)
        .AC LIN 1 1000 1000
        .PRINT AC VM(1) VP(1) VM(2) VP(2)
        .END

Running this program yields for the working point

I1	V(1)	V(2)
10 μA	0.6847 V	5.073 V

For the AC signal we find

I2	VM(1)	VP(1)	VM(2)	VP(2)
1 μA	3.024 mV	0°	472.7 mV	180°

Agreement with the graphically derived results of Examples 13.4 is satisfactory.

15.13 ◆ Subcircuits

Circuits often contain several components of identical type: for example, all the transistors may be of the same type. This is the reason for including a model name in the component description as well as the component label. The model name is then given a full description in the .MODEL statement. SPICE makes a similar provision for repeats of a particular configuration within the main circuit. The repeated configuration is termed a subcircuit, and can be given a separate name and description. The general format for a subcircuit entry is as follows:

        .SUBCKT   name   external nodes
        circuit description
        .ENDS   name

Reference in the description of the main circuit then takes the form

        X. . .   name   nodes

in which the initial letter X is the identifier for a subcircuit. The procedure will be illustrated in the following example.

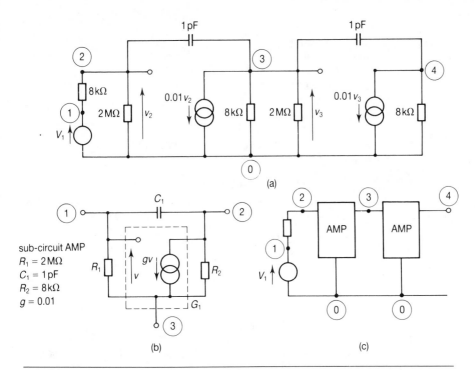

Figure 15.25 Circuit for Example 15.22.

EXAMPLE 15.22 In Figure 15.25(a) is shown the small-signal equivalent circuit of a two-stage FET amplifier. Determine the gain–frequency response over the range 1 kHz–10 MHz.

The circuit contains two repeats of the subcircuit shown in Figure 15.25(b), given the name AMP. This subcircuit is described in a SPICE file as follows:

.SUBCKT AMP 1 2 3
R1 1 3 2MEG
R2 2 3 8K
C1 1 2 1E-12
G1 2 3 1 3 0.01
.ENDS AMP

The component labels and the node numbering are independent of the main circuit, but these external nodes must not include the reference node 0, although this can be used internally in the subcircuit. Only the nodes identified in the first line can be accessed in the main circuit description. Using this subcircuit, the circuit of Figure 15.25(a) can be redrawn in the form shown in Figure 15.25(c). A suitable text file is given below.

File X15-22.CIR

Example 15.22
.SUBCKT AMP 1 2 3
R1 1 3 2MEG
R2 2 3 8K

C1 1 2 1E-12
G1 2 3 1 3 0.01
.ENDS AMP
R1 1 2 8K
X1 2 3 0 AMP
X2 3 4 0 AMP
V1 1 0 AC 1 0
.AC DEC 3 1000 10000000
.PRINT AC VM(2) VP(2) VM(3) VP(3) VM(4) VP(4)
.END

The results obtained on running this program are given below.

FREQ	VM(2)	VP(2)	VM(3)	VP(3)	VM(4)	VP(4)
1.000 kHz	0.9960	−0.231	79.36	179.5	6349	−0.4696
2.154	0.9959	−0.499	79.35	179.0	6348	−1.012
4.642	0.9955	−1.073	79.31	177.8	6345	−2.179
10.00	0.9936	−2.305	79.10	175.3	6328	−4.686
21.54	0.9849	−4.898	78.17	170.0	6254	−10.02
46.42	0.9488	−9.925	74.24	159.3	5939	−20.85
100.0	0.8363	−16.88	61.54	140.8	4923	−39.46
215.4	0.6618	−19.10	39.30	119.7	3143	−60.96
464.2	0.5532	−13.91	20.30	104.8	1623	−76.55
1.000 MHz	0.5181	−9.370	9.672	96.97	772.8	−85.94
2.154	0.5075	−9.206	4.516	93.20	359.1	−93.06
4.642	0.4954	−14.40	2.099	91.37	163.5	−101.9
10.00	0.4543	−26.97	0.9748	90.39	69.67	−116.7

The reliability of results of this sort is difficult to check. It is not too difficult to obtain an analytic solution to the problem, but a lot of complex arithmetic is involved in computing results for comparison. It is sometimes possible to use approximate analytical results as a check on plausibility. Single stages with the amplifier circuit of Figure 15.25(b) have been studied in Example 5.6 and Example 15.8. From the results obtained in those examples, the gain VM(4)/VM(3) would be expected to be close to $0.01 \times 8000 = 80$. The results above satisfy this criterion. The gain of the first stage will be modified by the input impedance of the second stage. The earlier analysis has shown this to be approximately $(80 + 1) \times 1$ pF. Thus it would be expected, at least for lower frequencies, that

$$\frac{V(3)}{V(2)} \approx \frac{-gR}{1 + j\omega CR}$$

where $g = 0.01$, $R = 8$ kΩ, $C = 81$ pF. The results also satisfy this criterion. In the same range of frequencies the input admittance of the first stage should be approximately

$$j\omega C \left(1 + \frac{gR}{1 + j\omega CR}\right)$$

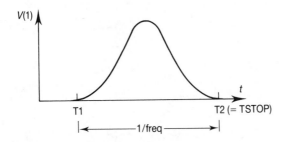

Figure 15.26 Schematic waveform indicating period (T2 − T1) = 1/FREQ assumed for Fourier analysis by the commands

.TRAN TSTEP T2 < TSTART < TMAX ≫ < UIC >

.FOUR FREQ V(1)

This expression can be used to check VM(2)/VM(1).

    This method of applying checks is precisely the method adopted to check the reliability of experimental measurements. It is equally applicable to results gained by simulation.

15.14 ◆ Fourier analysis: .FOUR

Fourier analysis was introduced in Chapter 11. In that chapter the determination of the coefficients in the Fourier series was primarily through analytic methods. SPICE includes in the transient analysis process an option allowing determination of a Fourier expansion for an output variable. Consider the situation illustrated in Figure 15.26, in which the waveform of an output variable V(1) is shown as determined using a .TRAN command. That part of the waveform between T1 and T2 can be treated as one period of a repetitive waveform and the Fourier coefficients obtained by the commands

    .TRAN TSTEP T2 ⟨ TSTART ⟨ TMAX⟩⟩ ⟨ UIC⟩
    .FOUR FREQ V(1)

in which FREQ = 1/(T2 − T1) is the fundamental frequency. The Fourier series is in the form given in Equation 11.11, Section 11.2.3:

$$v(t) = c_0 + \sum_{n=1}^{9} c_n \sin\left(\frac{2\pi}{T}(t - T1) + \theta_n\right) \qquad (15.1)$$

in which the time axis is taken as in Figure 15.26. The accuracy of the calculation is dependent on the calculation step length TMAX, and this should be made less than $T/100$.

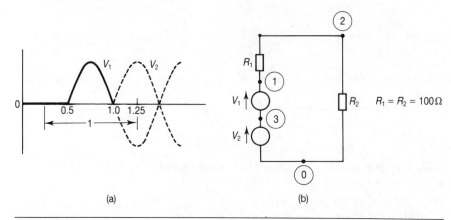

Figure 15.27 Example 15.23: (a) composition to give waveform; (b) circuit.

EXAMPLE 15.23 Determine a Fourier expansion of the waveform for which one period is given by

$$0 < t < 0.25 \qquad v = 0$$

$$0.25 < t < 0.75 \qquad v = \sin\left[2\pi(t - 0.25)\right]$$

$$0.75 < t < 1.0 \qquad v = 0$$

The waveform can be generated by using the SIN time-dependent source to produce two sine waves, each of period 1 but one of which is delayed by one half-period. Thus the addition of the waveforms represented by

> V1 SIN(0 2 1 0.5 0)
> V2 SIN (0 2 1 1 0)

will give the required waveshape with an amplitude of 2 and with an initial delay of 0.5. This is shown in Figure 15.27(a). A suitable circuit configuration is shown in Figure 15.27(b), in which the output V(2) should have the requisite waveform. A suitable text file appears below.

File X15-23. CIR

> Example 15.23
> R1 1 2 100
> R2 2 0 100
> V1 1 3 SIN(0 2 1 0.5 0)
> V2 3 0 SIN(0 2 1 1 0)
> .TRAN 0.1 1.25 0.25 0.01
> .FOUR 1 V(2)
> .PRINT TRAN V(2)
> .END

The segment analysed lies between 0.25 and 1.25. The output file will confirm that the waveform V(2) has the correct shape. The Fourier coefficients, to four decimal places, are given below.

n	f (Hz)	c_n	θ_n (deg)
0	0	0.3183	–
1	1.000	0.4999	−90.000
2	2.000	0.2122	90.000
3	3.000	0.0000	−90.541
4	4.000	0.0424	−90.000
5	5.000	0.0000	−89.023
6	6.000	0.0182	90.000
7	7.000	0.0000	91.529
8	8.000	0.0101	−89.999
9	9.000	0.0000	−89.279

Apart from the first, the coefficients for the odd harmonics are too small to be significant. Substitution in the series of Equation 15.1 gives

$$v(t) = c_0 + c_1 \sin(2\pi t - \pi/2) + c_2 \sin(4\pi t + \pi/2)$$
$$+ c_4 \sin(8\pi t - \pi/2) + c_0 \sin(12\pi t + \pi/2)$$
$$+ c_8 \sin(16\pi t - \pi/2)$$
$$= c_0 - c_1 \cos(2\pi t) + c_2 \cos(4\pi t) - c_4 \cos(8\pi t) + c_6 \cos(12\pi t)$$
$$- c_8 \cos(16\pi t) \qquad \textbf{(15.2)}$$

The waveform analysed in Example 11.2 is the same as the one now being considered apart from a time shift of one half-period. This can be accounted for by subtracting $\cos(2\pi t/T)$ from the series of Equation 11.9, just reversing the sign of the first harmonic term. With this modification and with $T = 1$, $\hat{V} = 1$, Equation 11.9 leads to Equation 15.2 with the following values for the non-zero coefficients:

$$c_0 = 1/\pi$$
$$c_1 = 0.5$$
$$c_2 = 2/3\pi = 0.2122$$
$$c_4 = 2/15\pi = 0.0424$$
$$c_6 = 2/35\pi = 0.0182$$
$$c_8 = 2/63\pi = 0.0101$$

The agreement is thus satisfactory to four decimal places.

EXAMPLE 15.24 Write a SPICE file to determine the output from the resistive divider in Figure 15.28(a) when the source V1 has the waveform shown in Figure 15.28(b). That part of the waveform in the range $0.2 < t < 1.2$ forms one period of a repetitive waveform. Use the analysis to determine a Fourier expansion for this waveform.

The waveform of Figure 15.28(b) can be simulated using the PULSE time-dependent source. A suitable text file appears below.

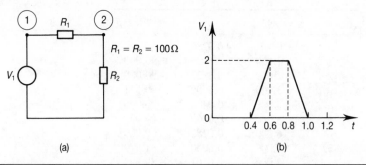

(a) (b)

Figure 15.28 Circuit for Example 15.24.

File X15-24.CIR

    Example 15.24
    R1 1 2 100
    R2 2 0 100
    V1 1 0 PULSE(0 2 0.4 0.2 0.2 0.2 1.2)
    .TRAN 0.05 1.2 0.2 0.01
    .FOUR 1 V(2)
    .PRINT TRAN V(2)
    .END

The output will confirm that $V(2)$ has the same shape as Figure 15.28(b) but of amplitude 1. The values for the Fourier coefficients are given below.

n	f (Hz)	c_n	θ_n (deg)
0	0	0.4000	–
1	1.000	0.5663	−90.000
2	2.000	0.1416	90.000
3	3.000	0.0629	90.000
4	4.000	0.0353	−90.000
5	5.000	0.0001	−89.997
6	6.000	0.0157	−90.000
7	7.000	0.0115	90.000
8	8.000	0.0089	90.000
9	9.000	0.0069	−90.000

The methods of Chapter 11 can be applied to show that, referred to a time origin at $t = 0.2$ in Figure 15.28(b),

$$v(t) = c_0 + \sum_{n=1}^{\infty} c_n \cos(2\pi nt)$$

in which

$$c_0 = 0.4, \quad c_n = (-1)^n \frac{5}{(\pi n)^2} \left[\cos(0.2\pi n) - \cos(0.6\pi n) \right]$$

Application of this formula will show that the SPICE results are in agreement.

It is instructive to sum the series of Equations 15.1 with the values of coefficients and the phases produced by a SPICE program. The result can then be compared with the original waveform. If the phases of the various harmonics are all zero, the summation can be effected by simulation of a number of current sources in parallel with a 1 Ω resistor, each current source having an appropriate SIN waveform to represent one term of the series. Otherwise it will be desirable to obtain a short computer program.

An alternative Fourier analysis facility is incorporated in PROBE. The discrete Fourier transform of the waveform covering the duration of the .TRAN statement is carried out and displayed graphically. The facility is called as an option on the X-menu.

15.15 ◆ The options statement: .OPTIONS

This statement can be used to vary certain aspects of the format of the output file and also the default values built into the program. The statement to be inserted in the text file takes the form

.OPTIONS OPT1 OPT2 . . .

in which OPT1, . . . stand for a number of the variations detailed in the SPICE *User's Guide*. Only three are likely to be relevant in the present context, NODE, NOMOD, NOPAGE.

The statement

.OPTIONS NODE

causes a table of node connections to be printed, sometimes useful in checking correct entry of the circuit to be simulated.

The statement

.OPTIONS NOMOD

inhibits printing of a table listing model parameters.

The statement

.OPTIONS NOPAGE

suppresses some of the page breaks when the output file is printed. For a small circuit the printed file can be unnecessarily long.

These statements can be run into one entry, as in

.OPTIONS NODE NOMOD NOPAGE

15.16 ◆ Design and computer-aided analysis

The foregoing sections have shown that it is possible to simulate circuits and to predict the response to various excitations. This process yields results which should be the same as those which would have been obtained

Table 15.8

Component	Entry	Comments
Vx–x	DC value Vx–x N+ N– AC Magnitude phase PULSE/SIN/EXP/PWL	PULSE, ..., carry parameters (see Section 15.7.3)
Ix–x	DC value Ix–x N+ N– AC Magnitude phase PULSE/SIN/EXP/PWL	PULSE, ... carry parameters (see Section 15.7.3)
Rx–x	Rx–x N1 N2 value	
Lx–x	Lx–x N+ N– value $\langle IC = i(0)\rangle$	
Cx–x	Cx–x N+ N– value $\langle IC = v(0)\rangle$	
Kx–x	Kx–x L1 L2 k L1 N1 N2 value L2 N3 N4 value	$k = M/(L1\ L2)^{1/2}$; node order important
Ex–x	Ex–x N3 N4 N1 N2 value(A)	
Gx–x	Gx–x N3 N4 N1 N2 value(g)	
Hx–x	Hx–x N3 N4 V0 value(r) V0 N1 N2	Abbreviated format for V0 permissible for a zero source
Fx–x	Fx–x N3 N4 V0 value(A) V0 N1 N2	
Xx–x	Xx–x N1 N2 N3 SNAME .SUBCKT SNAME NS1 + NS2 NS3 subcircuit description .ENDS	N1, ... refer to main circuit; NS1, ... refer to subcircuit definition ('0' not allowed)

Table 15.8 (*cont.*)

Component	Entry	Comments
N+ / N− Dx–x	Dx–x N+ N− MNAME .MODEL MNAME D(parameters)	
ND / NG / NS Jx–x	Jx–x ND NG NS MNAME .MODEL MNAME NJF(parameters)	
ND / NG / NS Jx–x	Jx–x ND NG NS MNAME .MODEL MNAME PJF(parameters)	
NC / NB / NE Qx–x	Qx–x NC NB NE MNAME .MODEL MNAME NPN(parameters)	
NC / NB / NE Qx–x	Qx–x NC NB NE MNAME .MODEL MNAME PNP(parameters)	

by measurement on a made-up circuit. Simulation is especially valuable in investigating the working of large circuits on which 'internal' measurements are difficult or even impossible. In either case the design process occurs at a prior stage: the simulation or experiment serves to confirm the validity of the design. The experiment will, of course, often show disagreement with expectations, thus revealing factors which have not been taken into account. This a simulation will not do and so in the final analysis is not a substitute for experiment. It can, however, in a complicated circuit carefully modelled confirm the general design process and provide a relatively painless way of trying modifications to aspects of the circuit which were difficult to otherwise calculate. The preliminary design process must be carried out using the algebraic methods of the previous chapters, making such simplifications as are necessary. The effect of these simplifications can then be checked by simulation.

Simulation can also help in understanding the operation of very fast circuits or high-frequency circuits. In such circuits connection of, for example, an oscilloscope for monitoring purposes can greatly perturb the operation of the circuit by addition of stray capacitance. A simulation can help the interpretation of such monitoring.

Table 15.9

.DC SRCNAME VSTART VSTOP +VINCR ⟨REPEAT⟩	Causes one or two sources to increment
.AC LIN NP FSTART FSTOP	Linear coverage with NP points
.AC DEC ND FSTART FSTOP	Logarithmic coverage with ND points per decade
.AC OCT NO FSTART FSTOP	Logarithmic coverage with NO points per octave
.TRAN TSTEP TSTOP ⟨TSTART +⟨TMAX⟩⟩ ⟨UIC⟩	TSTEP, TSTOP: printing TMAX: computation step UIC: initial conditions
.FOUR FREQ output variables	FREQ: fundamental frequency analysis over interval (TSTOP − period) to TSTOP
.PRINT DC/AC/TRAN output variables	
.PLOT DC/AC/TRAN output variables	
.OPTIONS NOPAGE/NODE/NOMOD	

In conclusion, simulation is no substitute for experiment but can reduce the number of experiments and help in interpretation.

15.17 ◆ Summary

SPICE performs three types of analyses, DC, AC and transient. In circuits containing semiconductors

- DC analysis determines the working points,
- AC analysis does a small-signal perturbation about the working points previously determined, and
- TRAN determines the response to specified time-varying sources.

Component labels: All components have a label consisting of the appropriate initial letter followed by up to seven alphanumeric characters denoted below by x–x:

Vx–x	independent voltage source
Ix–x	independent current source
Rx–x	resistor

Lx–x	inductor
Cx–x	capacitor
Kx–x	coupled inductors
Ex–x	voltage-controlled voltage source
Gx–x	voltage-controlled current source
Hx–x	current-controlled voltage source
Fx–x	current-controlled current source
Xx–x	sub-circuit

Sources: Independent sources, voltage or current, can be specified as DC, AC or TRAN. Transient sources are described in Section 15.7.3 and include

PULSE(V1 V2 TD TR TF PW PER), Figure 15.12
SIN(V0 VA FREQ TD THETA), Figure 15.13
EXP(V1 V2 TD1 TAU1 TD2 TAU2), Figure 15.14
PWL(T1 V1 T2 V2 . . .), Figure 15.15

Text file entries for sources and components are summarized in Table 15.8.

Control statements: These are summarized in Table 15.9.

Bibliography

PSpice is a registered trademark of the MicroSim Corporation, 20 Fairbanks, Irvine, CA 92718, USA. The United Kingdom agent is ARS Microsystems, Doman Road, Camberley, Surrey, GU15 3DF, UK.
    The User's guide referred to in the text is

Vladimirescu A., Zhang K., Newton A.R., Pederson D.O. and Sangiovanni-Vincentelli A. (1981) *SPICE version 2G User's Guide.* University of California

This document is widely available but has not been formally published.
    Apart from the reference manual for PSpice, the following text deals with PSpice in more detail and wider scope than appropriate in Chapter 15. The classroom version software can be obtained with this text.

Tuinenga P.W. (1988) SPICE. *A Guide to Circuit Simulation and Analysis Using PSpice.* Englewood Cliffs NJ: Prentice-Hall

PROBLEMS

Many of the end-of-chapter problems which have appeared earlier can be done using numerical simulation. A selection is given below. In some cases it will be necessary to modify the problem slightly, as for example in Chapters 4 and 7 in which reactances are specified by numbers rather than components and frequency. In such cases an inductance or capacitance and associated frequency must replace the reactance before SPICE can be used. The order of treatment in the present chapter follows roughly that of the separate chapters.

Chapter 3: Problems 3.4, 3.5, 3.8, 3.9, 3.11, 3.15(a).
Chapter 4: Problems 4.7, 4.9.
Chapter 5: Problems 5.2, 5.3, 5.10, 5.13, 5.15.
Chapter 6: Problem 6.11 gives practice in setting up a text file.
Chapter 7: All problems.
Chapter 8: Problems 8.2, 8.5, 8.9, 8.12.
Chapter 9: Problems 9.3, 9.9, 9.10, 9.11, 9.13, 9.15.
Chapter 11: Problems 11.7 (using $L = 25.3\ \mu H$, $C = 1\ nF$, $R = 15.9\ \Omega$), 11.8.
Chapter 12: Problems 12.1, 12.6 (the diode characteristic in Problem 12.1 is the 1N752; see Example 15.16, Section 15.10).
Chapter 13: Problems 13.3, 13.4, FET model NJF(BETA=5E-3 LAMBDA = 2E-2 VTO = −2.5), 13.5, 13.6, BJT model NPN(IS=1E-14 BF = 80 BR = 5 RC = 1 VAF = 60).

The following problems are somewhat in the nature of investigations, and can be developed by the interested student.

15.1 In Figure 15.29 is shown the small-signal equivalent circuit of a pulse amplifier. The pulse waveform is trapezoidal, of total duration 1 μs, rise and fall times each 0.2 μs. The capacitance C is set by circuit limitations at 15 pF. The design process suggested is as follows:

(i) choose R so that without L the gain would be $1/\sqrt{2}$ of its maximum value at the frequency of 1 MHz (reciprocal of the pulse length);

(ii) with this value of R, choose L so that the gain of the complete circuit is the same at 1 MHz as at DC.

Having calculated the values of R and L, use SPICE

(a) to make graphs of the gain and phase as functions of frequency, and

(b) to determine the waveform of the response to the pulse. Compare the 10–90% rise time of the output with that of the input.

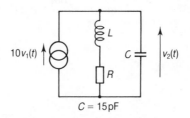

Figure 15.29

15.2 The circuit shown in Figure 15.30 can be used to delay a pulse.

(i) Make graphs of gain and phase over the frequency range 10 kHz–5 MHz. Estimate a value for $d\phi/d\omega$ over the low-frequency linear region, and estimate the delay expected (cf. Section 11.4).

(ii) Make a direct estimate of delay by taking as the input V1 a trapezoidal pulse of total duration 0.6 μs, linear rise and fall times each 0.15 μs.

15.3 All transistors have built-in capacitance which limit performance at high frequencies. For a field-effect transistor the important ones are the gate–source and gate–drain capacitances. These components can be incorporated in the small-signal equivalent circuit, or appear in the model statement.

(i) Obtain a set of DC characteristics for the FET with model statement

$$\text{NJF(BETA}=5\text{E-4 LAMBDA} = 1\text{E-2 CGS}=5\text{E-12 CGD}=1\text{E-12)}$$

(ii) This transistor is used in the amplifier circuit of Figure 15.31. Use SPICE to perform a small-signal AC analysis over the range 10 kHz–10 MHz, determining the gain and phase as functions of frequency. Record the gain at mid-frequencies and the frequencies at which the gain is $1/\sqrt{2}$ of this value. Check the results against approximate theory.

15.4 The fact that the transistor of Problem 15.3 is non-linear means that the output waveform is not an exact replica of the input. For a sinusoidal input this means that the output will contain harmonic frequency components. These can be detected by a Fourier analysis of the output.

The source V1 in the circuit of Figure 15.31 is a sine wave of 1 V peak-to-peak amplitude and 1 kHz frequency (this can be realized using the SIN time-dependent source). Carry out a transient analysis over a period of 2 cycles of the input, incorporating the .FOUR statement. Determine the ratios of the second and third harmonic components to the fundamental.

15.5 In Figure 15.32 is shown the circuit of a full-wave rectifier with inductance smoothing. The sources are 50 Hz AC, peak amplitude 75 V; the diodes are described by the model statement D(IS=1E-6 RS=1). To simulate the steady-state circuit behaviour it would be necessary to follow many cycles. The number can be reduced by allowing an initial current in the inductor: assuming smoothing is good,

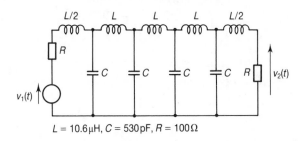

$L = 10.6\,\mu\text{H}, C = 530\,\text{pF}, R = 100\,\Omega$

Figure 15.30

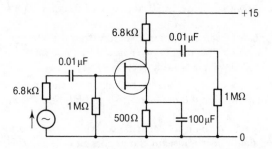

Figure 15.31

the average current $2E/\pi R$ is suitable. With this value carry out a transient analysis over three cycles. Check that periodicity is approximately reached. Estimate the average output voltage and the peak-to-peak ripple. (Comparison should be made with Problem 12.13.)

15.6 Determine the behaviour of the current of Problem 15.5 when one diode goes permanently open circuit. (The current is now discontinuous, so that the initial inductor current will be zero.) Find the conduction angle. (Compare with Problem 12.14.)

15.7 In Figure 15.33 is shown a twin-T notch filter. By means of a star–delta transformation, find the frequency of zero transmission. Simulate the action of the filter and make graphs of gain and phase over a range two decades either side of the notch frequency.

15.8 The circuit shown in Figure 15.34(a) uses the notch filter of Problem 15.7 to make a frequency-selective amplifier. The small-signal equivalent is shown in Figure 15.34(b). By simulation investigate the gain–frequency response of this amplifier.

15.9 An operational amplifier has the small-signal equivalent circuit shown in Figure 15.35(a). The gain A varies with frequency according to

$$A = \frac{A_0}{(1 + j\omega T_1)(1 + j\omega T_2)}$$

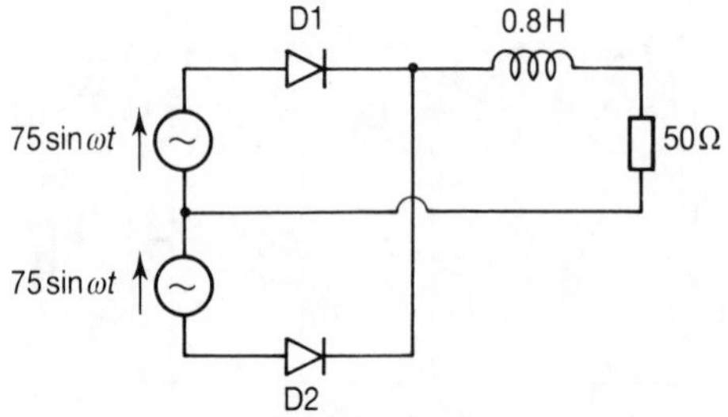

Figure 15.32

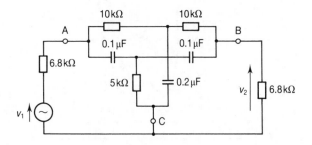

Figure 15.33

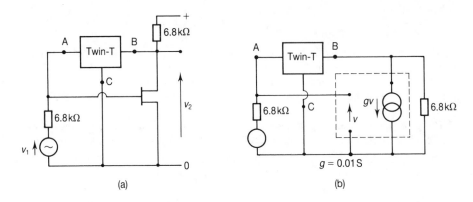

Figure 15.34

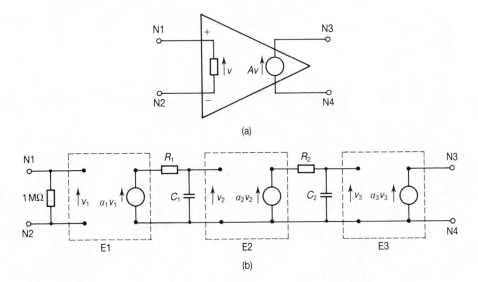

Figure 15.35

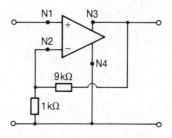

Figure 15.36

in which

$$A_0 = 10^5, \ T_1 = 10^{-4}, \ T_2 = 2 \times 10^{-5}$$

Simulate this amplifier by a subcircuit of the form shown in Figure 15.35(b). Make graphs of gain and phase as functions of frequency over the range 1 kHz–10 MHz. (To do this, join nodes 2 and 4 to the common node, put a source between nodes 1 and 0, and a resistance between nodes 3 and 0. Use decibel and logarithmic scales.)

15.10 The operational amplifier of Problem 15.9 is used in the circuit shown in Figure 15.36, designed to give an overall gain of ×10. Simulate the operation of this circuit over the range 0.1–10 MHz.

◆ *Appendix A:*
Network Theorems

A.1 Linearity	A.4 Reciprocity
A.2 Superposition	A.5 Conservation of
A.3 Theorems of Thévenin	volt–amps reactive
and Norton	

In the course of the book a number of theorems, such as those of Thévenin and Norton, have been introduced and demonstrated by examples. Formal proof depends on the formulation of equations for a general network given in Chapter 6, and is provided in the following sections. It is assumed that nodal equations have been set up and solved in the form given in Section 6.5:

$$\sum_{q=1}^{n-1} Y_{pq}V_q = I_p, \quad p = 1, 2, 3, \ldots, n-1 \tag{A.1}$$

$$V_p = \sum_{q=1}^{n-1} M_{pq}I_q \quad p = 1, 2, 3, \ldots, n-1 \tag{A.2}$$

In Equations A.1 it is assumed that node n is the reference node, the nodal currents I_p are prescribed, and the nodal voltages V_p are to be found. The coefficients Y_{pq} may or may not be symmetrical, depending on the absence or presence of dependent sources. The coefficients M_{pq} in Equation A.2 are given in terms of Y_{pq} by

$$M_{pq} = \Delta_{pq}/\Delta \tag{A.3}$$

in which

$$\Delta = \begin{vmatrix} Y_{11} & Y_{12} & Y_{13} & \cdots & Y_{1,n-1} \\ Y_{21} & Y_{22} & Y_{23} & \cdots & Y_{2n-1} \\ \vdots & \vdots & \vdots & & \\ Y_{n-1,1} & Y_{n-1,2} & & & \end{vmatrix} Y_{n-1,n-1}, \tag{A.4}$$

and Δ_{pq} is the cofactor of Y_{pq}. If $Y_{pq} = Y_{qp}$ for all p, q, then the same is true of M_{pq}.

A.1 ◆ Linearity

If a network is excited with a set of nodal currents I_p for which the nodal voltages are V_p, then it follows directly from Equation A.2 that excitation by the set kI_p, k being any constant, positive or negative, results in nodal voltages kV_p.

A.2 ◆ Superposition

Superposition relates to excitation of the network by firstly two different sets of nodal currents and secondly by a set equal to the sum of the first two sets.

Let the set of nodal currents $I_p^{(1)}$ result in nodal voltages $V_p^{(1)}$, so that from Equations A.2

$$V_p^{(1)} = \sum_{q=1}^{n-1} M_{pq} I_q^{(1)}, \quad p = 1, 2, 3, \ldots, n-1$$

In a similar fashion, let the set $I_p^{(2)}$ result in $V_p^{(2)}$ so that

$$V_p^{(2)} = \sum_{q=1}^{n-1} M_{pq} I_q^{(2)}, \quad p = 1, 2, 3, \ldots, n-1$$

Adding these two equations together gives

$$V_p^{(1)} + V_p^{(2)} = \sum_{q=1}^{n-1} M_{pq} (I_q^{(1)} + I_q^{(2)}), \quad p = 1, 2, 3, \ldots, n-1$$

Thus when the excitation currents are added, the nodal voltages may be found by adding together the results of the two separate excitations.

A very powerful practical test of linearity is to excite the network simultaneously at two frequencies and to examine the spectrum of the output: if the network is linear so that superposition applies, only the two input frequencies will be present. Any non-linearity will produce sum and difference frequencies.

A.3 ◆ Theorems of Thévenin and Norton

These theorems concern the connection of two free terminals of a network to external components, the free terminals being regarded as equivalent to a source. Take the reference node, n, and node 1 to be the free terminals. Currents $I_2, I_3, \ldots, I_{n-1}$ are the fixed excitation of the network, $-I_1$ is the current provided by the equivalent source and V_1 is the voltage at the source terminals. The first of Equations A.2 is

$$V_1 = M_{11} I_1 + \sum_{q=2}^{n-1} M_{1q} I_q$$

The summation term is constant by supposition, so that the equation can be written in the form

$$V_1 = M_{11}I_1 + E$$

or, remembering the direction of I_1,

$$V_1 = E - M_{11}(-I_1)$$

This is the form for Thévenin's theorem; Norton's theorem follows by rearrangement.

A.4 ◆ Reciprocity

A network is said to be reciprocal when the nodal admittance matrix is symmetrical, $Y_{pq} = Y_{qp}$. Consider two sets of excitation as in Section A16.1.2. Then using Equations A.1 we have

$$\sum_{p=1}^{n-1} V_p^{(1)} I_p^{(2)} = \sum_{p=1}^{n-1} V_p^{(1)} \sum_{q=1}^{n-1} Y_{pq} V_q^{(2)}$$

$$= \sum_{p=1}^{n-1} \sum_{q=1}^{n-1} Y_{pq} V_p^{(1)} V_q^{(2)}$$

$$= \sum_{q=1}^{n-1} V_q^{(2)} \sum_{p=1}^{n-1} Y_{pq} V_p^{(1)}$$

If the condition $Y_{pq} = Y_{qp}$ is satisfied for all p,q, then

$$\sum_{p=1}^{n-1} Y_{pq} V_p^{(1)} = \sum_{p=1}^{n-1} Y_{qp} V_p^{(1)} = I_q$$

Hence

$$\sum_{p=1}^{n-1} V_p^{(1)} I_p^{(2)} = \sum_{q=1}^{n-1} V_q^{(2)} I_q^{(1)}$$

This is the condition for reciprocity expressed in terms of the observable voltage and current variables.

A.5 ◆ Conservation of volt–amps reactive

We consider a network made up of passive components. We denote by y_p the admittance of the component connected between node p and the reference node, and by y_{pq} the admittance of the component between nodes p and q. We may then, as shown in Chapter 6, express the coefficients Y_{pq} in Equations A.1 in the form

$$Y_{pp} = y_p + \sum_{q \neq p} y_{pq}$$

$$Y_{pq} = Y_{qp} = -y_{pq}$$

Hence from Equations A.1

$$I_p = y_p V_p + V_p \sum_{q \neq p} y_{pq} - \sum_{q \neq p} y_{pq} V_q$$

$$= y_p V_p + \sum_q y_{pq}(V_p - V_q)$$

In this last summation the restriction $p \neq q$ can be dropped, since the term vanishes. We then form

$$P + jQ = \sum_p V_p I_p^* = \sum_p \left[V_p y_p^* V_p^* + V_p \sum_q y_{pq}^*(V_p - V_q)^* \right]$$

The term $V_p y_p^* V_p^*$ represents VI^* for the component connected between node p and reference. The double summation

$$\sum_p \left[V_p \sum_q y_{pq}^*(V_p - V_q)^* \right]$$

when expanded takes the form

$$V_1 \left[y_{12}^*(V_1 - V_2)^* + y_{13}^*(V_1 - V_3)^* + \ldots \right]$$

$$+ V_2 \left[y_{21}^*(V_2 - V_1)^* + y_{23}^*(V_2 - V_3)^* + \ldots \right]$$

$$+ V_3 \left[y_{31}^*(V_3 - V_1)^* + y_{32}^*(V_3 - V_2)^* + \ldots \right]$$

$$+ \ldots$$

$$+ \ldots$$

$$+ V_{n-1} \left[\ldots \right]$$

It will be seen that since y_{pq} and y_{qp} represent the same component the terms in the first three lines pair up to give

$$y_{12}^*(V_1 - V_2)(V_1 - V_2)^* + y_{13}^*(V_1 - V_3)(V_1 - V_3)^*$$
$$+ y_{23}^*(V_2 - V_3)(V_2 - V_3)^*$$

These are the correct expressions for VI^* associated with y_{12}, y_{13} and y_{23} respectively. All the terms in the double summation can be paired in the same way, so that we finally have

$$P + jQ = \sum (P + jQ)_{pq}$$

in which the sum is taken over all components. Thus both power and VAR are conserved.

◆ *Appendix B:*
Proof That $M^2 \leqslant L_1 L_2$

It was shown in Section 1.6.3 that the energy stored in a pair of magnetically coupled inductors L_1, L_2 carrying currents i_1, i_2 respectively is given by

$$W = \tfrac{1}{2}(L_1 i_1^2 + L_2 i_2^2 + 2M i_1 i_2)$$

Writing $\theta = i_1/i_2$ we have

$$W = \tfrac{1}{2}L_1 i_2^2[\theta^2 + 2(M/L_1)\theta + L_2/L_1]$$

The quadratic expression can be factorized to

$$[\theta^2 + 2(M/L_1)\theta + L_2/L_1] = (\theta - \theta_1)(\theta - \theta_2)$$

in which

$$\theta_1, \theta_2 = -\frac{M}{L_1} \pm \left[\left(\frac{M}{L_1}\right)^2 - \frac{L_2}{L_1}\right]^{1/2}$$

If these roots are real then the quadratic expression, and hence W, can be either positive or negative depending on the value of the ratio i_1/i_2, which may have any value. Accepting that the energy stored can never be negative we conclude that θ_1, θ_2 cannot be real and must be complex conjugates, requiring that

$$M^2 \leqslant L_1 L_2$$

◆ *Appendix C:*
Integrals Arising from Fourier Series

The results of certain integrations were used in Section 11.2.3. These results are obtained here.

Average of $\cos{(2\pi mt/T)}\cos{(2\pi nt/T)}$
The average is given by

$$I_1 = \frac{1}{T}\int_0^T \cos{(2\pi mt/T)}\cos{(2\pi nt/T)}\,\mathrm{d}t$$

Changing the variable to

$$\theta = 2\pi t/T$$

$$I_1 = \frac{1}{2\pi}\int_0^{2\pi} \cos{(m\theta)}\cos{(n\theta)}\,\mathrm{d}\theta$$

$$= \frac{1}{4\pi}\int_0^{2\pi} \{\cos{[(n-m)\theta]} + \cos{[(n+m)\theta]}\}\,\mathrm{d}\theta$$

If $n \neq m$

$$I_1 = \frac{1}{4\pi}\left[\frac{\sin{[(n-m)\theta]}}{n-m} + \frac{\sin{[(n+m)\theta]}}{n+m}\right]_0^{2\pi}$$

$$= 0$$

If $n = m \neq 0$

$$I_1 = \frac{1}{4\pi}\int_0^{2\pi} [1 + \cos{(2m\theta)}]\,\mathrm{d}\theta$$

$$= \frac{1}{2}$$

If $m = n = 0$

$$I_1 = 1$$

Average of $\sin{(2\pi mt/T)}\sin{(2\pi nt/T)}$
The average is given by

$$I_2 = \frac{1}{T} \int_0^T \sin\left(2\pi mt/T\right) \sin\left(2\pi nt/T\right) dt$$

Changing the variable to

$$\theta = 2\pi t/T$$

$$I_2 = \frac{1}{2\pi} \int_0^{2\pi} \sin\left(m\theta\right) \sin\left(n\theta\right) d\theta$$

$$= \frac{1}{4\pi} \int_0^{2\pi} \{\cos\left[(n-m)\theta\right] - \cos\left[(n+m)\theta\right]\} d\theta$$

If $n \neq m$

$$I_1 = \frac{1}{4\pi} \left[\frac{\sin\left[(n-m)\theta\right]}{n-m} - \frac{\sin\left[(n+m)\theta\right]}{n+m} \right]_0^{2\pi}$$

$$= 0$$

If $n = m \neq 0$

$$I_1 = \frac{1}{4\pi} \int_0^{2\pi} \left[1 - \cos\left(2m\theta\right)\right] d\theta$$

$$= \frac{1}{2}$$

If $m = 0$ or $n = 0$

$$I_2 = 0$$

Average of $\cos\left(2\pi mt/T\right) \sin\left(2\pi nt/T\right)$

The average is given by

$$I_3 = \frac{1}{T} \int_0^T \cos\left(2\pi mt/T\right) \sin\left(2\pi nt/T\right) dt$$

Changing the variable to

$$\theta = 2\pi t/T$$

$$I_3 = \frac{1}{2\pi} \int_0^{2\pi} \cos\left(m\theta\right) \sin\left(n\theta\right) d\theta$$

$$= \frac{1}{4\pi} \int_0^{2\pi} \left(\sin\left[(n+m)\theta\right] + \sin\left[(n-m)\theta\right]\right) d\theta$$

If $n = m$

$$I_3 = -\frac{1}{4\pi} \left[\frac{\cos\left[(n+m)\theta\right]}{n+m} \right]_0^{2\pi}$$

$$= 0$$

If $n \neq m$

$$I_3 = -\frac{1}{4\pi}\left[\frac{\cos[(n+m)\theta]}{n+m} + \frac{\cos[(n-m)\theta]}{n-m}\right]_0^{2\pi}$$

$$= 0$$

If $n = 0$

$$I_3 = 0$$

If $n = 0, n \neq 0$

$$I_3 = -\frac{1}{4\pi}\left[2\frac{\cos(n\theta)}{n}\right]_0^{2\pi}$$

$$= 0$$

C.1 ◆ Integral of Example 11.2

The required integral was

$$I = \int_{-\pi/2}^{\pi/2} \cos\theta \cos(m\theta)\,d\theta$$

If $m = 0$

$$I = [\sin\theta]_{-\pi/2}^{\pi/2} = 2$$

If $m = 1$

$$I = \int_{-\pi/2}^{\pi/2} \cos^2\theta\,d\theta$$

$$= \frac{1}{2}\int_{-\pi/2}^{\pi/2}[1 + \cos(2\theta)]\,d\theta$$

$$= \frac{\pi}{2}$$

Otherwise, for $m \geqslant 2$

$$I = \frac{1}{2}\int_{-\pi/2}^{\pi/2}\{\cos[(m-1)\theta] + \cos[(m+1)\theta]\}\,d\theta$$

$$= \int_0^{\pi/2}\{\cos[(m-1)\theta] + \cos[(m+1)\theta]\}\,d\theta$$

$$= \frac{\sin[(m-1)\pi/2]}{m-1} + \frac{\sin[(m+1)\pi/2]}{m+1}$$

If m is an odd integer, both terms vanish. For m even we write $m = 2p$, in which case

$$I = \frac{\sin(p\pi - \pi/2)}{2p - 1} + \frac{\sin(p\pi + \pi/2)}{2p + 1}$$

$$= \frac{-\cos(p\pi)}{2p - 1} + \frac{\cos(p\pi)}{2p + 1}$$

$$= -(-1)^p \frac{2}{4p^2 - 1}, \quad p = 1, 2, \ldots$$

These are the results used in Example 11.2.

◆ *Answers to End-of-chapter Problems*

Below are given answers to numerical problems, together with expressions for the problems of Chapter 10.

Chapter 1

1.1 25.5 mA, 0.306 W; 6.82 mA, 0.102 W; 0.21 mA, 2.57 mW.
1.2 79.9 V, 1.36 mW.
1.3 $i = Vt/L$. 0.3 A; 40 ms.
1.4 2.25 J; 90 mJ.
1.5 $-50 \exp(-1000t)$
1.6 0.45 C, 10.1 J; 0.021 C, 0.476 J; 99 μC, 2.23 mJ.
1.7 $V = It/C$. 0.84 V.
1.8 28 Ω.
1.9 13.2 μA in parallel with 680 kΩ.
1.10 5 V; 100 μC, 50 μC.

Chapter 2

2.1 (i) $\hat{I}/\pi$, $\hat{I}/2$; (ii) $2\hat{I}/\pi$, $\hat{I}/\sqrt{2}$; (iii) $0.632\hat{I}$, $0.658\hat{I}$.
2.2 1.08 A; 0.772 A.
2.3 198 V.
2.4 153 cos α; (i) 49°; (ii) 79°.
2.5 $<i> = 0.244$ A, $i_{RMS} = 0.416$ A. Power, 3.015 W; resistance, 2.928 Ω.
2.6 (i) 3.36 J; (ii) 0.24 A; (iii) 1.68 kW. 0.672 MW.
2.7 44.7 V.
2.8 83 μJ, 167 μ, 1.25 W.
2.9 4.2 W.
2.10 275 W, 225 W.
2.11 55 W, 445 W.
2.12 ≈ 500 C, 2 kJ.
2.13 67 MJ.
2.14 6.7 A, 1.5 kW.
2.15 $A = 207$, $B = 122$.

Chapter 3

3.1 6.23 V.
3.2 5.91 V, 1.31 A, 0.47 A, 1.73 A.
3.3 $R_1 = 577$ Ω, $R_2 = 306$ Ω.
3.4 5 Ω, 4.73 A; 4 Ω, 6.59 A; 5 Ω, 1.86 A; 5 Ω, 7.14 A. 236 W, 321 W.

3.5   (i) 14.3 V; (ii) 4.85 A; 2.40 A.
3.6 $e_0 = 0.015$ V; $R_0 = 1$ kΩ; 10 μA.
3.7 $R_1 = 117$ Ω; $R_2 = 32$ Ω.
3.8   4.4 V.
3.9   1.55 mA, 7.7 V.
3.10 $A/(R + R_0 + AR)$; 0.01%.
3.11 1.74 MΩ
3.12 $R_1 = 8.6$ Ω, $R_2 = 141.5$ Ω; 0.741 V, 9.98 mW, 9.99 dBm.
3.15 (a) 0.957 A, 0.971 A.

Chapter 4

4.1   94.2 Ω, 754 Ω; 482 Ω, 60.3 Ω; 339 Hz.
4.2   1000 + j3047 Ω; 3207∠ 71.8°; 318 Hz, 1414 Ω.
4.3   97.2 − j296 μS; 10.3 kΩ, 0.554 H.
4.4   100 − j159 Ω = 188∠ −57.8°; 2.83 + j4.51 mS = 5.32∠ 57.8° mS;
          353 Ω, 0.718 μF.
4.5   (i) j461 Ω; (ii) −j2170 Ω. 285 − j2108 Ω
4.6   157∠ 90°, 398∠ −90°, 200∠ 0; 313∠ −50.3°.
4.7 $P = 965$ W, $Q = 539$ VAR.
4.8   (a) 27.6 A, 4.58 kW, 4.79 VAR, 0.69; (b) 265 μF; (c) 266 V.
4.9   28.2∠ −64.8°, 12∠ 0°, 6.39∠ 57.9°; 6.58 kW, 4.81 VAR.
4.10 5.07 μF.
4.11 16° < φ < 178°.
4.12 9.65∠ 92.0°.
4.13 $L_1 = 13.5$ mH, $C_2 = 56.2$ nF or $L_2 = 4.50$ mH, $C_1 = 18.8$ nF.

Chapter 5

5.1   0.3 μF; 25.8° leading.
5.2   3.18 kHz; 2.23∠ 21.8°.
5.3   31.8 kHz, 5 V; 8.5, 119 kHz.
5.4   353 Ω, 0.537 H.
5.5   (1/3) ∠ 90°; 72.7° output leading.
5.6   30.6 μF; −158.2°.
5.9   0.176 H, 1.42 kΩ.
5.10 2.25 MHz, 22.5, 0.1 MHz; 1.1 kHz below f_0.
5.11 (a) 1.04 μH; (b) 10.2; (c) 4.4 MHz; 17 mA.
5.12 15.6 nF, 11.3 kΩ.
5.13 74.4 kHz, 2.07; 1.47 ∠ −27.4°.
5.14 80 μH, 10 pF; 119, 95, 80.
5.15 √ 10 Hz, 4.52 (13.1 dB).

Chapter 7

7.1   (a) 4.8 A, 3.44 kW; (b) 14.4 A, 10.3 kW.
7.2   8.91 A, 5.94 kW, 0.928.
7.3   8.24 A, 5.18 kW, 0.874.
7.4   17.1 A, 11.1 kW, 0.904.
7.5   8.02 A, 3.86 kW, 0.126.
7.6   3.33 A, 3.86 kW, 0.304.

7.7   6.37 μF.
7.8   12.9 kW, 12.9 kW.
7.9 $\pm$ 7.46 kW.
7.10   4.04, 7.07 kW
7.11   547 A; 139 kV, 1.80 MW.
7.12   2.57 μF; 136 kV, 1.27 MW.
7.13   16.1, 16.1, 8.3 A.
7.14   24 V.
7.15   10.5 A, 3.31 kW; 39.1 V, 11.8 A, 9.4 A, 12.4 A.

Chapter 8
8.1 $M_{max} = 0.632$ mH, $k = 0.791$.
8.2   0.375 mH.
8.4   69.3 μH, 0.866.
8.5   (a) $120 + j352$ Ω; (b) $j634$ Ω.
8.6   (a) 3.39 kHz; (b) 4.75 kHz.
8.7   100 μH, 10 μH, 0.632.
8.9   2.18 V, 0.95.
8.10   255 V.
8.11   104 V, 602 W, 14.0 A.
8.12   114 V; 91.8%.

Chapter 9
9.1   (a) 3.1 s; (b) 190 s.
9.2   (a) 1.0 s; (b) 5.1 s; 8.3 A s^{-1}.
9.3   125 V, 30 ms.
9.5   20 A, 11.5 s.
9.6   (a) 1.7 s; (b) 11.8 s.
9.9   6.93 ms, 12.5 mA.
9.11   10%, 0.096 μs; 90%, 1.071 μs; maximum 2.23 μs, 15.8%.
9.12   120 kHz, 140.
9.13   5.23 kV, 78 μs.

Chapter 10
10.1 $v(t) = ARC\,[(t/RC) - 1 + \exp(-t/RC)]$.
10.2 $v(t) = \hat{E}\,[C_1(C_1 + C_2)]\cos(\phi)\,[\sin(\omega t + \phi) - \alpha\exp(-\alpha t)]$ in which
$\alpha = 1/R_1\,(C_1 + C_2)$, $\phi = \tan^{-1}(\alpha/\omega)$.
10.3 $v_2(t) = (E/\sqrt{2})\,[\exp(-\alpha_1 t) - \exp(-\alpha_2 t)]$, $\alpha_1 = 2 - \sqrt{2}$, $\alpha_2 = 2 + \sqrt{2}$.
10.4 $v(t) = -\,[IR^2/(R + r)]\exp[-(2R + r)t/L]$.
10.7 $i_1(t) = (E/R)\,\{1 - (1/\sqrt{5})\,[\exp(-\alpha_1 t) - \exp(-\alpha_2 t)]\}$ and
$i_2(t) = (E/R)\,\{1 - (1/2\sqrt{5})\,[(\sqrt{5} + 1)\exp(-\alpha_1 t) + (\sqrt{5} - 1)\exp(-\alpha_2 t)]\}$.
10.8 $i(t) = 0.05[\exp(-100t) - \exp(200t)]$ and
$v(t) = 50[2\exp(-100t) - \exp(-200t)]$.
10.9 $v(t) = (50/3)\,[4\exp(-100t) - \exp(-200t)]$.
10.11 $v(t) = (E/2)\,[1 - \exp(-t) - (2/\sqrt{3})\exp(-t/2)\sin(\sqrt{3}t/2)]$.
10.13 $v_4(t) = -(gR)^3\,E\{1 - \exp(-\alpha t)\,[1 + \alpha t + (\alpha t)^2\,/2]\}$;
10%, 1.10; 90%, 5.30; 0.3 μs.
10.14 $0.7Q$.

10.15 $V_2(s)/E = As^2 R_1 R_2 C_1 C_2/[s^2 R_1 R_2 C_1 C_2$
$+ s(R_1 C_1 + R_2 C_2 + R_1 C_2 - AR_2 C_2) + 1]$
and $v_2(t) = (E/6) [4 \exp(-2000t) - \exp(-500t)]$.

Chapter 11

11.1 $a_0 = \hat{V}\tau/T; a_n = (2\hat{V}/\pi n) \sin(n\pi\tau/T)$.
11.2 $a_0 = \hat{V}\tau/T; a_n = (2\hat{V}/\pi n) \sin(n\pi\tau/T) \cos(n\pi\tau/T)$;
$b_n = (2\hat{V}/\pi n) \sin^2(n\pi\tau/T)$.
11.3 $a_0 = 2\hat{V}/\pi; a_n = -4\hat{V}/\pi(4n^2 - 1); b_n = 0$.
11.4 With a_n of problem 11.3,

$$i(t) = a_0/R + \sum a_n |\mathbf{Z}_n|^{-1} \cos(2\pi nt/T - \angle \mathbf{Z}_n),$$

$\mathbf{Z}_n = R + jL(2\pi n/T)$
11.5 4.2 H.
11.7 10.0, 0.33, 0.12 V; 3.3%, 1.2%.
11.8 $a_0 = \hat{V}/4; a_2 = \hat{V}/4; n = 4, 6, 8, \ldots, a_n = 0, n = 1, 3, 5, \ldots,$
$a_n = -4\hat{V} \sin(n\pi/2)/\pi n(n^2 - 4); 1.698, 2, 0.340, 0, -0.485$.
11.9 $a_0 = 0.25, a_1 = 0.427, a_2 = 0.25, a_3 = 0.073, a_4 = 0$.
11.10 2 MHz.
11.11 $(j\hat{V}T/\pi) \sin(2\pi NfT)/(1 - f^2 T^2)$

Chapter 12

12.1 (a) 0.26 V, 4.1 mA, (b) 0.42 V, 23.3 mA; (c) 0.27 V, 4.7 mA
12.3 (a) 0.28 V, 4.1 mA; (b) 0.41 V, 23.5 mA; (c) 0.28 V, 4.4 mA.
12.5 (ii) maximum 1.20, edge 1.15.
12.6 Maximum 1.40, edge 1.22.
12.7 106°, 1.2 A, 0.23 A.
12.8 0.83 A, 0.15 A.
12.9 354 V RMS, 2 μF.
12.10 191 V, 14 V.
12.13 48 V, 6 V.
12.14 280°.
12.15 15 V, 1.5 V.

Chapter 13

13.1 $v = 0.44$ V, $i = 7.0$ mA; 7 Ω.
13.2 $v = 0.443$ V, $i = 7.05$ mA; 7.1 Ω.
13.3 -0.75 V, 8.8 V, 18 mA, 21 mA V^{-1}, 3.3 kΩ.
13.4 4.6 V.
13.5 $V_{BE} = 0.74$V, $V_{CE} = 5.1$V, $I_B = 0.3$ mA, $I_C = 26$ mA $h_{ie} = 8700$, $h_{fe} = 86$,
$h_{oe} = 4 + 10^{-4}$S.
13.6 0.71 mV.
13.7 384 Ω, 0.38 V.
13.8 $R_1 = 42 \Omega$, $R_2 = 302 \Omega$; 1.6 V.
13.9 $g_m = 2\beta(V_{GS} - V_T)(1 + \lambda V_{DS}); r_{ds}^{-1} = \beta\lambda(V_{GS} - V_T)^2$
13.10 $g_m = 21$ mA V^{-1}, $r_{ds} = 3.3$ kΩ.

✦ *Index*